Günther Nürnberger

Approximation by Spline Functions

Springer-Verlag Berlin Heidelberg GmbH

Prof. Dr. Günther Nürnberger
Fakultät für Mathematik und Informatik
Universität Mannheim, Seminargebäude A 5
D-6800 Mannheim, Federal Republic of Germany

Mathematics Subject Classification (1980): 41-02, 41A05, 41A10, 41A15, 41A25, 41A30, 41A50, 41A52, 41A55, 41A63, 65D05, 65D07, 65D10, 65D15, 65D32, 65L05, 65L10

ISBN 978-3-642-64799-4   ISBN 978-3-642-61342-5 (eBook)
DOI 10.1007/978-3-642-61342-5

Library of Congress Cataloging-in-Publication Data.
Nürnberger, G. (Günther), 1948–.
Approximation by spline functions / Günther Nürnberger. p. cm.
Includes bibliographical references.

1. Spline theory. 2. Approximation theory. I. Title.   QA224.N87   1989   511'.42 – dc20   89-21710

© Springer-Verlag Berlin Heidelberg 1989
Originally published by Springer-Verlag Berlin Heidelberg New York in 1989
Softcover reprint of the hardcover 1st edition 1989

2141/3140-543210 – Printed on acid-free paper

Preface

The *aim of this book* is to analyse in a unified approach basic theoretical and numerical aspects of interpolation and best approximation by polynomial splines in one variable. We begin by examining approximation–theoretic properties of spaces of polynomials (as prototypes of Chebyshev spaces) which are important for investigating the more complex structure of spline spaces (as prototypes of weak Chebyshev spaces). In the appendix we give a brief introduction to splines with free knots, to splines in two variables and to spline collocation for differential equations. A large portion of the results and methods was developed and completed in the last decade, and cannot be found in earlier books on splines.

Since the amount of available material is so large, the results are presented in their standard form for polynomial splines. On the other hand, the researcher will find several references to the recent literature concerning more general aspects. The book can be used for graduate courses on splines or approximation theory. Students can read the book with a basic knowledge of analysis and linear algebra. Engineers will find various practice interpolation and approximation methods; in particular, an algorithm for computing spline approximations with free knots of high accuracy.

In the following we describe the *topics* which are considered in this book.

Functions are the basic mathematical tools for describing objects and processes from the real world. A fundamental problem in Applied Mathematics is to approximate functions which are given explicitly or only implicitly by operator equations, e.g. differential equations, by functions of a simple structure. In practice the functions to be approximated may be defined by only finitely many data. To insure a unified approach, we consider approximation of functions only on an interval. However, the methods described in this book can also be applied to functions on a finite set.

For applications the approximating functions should be simple enough to be easily manipulated on a computer and flexible enough to provide efficient approximations.

A class of functions with a simple structure is the space

$$P_m = \{p : [a, b] \to \mathbf{R} : p(t) = \sum_{i=0}^{m} a_i\, t^i,\ a_0, \ldots, a_m \text{ real} \}$$

of polynomials of degree m. Although polynomials possess optimal structural properties, they do not provide small approximation errors for functions which fail to be extremely smooth. Polynomials are not flexible enough to approximate such functions efficiently on a relatively large interval.

To obtain better approximations the considered interval can be divided into several subintervals and we can work with polynomials on these smaller subintervals. Since in practice the approximating functions should be differentiable, this leads to the space

$$S_m(x_1, \ldots, x_k) = \{s \in C^{m-1}[a, b] : s|_{[x_i, x_{i+1}]} \in P_m, i = 0, \ldots, k\}$$

of polynomial splines of degree m with k fixed knots $a = x_0 < x_1 < \cdots < x_k < x_{k+1} = b$. Splines possess high flexibility, and can approximate even nonsmooth functions efficiently.

Spaces of polynomials are prototypes of Chebyshev spaces while spaces of splines are prototypes of the more general class of weak Chebyshev spaces. We discuss the theory of these spaces only as far as applications to approximation by polynomials and splines are concerned. Several results on Chebyshev spaces given in Chapter I are used for proving various results from Chapter II.

The standard methods for approximating functions by polynomials or splines are interpolation and best approximation. The algorithms described in this book are practice methods for constructing such approximations using a modern computer.

Chebyshev spaces possess desirable approximation–theoretic properties. In particular, interpolation at arbitrary points is always possible and best uniform approximations from these spaces are always unique. These properties are lost when we pass to weak Chebyshev or spline spaces. However, as we will see, many nice structural properties are still preserved.

In *Section 1* of *Chapter I* we investigate the unique solvabiltiy of interpolation problems for various types of Chebyshev spaces and derive further properties of these spaces. *Section 2* is devoted to the construction of interpolating polynomials and the choice of nearly optimal interpolation points.

Sections 3–6 deal with best approximation by functions from Chebyshev spaces in the uniform norm, L_1–norm, and L_2– norm. We give results on the characterization, unicity and strong unicity of best approximations. Moreover, we describe an algorithm for computing best uniform approximations from Chebyshev spaces. The connection between best L_1–approximation and interpolation, and the relationship between best one–sided L_1–approximation and quadrature formulas are investigated.

Chapter II begins with a section on properties of weak Chebyshev spaces. It is shown that weak Chebyshev spaces can be "approximated" by Chebyshev spaces and therefore, certain approximation properties of Chebyshev spaces pass over to weak Chebyshev spaces. In *Section 2* we show that B–splines form a basis for spline spaces and discuss some numerical aspects concerning this type of splines. *Section 3* is devoted to interpolation and quasi–interpolation by splines. We give a characterization of those sets of points for which Hermite interpolation is possible. Moreover, it is shown that interpolating splines which satisfy special boundary conditions possess certain optimality properties.

Sections 4–7 deal with best approximation by functions from weak Chebyshev and spline spaces in the uniform norm and the L_1-norm. We investigate the characterization, unicity, and strong unicity of best approximations, especially from spline spaces. Moreover, we examine continuity properties of the metric projection onto spline spaces. A result on the approximation power of splines is established. Concerning numerical purposes we develop an algorithm for computing best uniform approximations from spline splaces. Moreover, we describe numerical methods for best piecewise polynomials with free knots and good spline approximations with free knots. Finally, the relationship between best L_1-approximation and interpolation is examined. In the section on best one–sided L_1-approximation the existence and uniqueness of Gauss quadrature formulas for spline spaces is proved.

In *Section 8* we investigate the approximation of linear functionals by linear combinations of point functionals which possess certain optimality properties. It is shown that these optimal functionals are defined by various types of splines. In *Section 9* we point out that most of the standard results on polynomial splines have analogous versions for more general classes of splines.

In contrast to splines with fixed knots, there is no fully developed theory in the case of splines with free knots and splines in two variables. In the *Appendix* we give an illustration of how far these domains are developed. Without going into details, we finally decribe some basic ideas concerning the use of splines in the numerical solution of differential equations in connection with collocation methods. The aim of the appendix is to give the reader a first impression of subjects which are not considered in the main part of the book and to illustrate that despite the great progress of the last twenty years in spline theory several deep problems are still unsolved at present.

Finally, I would like to *thank* Professors H. Berens, H.-P. Blatt, D. Braess, B. Brosowski, L. Collatz, F. Deutsch, D. Kölzow, G. Meinardus, L.L. Schumaker, I. Singer, M. Sommer and H. Strauß with whom I had the opportunity to carry out joint scientific projects. A lot of material which resulted from this cooperation is contained in this book. My special thanks is due to Professor G. Meinardus for his strong encouragement to write this book.

Erlangen, Spring 1989   Günther Nürnberger

Contents

Chapter II. Splines and Weak Chebyshev Spaces

Appendix

Chapter I. Polynomials and Chebyshev Spaces

1. Interpolation by Chebyshev Spaces

Interpolation is a standard method for approximating functions. We consider Lagrange and Hermite interpolation by arbitrary finite–dimensional spaces. The so-called Chebyshev and extended Chebyshev spaces play an important role in connection with these interpolation problems. Prototypes of such spaces are spaces of polynomials and exponentials.

1.1. Lagrange Interpolation by Chebyshev Spaces

Lagrange interpolation is a well-known procedure to approximate a given function by functions from a finite–dimensional space. The Chebyshev spaces introduced in this section have the property that Lagrange interpolation is possible at arbitrarily chosen points.

We first introduce some basic notations. Let $[a, b]$ be a closed bounded interval of the real line. The space of continuous functions on $[a, b]$ is defined by

$$C[a, b] = \{f : [a, b] \to \mathbf{R} : f \text{ continuous}\},$$

where $\mathbf{R}$ denotes the set of real numbers. If r is a positive integer, then the space of r-times continuously differentiable functions on $[a, b]$ is defined by

$$C^r[a, b] = \{f : [a, b] \to \mathbf{R} : f^{(r)} \in C[a, b]\}.$$

In order to approximate a given function $f \in C[a, b]$ by functions of a simple structure it is natural to require that the approximation coincides with f at certain points of the interval $[a, b]$. This leads to the following interpolation problem.

Definition 1.1. Let an n–dimensional subspace G of $C[a, b]$, a function $f \in C[a, b]$ and points $t_1 < ... < t_n$ in $[a, b]$ be given. The *Lagrange interpolation problem* is to determine a function $g \in G$ such that

$$g(t_j) = f(t_j), \quad j = 1, ..., n. \tag{1.1}$$

If $\{g_1, \ldots, g_n\}$ is a basis of G, then we have to find coefficients $a_1, \ldots, a_n$ such that

$$\sum_{i=1}^{n} a_i g_i(t_j) = f(t_j), \quad j = 1, \ldots, n.$$

This is equivalent to the following system of linear equations.

$$\begin{pmatrix} g_1(t_1) & \cdots & g_n(t_1) \\ \vdots & & \vdots \\ g_1(t_n) & \cdots & g_n(t_n) \end{pmatrix} \begin{pmatrix} a_1 \\ \vdots \\ a_n \end{pmatrix} = \begin{pmatrix} f(t_1) \\ \vdots \\ f(t_n) \end{pmatrix}.$$

We denote the corresponding determinant by

$$D \begin{pmatrix} g_1, \ldots, g_n \\ t_1, \ldots, t_n \end{pmatrix} = \begin{vmatrix} g_1(t_1) & \cdots & g_n(t_1) \\ \vdots & & \vdots \\ g_1(t_n) & \cdots & g_n(t_n) \end{vmatrix}. \tag{1.2}$$

A standard result in Linear Algebra says that the problem has a unique solution if and only if

$$D \begin{pmatrix} g_1, \ldots, g_n \\ t_1, \ldots, t_n \end{pmatrix} \neq 0.$$

This observation leads to the following definition.

Definition 1.2. An n–dimensional subspace G of $C[a, b]$ is called a *Chebyshev subspace* or a *Haar subspace* if there exists a basis $\{g_1, \ldots, g_n\}$ of G such that

$$D \begin{pmatrix} g_1, \ldots, g_n \\ t_1, \ldots, t_n \end{pmatrix} > 0 \tag{1.3}$$

for all $t_1 < \cdots < t_n$ in $[a, b]$.

Example 1.3. (i) The prototype of a Chebyshev subspace of $C[a, b]$ is

$$P_n = \left\{ p : [a, b] \to \mathbf{R} : \quad p(t) = \sum_{j=0}^{n} a_j t^j, \quad a_0, \ldots, a_n \in \mathbf{R} \right\}, \tag{1.4}$$

the subspace of *polynomials of degree n*. In fact it is easy to verify that

$$D \begin{pmatrix} 1, & t, & \cdots, & t^n \\ t_1, & t_2, & \cdots, & t_{n+1} \end{pmatrix} = \prod_{1 \leq i < j \leq n+1} (t_j - t_i) \tag{1.5}$$

which shows that the determinant is positive for all $t_1 < \cdots < t_{n+1}$ in $[a, b]$. This determinant is called the *Vandermonde determinant*.

(ii) Let $[a, b]$ be a subinterval of $[-\pi, \pi)$. We call

$$Q_n = \Big\{ f : [a, b] \to \mathbf{R} : \quad f(t) = a_0 + \sum_{j=1}^n (a_j \cos(jt) + b_j \sin(jt)),$$
$$a_0, a_1, \ldots, a_n, b_1, \ldots, b_n \in \mathbf{R} \Big\}$$

the subspace of *trigonometric polynomials* of degree n. Since

$$\cos(jt) + i\sin(jt) = e^{ijt},$$

every function $f \in Q_n$ can be written as

$$f(t) = \frac{1}{2} \sum_{j=0}^n ((a_j - ib_j)e^{ijt} + (a_j + ib_j)e^{-ijt}).$$

By setting $z = e^{it}$, we obtain $f(t) = z^{-n}p(z)$, where p is a polynomial of degree $2n$ with complex coefficients. Since every polynomial $p \neq 0$ has at most $2n$ distinct zeros, the function f has at most $2n$ distinct zeros in $[a, b]$. Therefore, it follows from Theorem 1.14 below that Q_n is a $(2n + 1)$-dimensional Chebyshev subspace of $C[a, b]$.

A basic question is the unique solvability of interpolation problems. The first result shows that the Chebyshev spaces are exactly those spaces for which Langrange interpolation problems always have a unique solution.

Theorem 1.4. *For an n-dimensional subspace G of $C[a, b]$, the following statements are equivalent:*
(i) *For all functions $f \in C[a, b]$ and all points $t_1 < \cdots < t_n$ in $[a, b]$ the Lagrange interpolation problem (1.1) has a unique solution from G.*
(ii) *G is a Chebyshev subspace.*

Proof. (ii) $\Rightarrow$ (i). This implication follows from the fact that all determinants which correspond to the Lagrange interpolation problems are different from zero for Chebyshev subspaces.

(i) $\Rightarrow$ (ii). Let $\{g_1, \ldots, g_n\}$ be a basis of G. It follows from (i) that

$$D\begin{pmatrix} g_1, \ldots, g_n \\ t_1, \ldots, t_n \end{pmatrix} \neq 0$$

for all $t_1 < \cdots < t_n$ in $[a, b]$. Since the function

$$(t_1, \ldots, t_n) \to D\begin{pmatrix} g_1, \ldots, g_n \\ t_1, \ldots, t_n \end{pmatrix}$$

is continous on the set $\{(t_1, \ldots, t_n) : t_1 < \cdots < t_n \text{ in } [a, b]\}$, either

$$D\begin{pmatrix} g_1, \ldots, g_n \\ t_1, \ldots, t_n \end{pmatrix} > 0$$

or

$$D\begin{pmatrix} -g_1, & g_2, \ldots, g_n \\ t_1, & t_2, \ldots, t_n \end{pmatrix} > 0$$

for all $t_1 < \cdots < t_n$ in $[a, b]$. This proves Theorem 1.4.

1.2. Hermite Interpolation by Extended Chebyshev Spaces

Hermite interpolation is a generalization of Lagrange interpolation in the sense that not only function values are interpolated, but also certain derivatives. The extended Chebysev spaces introduced in this section have the property that Hermite interpolation is possible at arbitrarily chosen points.

We first state the Hermite interpolation problem.

Definition 1.5. Let an n–dimensional subspace G of $C[a, b]$, a function $f \in C[a, b]$ and points $t_1 \leq \cdots \leq t_n$ in $[a, b]$ be given. We assume that the function $f \in C[a, b]$ and the functions in G are sufficiently differentiable. We set for all $j \in \{1, \ldots, n\}$,

$$d_j = \max \{i : t_j = \cdots = t_{j-i}\}. \tag{1.6}$$

The *Hermite interpolation problem* is to determine a function $g \in G$ such that

$$g^{(d_j)}(t_j) = f^{(d_j)}(t_j), \quad j = 1, \ldots, n. \tag{1.7}$$

Remark 1.6. To clarify what the Hermite interpolation problem means, we give a different formulation of the problem. For $t_1 \leq \cdots \leq t_n$ there exists points $u_1 < \cdots < u_p$ such that

$$\{u_1, \ldots, u_p\} = \{t_1, \ldots, t_n\}, \tag{1.8}$$

where the point u_j appears exactly r_j times in the set $\{t_1, \ldots, t_n\}$, $j = 1, \ldots, p$. Then the Hermite interpolation problem is to determine a function $g \in G$ such that

$$g^{(i)}(u_j) = f^{(i)}(u_j), \quad i = 0, \ldots, r_i - 1, \quad j = 1, \ldots, p. \tag{1.9}$$

Let $\{g_1, \ldots, g_n\}$ be a basis of G. The determinant which corresponds to the Hermite interpolation problem is defined by

$$D \begin{pmatrix} g_1, \ldots, g_n \\ t_1, \ldots, t_n \end{pmatrix} = \begin{vmatrix} g_1^{(d_1)}(t_1) & \cdots & g_n^{(d_1)}(t_1) \\ \vdots & & \vdots \\ g_1^{(d_n)}(t_n) & \cdots & g_n^{(d_n)}(t_n) \end{vmatrix}. \tag{1.10}$$

This leads to the following definition.

Definition 1.7. An n–dimensional subspace G of $C^{n-1}[a, b]$ is called an *extended Chebyshev subspace* if there exists a basis $\{g_1, \ldots, g_n\}$ of G such that

$$D \begin{pmatrix} g_1, \ldots, g_n \\ t_1, \ldots, t_n \end{pmatrix} > 0 \tag{1.11}$$

for all $t_1 \leq \cdots \leq t_n$ in $[a, b]$.

The following result is an extension of Theorem 1.4.

Theorem 1.8. *For an n-dimensional subspace G of $C^{n-1}[a, b]$, the following statements are equivalent:*
(i)   For all functions $f \in C^{n-1}[a, b]$ and all points $t_1 \leq \cdots \leq t_n$ in $[a, b]$ the Hermite interpolation problem (1.7) has a unique solution from G.
(ii) G is an extended Chebyshev subspace.

Proof. (ii) $\Rightarrow$ (i). This implication follows from the fact that all determinants which correspond to the Hermite interpolation problems are different from zero for extended Chebysev subspaces.

(i) $\Rightarrow$ (ii). Let $\{g_1, \ldots, g_n\}$ be a basis of G. It follows from (i) that

$$D \left(\begin{matrix} g_1, \ldots, g_n \\ t_1, \ldots, t_n \end{matrix} \right) \neq 0 \tag{1.12}$$

for all $t_1 < \cdots < t_n$ in $[a, b]$. Replacing g_1 by $-g_1$, if necessary, and using continuity arguments, by (1.12)

$$D \left(\begin{matrix} g_1, \ldots, g_n \\ t_1, \ldots, t_n \end{matrix} \right) > 0 \tag{1.13}$$

for all $t_1 < \cdots < t_n$ in $[a, b]$. Let $t_1 \leq \cdots \leq t_n$ in $[a, b]$ be given and let p be the number of distinct points in $\{t_1, \ldots, t_n\}$. We show by induction on p that

$$D \left(\begin{matrix} g_1, \ldots, g_n \\ t_1, \ldots, t_n \end{matrix} \right) \geq 0. \tag{1.14}$$

If $p = n$, then (1.14) follows from (1.13). Let the result be true for $p+1$. We will show that it is true for p. Let $u_1, \ldots, u_p$ be the distinct points as in Remark 1.6 which correspond to $t_1, \ldots, t_n$. We assume that $r_1 > 1$. The other cases follow analogously. The determinant

$$D \left(\begin{matrix} g_1, \ldots, g_n \\ t_1, \ldots, t_n \end{matrix} \right)$$

can be written as

$$D \left(\begin{matrix} g_1, \ldots, g_n \\ t_1, \ldots, t_n \end{matrix} \right) = \begin{vmatrix} g_1(u_1) & \cdots\cdots & g_n(u_1) \\ g_1'(u_1) & \cdots\cdots & g_n'(u_1) \\ \vdots & & \vdots \\ g_1^{(r_1-1)}(u_1) & \cdots\cdots & g_n^{(r_1-1)}(u_1) \\ \vdots & & \vdots \\ g_1(u_p) & \cdots\cdots & g_n(u_p) \\ g_1'(u_p) & \cdots\cdots & g_n'(u_p) \\ \vdots & & \vdots \\ g_1^{(r_p-1)}(u_p) & \cdots\cdots & g_n^{(r_p-1)}(u_p) \end{vmatrix}.$$

For sufficiently small $\varepsilon > 0$ we set

$$
D_\varepsilon = \begin{vmatrix}
g_1(u_1) & \ldots\ldots & g_n(u_1) \\
g_1'(u_1) & \ldots\ldots & g_n'(u_1) \\
\vdots & & \vdots \\
g_1^{(r_1-2)}(u_1) & \ldots\ldots & g_n^{(r_1-2)}(u_1) \\
g_1(u_1+\varepsilon) & \ldots\ldots & g_n(u_1+\varepsilon) \\
\vdots & & \vdots \\
g_1(u_p) & \ldots\ldots & g_n(u_p) \\
g_1'(u_p) & \ldots\ldots & g_n'(u_p) \\
\vdots & & \vdots \\
g_1^{(r_p-1)}(u_p) & \ldots\ldots & g_n^{(r_p-1)}(u_p)
\end{vmatrix}.
$$

Then by the induction hypothesis, $D_\varepsilon > 0$. It follows from Taylor's theorem that for each $j \in \{1, \ldots, n\}$ there exists a point $u_1 \leq y_j^{(\varepsilon)} \leq u_1 + \varepsilon$ such that

$$
g_j(u_1 + \varepsilon) = \sum_{i=0}^{r_1-2} \frac{g_j^{(i)}(u_1)\varepsilon^i}{i!} + \frac{g_j^{(r_1-1)}(y_j^{(\varepsilon)})\varepsilon^{r_1-1}}{(r_1-1)!} \ . \tag{1.15}
$$

Let $\tilde{D}_\varepsilon$ be the determinant which arises if we replace in D_ε the row

$$
g_1(u_1 + \varepsilon), \ldots, g_n(u_1 + \varepsilon)
$$

by the row

$$
g_1^{(r_1-1)}(y_1^{(\varepsilon)}), \ldots, g_n^{(r_1-1)}(y_n^{(\varepsilon)}).
$$

By using (1.15), a simple calculation shows that

$$
D_\varepsilon = \frac{\varepsilon^{r_1-1}}{(r_1-1)!}\tilde{D}_\varepsilon,
$$

and thus $\tilde{D}_\varepsilon > 0$. Since

$$
\tilde{D}_\varepsilon \to D\begin{pmatrix} g_1, \ldots, g_n \\ t_1, \ldots, t_n \end{pmatrix}
$$

for $\varepsilon \to 0$, it follows that

$$
D\begin{pmatrix} g_1, \ldots, g_n \\ t_1, \ldots, t_n \end{pmatrix} \geq 0.
$$

Then by (i) we get

$$
D\begin{pmatrix} g_1, \ldots, g_n \\ t_1, \ldots, t_n \end{pmatrix} > 0.
$$

This shows that (i) $\Rightarrow$ (ii) and proves Theorem 1.8.

1.3. Characterization of Extended Complete Chebyshev Spaces

We introduce extended complete Chebyshev spaces and give characterizations of these spaces. The results can be used to construct spaces of this type. It is shown that spaces of polynomials and exponentials have this property.

We use the following notation. If $g_1, \ldots, g_m$ are linearly independent functions in $C[a, b]$, then we define the *linear hull* of $g_1, \ldots, g_m$ by

$$\text{span}\{g_1, \ldots, g_m\} = \Big\{\sum_{i=1}^{m} a_i g_i : \quad a_1, \ldots, a_m \in \mathbf{R}\Big\}.$$

Definition 1.9. An n–dimensional subspace G of $C^{n-1}[a, b]$ is called a *complete* (respectively *extended complete*) *Chebyshev subspace* if there exists a basis $g_1, \ldots, g_n$ of G such that for all $m \in \{1, \ldots, n\}$ the subspace $\text{span}\{g_1, \ldots, g_m\}$ is an m–dimensional Chebyshev (respectively extended Chebyshev) subspace.

The first result shows that extended complete Chebyshev spaces can be completely characterized by properties of the Wronskian determinant (see Karlin & Studden [1966]).

Definition 1.10. Let a set $\{g_1, \ldots, g_m\}$ of functions in $C^{m-1}[a, b]$ be given. The *Wronskian determinants* of $\{g_1, \ldots, g_m\}$ are defined by

$$W(g_1, \ldots, g_m)(t) = \begin{vmatrix} g_1(t) & \cdots & g_m(t) \\ g_1'(t) & \cdots & g_m'(t) \\ \vdots & & \vdots \\ g_1^{(m-1)}(t) & \cdots & g_m^{(m-1)}(t) \end{vmatrix} \tag{1.16}$$

for all $t \in [a, b]$.

Theorem 1.11. *For an n–dimensional subspace G of $C^{n-1}[a, b]$, the following statements are equivalent:*
(i) G is an extended complete Chebyshev subspace.
(ii)   There exists a basis $\{g_1, \ldots, g_n\}$ of G such that for all $m \in \{1, \ldots, n\}$ and all $t \in [a, b]$,

$$W(g_1, \ldots, g_m)(t) > 0. \tag{1.17}$$

Proof. The implication (i) $\Rightarrow$ (ii) follows directly from the definition of extended complete Chebyshev subspaces.
We prove the implication (ii) $\Rightarrow$ (i) by induction on n. The result is obvious for $n = 1$. Suppose that the result is true for n. We set for all $i \in \{1, \ldots, n-1\}$,
$\tilde{g}_i = \Big(\dfrac{g_{i+1}}{g_1}\Big)'$. By using the well-known Leibniz rule on differentiation

$$\left(\frac{f_1}{f_2}\right)^{(r)} = \sum_{i=0}^{r} \binom{r}{i} f_1^{(r-i)} \left(\frac{1}{f_2}\right)^{(i)}, \tag{1.18}$$

it is easy to verify that for all $m \in \{2, \ldots, n\}$,

$$W(g_1, \ldots, g_m) = g_1^m \, W(\tilde{g}_1, \ldots, \tilde{g}_{m-1}). \tag{1.19}$$

In fact, given a point $t \in [a, b]$ we factor $g_1(t)$ out of each row of $W(g_1, \ldots, g_m)(t)$. Then starting with the last row we can add to each row a suitable linear combination of the preceding row to obtain by using (1.18)

$$W(g_1, \ldots, g_m)(t) = g_1^m \, W(1, \frac{g_2}{g_1}, \ldots, \frac{g_m}{g_1})(t). \tag{1.20}$$

Expanding $W(1, \frac{g_2}{g_1}, \ldots, \frac{g_m}{g_1})(t)$ along the first column we obtain (1.19).

We now continue as follows. Let an index $m \in \{2, \ldots, n\}$ and points $t_1 \leq \cdots \leq t_m$ in $[a, b]$ be given. Then there exist points $u_1 < \cdots < u_p$ such that

$$\{u_1, \ldots, u_p\} = \{t_1, \ldots, t_m\},$$

where the point u_j appears exactly r_j times in the set $\{t_1, \ldots, t_m\}$, $j = 1, \ldots, p$. We now consider the determinant

$$D\left(\begin{array}{c} g_1, \ldots, g_m \\ t_1, \ldots, t_m \end{array} \right).$$

Analogously as above for each $i \in \{1, \ldots, p\}$ we factor $g_1(u_j) > 0$ out of the r_j rows which contain u_j and then transform them into the first r_j rows of

$$W(1, \frac{g_2}{g_1}, \ldots, \frac{g_m}{g_1})(u_j).$$

Then we only consider those rows which have 1 in its first place. Starting with the last row we subtract from each such row the preceding row of the same type. Then applying the mean value theorem to these rows, it follows that there exist points $v_j \in (u_j, u_{j+1})$ such that (up to some factor $u_{j+1} - u_j > 0$) we obtain a new row

$$(0, \left(\frac{g_2}{g_1}\right)' (v_j), \ldots, \left(\frac{g_m}{g_1}\right)' (v_j)), \qquad j = 1, \ldots, p - 1.$$

There exist points $\tilde{t}_1 \leq \cdots \leq \tilde{t}_{m-1}$ such that

$$\{\tilde{t}_1, \ldots, \tilde{t}_{m-1}\} = \{u_1, v_1, \ldots, u_{p-1}, v_{p-1}, u_p\}$$

and the point u_j appears exactly $r_j - 1$ times in the set $\{\tilde{t}_1, \ldots, \tilde{t}_{m-1}\}$, $j = 1, \ldots, p$. The above arguments show that

$$\text{sgn } D\left(\begin{array}{c} g_1, \ldots, g_m \\ t_1, \ldots, t_m \end{array} \right) = \text{sgn } D\left(\begin{array}{c} \tilde{g}_1, \ldots, \tilde{g}_{m-1} \\ \tilde{t}_1, \ldots, \tilde{t}_{m-1} \end{array} \right). \tag{1.21}$$

Finally, it follows from (ii) and (1.19) that for all $m \in \{2, \ldots, n\}$ and all $t \in [a, b]$,

$$W(\tilde{g}_1, \ldots, \tilde{g}_{m-1})(t) > 0.$$

Therefore, by the induction hypothesis, for all $m \in \{2, \ldots, n\}$ and all points $\tilde{t}_1 \leq \cdots \leq \tilde{t}_{m-1}$,

$$D \left(\begin{array}{c} \tilde{g}_1, \ldots, \tilde{g}_{m-1} \\ \tilde{t}_1, \ldots, \tilde{t}_{m-1} \end{array} \right) > 0.$$

Then using (1.21) we get that for all $m \in \{1, \ldots, n\}$ and all points $t_1 \leq \cdots \leq t_m$,

$$D \left(\begin{array}{c} g_1, \ldots, g_m \\ t_1, \ldots, t_m \end{array} \right) > 0.$$

This proves Theorem 1.11.

We now give a further characterization of extended complete Chebyshev spaces which can be used to construct spaces of this type by integrating strictly positive functions (see Karlin & Studden [1966]).

Theorem 1.12. *For an n–dimensional subspace G of $C^{n-1}[a, b]$, the following statements are equivalent:*
(i) G is an extended complete Chebyshev subspace.
(ii) There exists a basis $\{g_1, \ldots, g_n\}$ of G of the following form:

$$
\begin{aligned}
g_1(t) &= w_1(t) \\
g_2(t) &= w_1(t) \int_a^t w_2(y_2)\, dy_2 \\
&\ \vdots \\
g_n(t) &= w_1(t) \int_a^t w_2(y_2) \int_a^{y_2} w_3(y_3) \ldots \int_a^{y_{n-1}} w_n(y_n)\, dy_n \ldots dy_3\, dy_2,
\end{aligned}
\tag{1.22}
$$

where $w_i \in C^{n-i}[a, b]$ are strictly positive functions, $i = 1, \ldots, n$.

Proof. (ii) $\Rightarrow$ (i). Suppose that there exists a basis $\{g_1, \ldots, g_n\}$ of G as in (ii). We show that for all $m \in \{1, \ldots, n\}$,

$$W(g_1, \ldots, g_n) = w_1^m w_2^{m-1} \cdots w_m > 0. \tag{1.23}$$

Then it follows from Theorem 1.11 that (i) holds. We prove (1.23) by induction on m. Obviously (1.23) holds for $m = 1$. Suppose that (1.23) holds for $m - 1$. We set for all $i \in \{1, \ldots, n-1\}$, $\tilde{g}_i = \left(\dfrac{g_{i+1}}{g_1} \right)'$. Using (ii) we have

$$
\begin{aligned}
\tilde{g}_1(t) &= w_2(t) \\
\tilde{g}_2(t) &= w_2(t) \int_a^t w_3(y_3)\, dy_3 \\
&\ \vdots \\
\tilde{g}_{n-1}(t) &= w_2(t) \int_a^t w_3(y_3) \int_a^{y_3} w_4(y_4) \ldots \int_a^{y_{n-1}} w_n(y_n)\, dy_n \ldots dy_3.
\end{aligned}
$$

As shown in the proof of Theorem 1.11,

$$W(g_1, \ldots, g_m) = w_1^m W(\tilde{g}_1, \ldots, \tilde{g}_{m-1}).$$

By induction hypothesis we have

$$W(\tilde{g}_1, \ldots, \tilde{g}_{m-1}) = w_2^{m-1} w_3^{m-2} \cdots w_m.$$

This implies that $W(g_1, \ldots, g_m) = w_1^m w_2^{m-1} \ldots w_m > 0$ and shows (ii) $\Rightarrow$ (i).

(i) $\Rightarrow$ (ii). Suppose that (i) holds. Then by Theorem 1.11 there exists a basis $\{h_1, \ldots, h_n\}$ of G such that for all $m \in \{1, \ldots, n\}$,

$$W(h_1, \ldots, h_m) > 0. \tag{1.24}$$

It is easy to verify that there exists a basis $\{g_1, \ldots, g_n\}$ of G such that

$$W(g_1, \ldots, g_m) > 0, \quad m = 1, \ldots, n,$$

and   (1.25)

$$g_j^{(i)}(a) = 0, \qquad i = 0, \ldots, j-2; \quad j = 2, \ldots, n.$$

In fact, by setting $g_1 = h_1$ and $g_2 = h_2 - \frac{h_2(a)}{g_1(a)} g_1$, we have $g_2(a) = 0$. Suppose now that the functions $g_1, \ldots, g_{m-1}$ are constructed. Then it is easy to see that there exist real numbers $a_1, \ldots, a_{m-1}$ such that

$$g_m = h_m - \sum_{i=1}^{m-1} a_i g_i$$

has the desired property. Moreover, for all $m \in \{1, \ldots, n\}$,

$$W(g_1, \ldots, g_m) = W(h_1, \ldots, h_m).$$

We now show by induction on n that (1.25) $\Rightarrow$ (ii). This implication is obvious for $n = 1$. Suppose that this implication holds for $n - 1$. Moreover, suppose that (1.25) holds for n. Since

$$W(g_1, \ldots, g_m) = g_1^m \, W(\tilde{g}_1, \ldots, \tilde{g}_{m-1}),$$

we have

$$W(\tilde{g}_1, \ldots, \tilde{g}_{m-1}) > 0 \quad , \qquad m = 2, \ldots, n.$$

Moreover, it follows from (1.25) that

$$\tilde{g}_j^{(i)}(a) = 0, \qquad i = 0, \ldots, j-2; \quad j = 2, \ldots, n-1.$$

Therefore, by the induction hypothesis there exist strictly positive functions $w_i \in C^{n-i}[a, b]$, $i = 2, \ldots, n$, such that

$$\tilde{g}_1(t) = w_2(t)$$

$$\tilde{g}_2(t) = w_2(t) \int_a^t w_3(y_3) \, dy_3$$

$$\vdots$$

$$\tilde{g}_{n-1}(t) = w_2(t) \int_a^t w_3(y_3) \int_a^{y_3} w_4(y_4) \ldots \int_a^{y_{n-1}} w_n(y_n) \, dy_n \ldots dy_3.$$

We set $w_1 = g_1$. Then we have

$$\int_a^t \left(\frac{g_i}{g_1}\right)' (y_2)\, dy_2 = \int_a^t \tilde{g}_i(y_2)\, dy_2, \qquad i = 2, \dots, n,$$

and therefore,

$$g_i(t) = w_1(t) \int_a^t \tilde{g}_i(y_2)\, dy_2 + \frac{g_j(a)}{w_1(a)}\, w_1(t), \qquad i = 2, \dots, n.$$

It follows from (1.25) that

$$g_i(t) = w_1(t) \int_a^t \tilde{g}_i(y_2)\, dy_2, \qquad i = 2, \dots, n,$$

and therefore, the functions $g_1, \dots, g_n$ satisfy (ii). This shows that (i) $\Rightarrow$ (ii) and proves Theorem 1.12.

Example 1.13. (i) The prototype of an extended complete Chebyshev subspace of $C^n[a, b]$ is P_n, the subspace of polynomials of degree n. We set $w_1(t) = 1$ and $w_i(t) = i - 1$, $i = 2, \dots, n + 1$. Then the functions $g_1, \dots, g_{n+1}$ as defined in Theorem 1.12 are

$$g_i(t) = (t - a)^{i-1}, \qquad i = 1, \dots, n + 1.$$

Since these functions form a basis of P_n, it follows from Theorem 1.12 that P_n is an extended complete Chebyshev subspace of $C^n[a, b]$.

(ii) Let real numbers $\alpha_1 < \cdots < \alpha_n$ be given. We call

$$E_n = \left\{ f : [a, b] \to \mathbf{R} : f(t) = \sum_{i=1}^n a_i\, e^{\alpha_i t},\ a_1, \dots, a_n \in \mathbf{R} \right\} \qquad (1.26)$$

the subspace of *exponentials.*

We set $w_1(t) = e^{\alpha_1 t}$ and $w_i(t) = e^{(\alpha_i - \alpha_{i-1})t}$, $i = 2, \dots, n$. Then the functions $g_1, \dots, g_n$ as defined in Theorem 1.12 can be written as

$$g_i(t) = \sum_{j=1}^i a_{ij}\, e^{\alpha_j t}, \qquad i = 1, \dots, n,$$

where $a_{ii} \neq 0$, $i = 1, \dots, n$. Since these functions form a basis of E_n, it follows from Theorem 1.12 that E_n is an extended complete Chebyshev subspace of $C^n[a, b]$.

1.4. Further Properties of Chebyshev Spaces

So far we have investigated the role of various types of Chebyshev spaces in the interpolation of functions and the construction of such spaces. We now derive further properties of Chebyshev spaces which will be needed in subsequent sections.

Let functions $g_1, \ldots, g_n$ in $C[a, b]$ and arbitrary points $t_1, \ldots, t_n$ in $[a, b]$ be given. Then we set

$$\overline{D} \left(\begin{array}{c} g_1, \ldots, g_n \\ t_1, \ldots, t_n \end{array} \right) = \left| \begin{array}{ccc} g_1(t_1) & \cdots & g_n(t_1) \\ \vdots & & \vdots \\ g_1(t_n) & \cdots & g_n(t_n) \end{array} \right|. \tag{1.27}$$

This notation is used to avoid confusions with the definition of the determinant

$$D \left(\begin{array}{c} g_1, \ldots, g_n \\ t_1, \ldots, t_n \end{array} \right)$$

for coinciding points (see (1.10)).

Chebyshev spaces can be characterized by zero properties and by the existence of certain functions from these spaces.

Theorem 1.14. *For an n–dimensional subspace G of $C[a, b]$, the following statements are equivalent:*

(i)   *G is a Chebyshev subspace.*

(ii) *Every nontrivial function $g \in G$ has at most $n - 1$ distinct zeros.*

(iii) *For all points $a = t_0 \le t_1 < \cdots < t_{n-1} \le t_n = b$, there exists a function $g \in G$ such that*

$$\left. \begin{array}{l} g(t) = 0, \quad t \in \{t_1, \ldots, t_{n-1}\}, \\[4pt] g(t) \ne 0, \quad t \notin \{t_1, \ldots, t_{n-1}\}, \\[4pt] (-1)^i g(t) > 0, \quad t \in (t_{i-1}, t_i), \quad i = 1, \ldots, n. \end{array} \right\} \tag{1.28}$$

Proof. By the proof of Theorem 1.4, statement (i) is equivalent to the following condition:

(iv) There exists a basis $\{g_1, \ldots, g_n\}$ of G such that

$$D \left(\begin{array}{c} g_1, \ldots, g_n \\ t_1, \ldots, t_n \end{array} \right) \ne 0$$

for all $t_1 < \cdots < t_n$ in $[a, b]$. Moreover, it is easy to verify that (iv) $\Leftrightarrow$ (ii).

(i) $\Rightarrow$ (iii). Suppose that (i) holds, i.e. there exists a basis $\{g_1, \ldots, g_n\}$ of G such that

$$D \left(\begin{array}{c} g_1, \ldots, g_n \\ u_1, \ldots, u_n \end{array} \right) > 0$$

for all $u_1 < \cdots < u_n$ in $[a, b]$. Moreover, let points $a = t_0 \le t_1 < \cdots < t_{n-1} \le t_n = b$ be given. We define $g \in G$ by

$$g(t) = \overline{D} \left(\begin{array}{c} g_1, \ldots, g_{n-1}, g_n \\ t_1, \ldots, t_{n-1}, t \end{array} \right)$$

for all $t \in [a, b]$. Then by (i) the function g or $-g$ has the desired property.

(iii) $\Rightarrow$ (iv). Suppose that (iii) holds. Let $\{g_1, \ldots, g_n\}$ be a basis of G and let points $a \leq u_1 < \cdots < u_n \leq b$ be given. It follows from (iii) that for all $j \in \{1, \ldots, n\}$ there exists a $\tilde{g}_j \in G$ such that $\tilde{g}_j(u_i) = 0$, $i = 1, \ldots, n$, $i \neq j$, and $\tilde{g}_j(u_j) \neq 0$. This implies that

$$D \left(\begin{array}{c} \tilde{g}_1, \ldots, \tilde{g}_n \\ u_1, \ldots, u_n \end{array} \right) \neq 0.$$

Since $\{\tilde{g}_1, \ldots, \tilde{g}_n\}$ is a basis of G, there exists a matix M such that $\det M \neq 0$ and

$$\left(\begin{array}{c} g_1(t) \\ \vdots \\ g_n(t) \end{array} \right) = M \left(\begin{array}{c} \tilde{g}_1(t) \\ \vdots \\ \tilde{g}_n(t) \end{array} \right)$$

for all $t \in [a, b]$. Then we have

$$D \left(\begin{array}{c} g_1, \ldots, g_n \\ u_1, \ldots, u_n \end{array} \right) = (\det M) D \left(\begin{array}{c} \tilde{g}_1, \ldots, \tilde{g}_n \\ u_1, \ldots, u_n \end{array} \right) \neq 0.$$

This proves Theorem 1.14.

By using Theorem 1.14, we obtain the following result on the existence of certain functions in Chebyshev spaces.

Corollary 1.15. *Let G be a n–dimensional Chebyshev subspace of $C[a, b]$. Then for all integers $m \in \{1, \ldots, n\}$ and all points $a = t_0 < t_1 < \cdots < t_{m-1} < t_m = b$, there exists a nontrivial function $g \in G$ such that*

$$(-1)^i g(t) \geq 0, \quad t \in [t_{i-1}, t_i], \quad i = 1, \ldots, m. \tag{1.29}$$

Proof. Let points $a = t_0 < t_1 < \cdots < t_{m-1} < t_m = b$ be given. If $m = n$, then the claim follows directly from Theorem 1.14. Therefore, let $m < n$. For all $i \in \{m, \ldots, n-1\}$ we choose a sequence $(t_{i,p})$ such that $\lim_{p \to \infty} t_{i,p} = b$ and $t_{m-1} < t_{m,p} < \cdots < t_{n-1,p} < t_{n,p} = b$ for all p. Moreover, for all p and all $i \in \{0, \ldots, m-1\}$ we set $t_{i,p} = t_i$. It follows from Theorem 1.14 that for all p there exists a nontrivial function $g_p \in G$ such that

$$(-1)^i g_p(t) \geq 0, \quad t \in [t_{i-1,p}, t_{i,p}], \quad i = 1, \ldots, n. \tag{1.30}$$

Multiplying every function with $1/\|g_p\|_\infty$ we may assume that $\|g_p\|_\infty = 1$. (The uniform norm will be defined in (2.18).) Since G is a finite dimensional subspace, by choosing a subsequence we may assume that $\lim_{p \to \infty} g_p = g$, where $g \in G$ and $\|g\|_\infty = 1$. Taking limits it follows that

$$(-1)^i g(t) \geq 0, \quad t \in [t_{i-1}, t_i], \quad i = 1, \ldots, m.$$

This proves Corollary 1.15.

Extended Chebyshev spaces can also be characterized by zero properties, provided we consider multiple zeros.

Definition 1.16. A function $f \in C^r[a, b]$ is said to have a *zero of multiplicity* (at least) r at a point $t \in [a, b]$, if

$$f^{(i)}(t) = 0, \quad i = 0, \ldots, r - 1.$$

Theorem 1.17. *For an n-dimensional subspace G of $C^{n-1}[a, b]$, the following statements are equivalent:*
(i) G is an extended Chebyshev subspace.
(ii) Every nontrivial function $g \in G$ has at most $n - 1$ zeros (counting multiplicities)

Proof. (i) $\Rightarrow$ (ii). Suppose that (ii) fails, i.e. there exists a nontrivial function $g_1 \in G$ with zeros $u_1 < \cdots < u_p$ of multiplicities $r_1, \ldots, r_p$ such that $\sum_{i=1}^{p} r_i = n$. Then the Hermite interpolation problem

$$g_1^{(i)}(u_j) = 0, \quad i = 0, \ldots, r_j - 1, \quad j = 1, \ldots, p,$$

has two solutions g_1 and $g_2 = 0$ from G. It follows from Theorem 1.8 that G is not an extended Chebyshev space.

(ii) $\Rightarrow$ (i). Suppose that G is not an extended Chebyshev space. Then by Theorem 1.8 there exist points $t_1 \leq \cdots \leq t_n$ in $[a, b]$ such that the Hermite interpolation problem (1.7) has at least two distinct solutions $g_1, g_2 \in G$. This implies that the function $g_1 - g_2 \in G$ has at least n zeros at $t_1, \ldots, t_n$. This proves Theorem 1.17.

The next result shows that as in Theorem 1.14, Chebyshev spaces can also be characterized by zero properties, where we consider double zeros.

Definition 1.18. Let a function $f \in C[a, b]$ and a point $t \in (a, b)$ be given with $f(t) = 0$. We say that f *changes sign at t* if for all sufficiently small $\varepsilon > 0$,

$$f(t - \varepsilon) f(t + \varepsilon) < 0.$$

The point t is called a *double zero* of f, if f does not change sign at t.

Theorem 1.19. *For an n-dimensional subspace G of $C[a, b]$, the following statements are equivalent:*
(i) G is a Chebyshev subspace.
(ii) Every nontrivial function $g \in G$ has at most $n - 1$ zeros (counting double zeros twice).

Proof. (ii) $\Rightarrow$ (i). This implication follows from Theorem 1.14.

(i) $\Rightarrow$ (ii). Suppose that (i) holds, but (ii) fails, i.e. there exists a nontrivial $g \in G$ which has at least n zeros (counting double zeros twice). We denote the zeros of g by $a \leq u_1 \cdots < u_p \leq b$. Let $\varepsilon > 0$ be given. We add to this set the point $u_j - \varepsilon$, if u_j is the first double zero of f, and the point $u_i + \varepsilon$, if u_i is an arbitrary double zero of f. The resulting set is denoted by $a \leq t_1 \cdots < t_q \leq b$. Then it follows that $q \geq n+1$ and that there exists a sign $\sigma \in \{-1, 1\}$ such that

$$\sigma (-1)^{i-1} g(t_i) \geq 0, \qquad i = 1, \ldots, q.$$

Moreover, we have

$$\begin{aligned}
0 &= D \left(\begin{array}{c} g, g_1, \ldots, g_n \\ t_1, \ldots \ldots, t_{n+1} \end{array} \right) \\
&= \sum_{i=1}^{n+1} (-1)^{i-1} g(t_i) D \left(\begin{array}{ccc} g_1, & \cdots\cdots & , g_n \\ t_1, \ldots, t_{i-1}, t_{i+1}, \ldots, t_{n+1} \end{array} \right),
\end{aligned}$$

where $\{g_1, \ldots, g_n\}$ is a basis of G. Since G is a Chebyshev space, it follows that $g(t_i) = 0, i = 1, \ldots, n+1$, which contradicts Theorem 1.14. This proves Theorem 1.19.

1.5. Variation Diminishing Property of Order Complete Chebyshev Spaces

In this section we show that order complete Chebyshev spaces have a basis which is variation diminishing.

Definition 1.20. An n–dimensional subspace G of $C[a, b]$ is called an *order complete Chebyshev subspace* or a *Descartes subspace* if there exists a basis $\{g_1, \ldots, g_n\}$ of G such that for all $m \in \{1, \ldots, n\}$ and all $i_1 < \cdots < i_m$ in $\{1, \ldots, n\}$ the subspace $\mathrm{span}\{g_{i_1}, \ldots, g_{i_m}\}$ is an m–dimensional Chebyshev subspace.

In the next result we use the following convention. The number of *sign changes* in a finite sequence of real numbers $a_1, \ldots, a_n$ is the number of indices $i \in \{1, \ldots, n-1\}$ for which $a_i a_{i+1} < 0$.

An important property of order complete Chebyshev subspaces G is that there exists a basis of G which is *variation diminishing* in the following sense (Gantmacher & Krein [1950]).

Theorem 1.21. *Let G be an n–dimensional order complete Chebyshev subspace of $C[a, b]$. Then there exists a basis $\{g_1, \ldots, g_n\}$ of G with the following property: For every nontrivial function $g = \sum_{i=1}^{n} a_i g_i$ in G, the number of distinct zeros of g is less than or equal to the number of sign changes in the sequence $a_1, \ldots, a_n$.*

Proof. Let the number of sign changes in $a_1, \ldots, a_n$ be $p - 1$. Then there exist indices $N_2, \ldots, N_p$ and $N_1 = 0, N_{p+1} = n$ such that the sequence

$$a_1, \ldots, a_{N_2}, a_{N_2+1}, \ldots, a_{N_3}, \ldots, a_{N_p+1}, \ldots, a_n$$

has the property that for every $j \in \{1, \ldots, p\}$ the coefficients

$$a_{N_j+1}, \ldots, a_{N_{j+1}}$$

have the same sign and at least one of them is nonzero. We now set for every $j \in \{1, \ldots, p\}$,

$$\tilde{g}_j = \sum_{i=N_j+1}^{N_{j+1}} |a_i|\, g_i.$$

Then $\tilde{G} = \text{span}\{\tilde{g}_1, \ldots, \tilde{g}_p\}$ is a Chebyshev subspace. In fact, for all $t_1 < \cdots < t_p$ in $[a, b]$ we have

$$D\left(\begin{matrix} \tilde{g}_1, \ldots, \tilde{g}_p \\ t_1, \ldots, t_p \end{matrix} \right) = \sum_{i_1=N_1+1}^{N_2} \cdots \sum_{i_p=N_p+1}^{N_{p+1}} |a_{i_1}| \cdots |a_{i_p}|\, D\left(\begin{matrix} g_{i_1}, \ldots, g_{i_p} \\ t_1, \ldots, t_p \end{matrix} \right).$$

This determinant is positive since G is an order complete Chebyshev subspace. Let σ be the sign of the coefficients $a_1, \ldots, a_{N_2}$. Then by the choice of the indices $N_2, \ldots, N_p$ we have

$$g = \sum_{i=1}^{n} a_i g_i = \sum_{j=1}^{p} \sigma(-1)^{j-1} \tilde{g}_j \in \tilde{G}.$$

Since $\tilde{G}$ is a p–dimensional Chebyshev subspace, it follows from Theorem 1.14 that g has at most $p - 1$ distinct zeros. This proves Theorem 1.21.

Example 1.22. (i) Let $[a, b]$ be an arbitrary closed subinterval of $(-\infty, \infty)$. Then the space E_n of exponentials is an order complete Chebyshev subspace of $C[a, b]$ (compare Example 1.13).

(ii) Let $[a, b]$ be a closed subinterval of $(0, \infty)$. Then the space P_n of polynomials of degree n is an order complete Chebyshev subspace of $C[a, b]$. In fact,

$$P_n = \text{span}\{e^{0u}, \ldots, e^{nu}\},$$

where $u = \log t$, and $\text{span}\{e^{0u}, \ldots, e^{nu}\}$ is an order complete Chebyshev subspace of $C[\log a, \log b]$.

Finally, we note that there is a vast literature on Chebyshev spaces. Further details concerning spaces of this type can be found e.g. in the books of Karlin & Studden [1966], Karlin [1968], Zielke [1979] and Schumaker [1981].

2. Interpolation by Polynomials and Divided Differences

In the previous section we have seen that the Hermite interpolation problem for polynomials has a unique solution. In this section it is shown that such polynomials can be computed by using divided differences. Moreover, we give results on the error in polynomial interpolation. In particular, Lagrange interpolation at the so–called Chebyshev points — in contrast to equidistant points — is nearly optimal.

2.1. Divided Differences

We introduce divided differences and show that they can be computed recursively. Moreover, a result on divided differences of products of functions and a continuity result for divided differences are given.

We begin with the notion of divided differences. In most books on Numerical Analysis divided differences for distinct points are defined recursively (see (2.2)). The following definition for coalescing points using quotients of determinants was given by Popoviciu [1959].

Definition 2.1. Let a sufficiently often differentiable function $f \in C[a, b]$ and points $t_1 \leq \cdots \leq t_{n+1}$ in $[a, b]$ be given. Moreover, let $p(t) = \sum_{i=0}^{n} a_i t^i$ be the unique solution of the corresponding Hermite interploation problem from P_n. The *divided difference* of order n of f with respect to the points $t_1, \ldots, t_{n+1}$ is defined by

$$f[t_1, \ldots, t_{n+1}] = a_n.$$

It is clear that

$$f[t_1, \ldots, t_{n+1}] = \frac{D\left(\begin{array}{c} 1, t, \ldots, t^{n-1}, f \\ t_1, \quad \ldots \ldots, \quad t_{n+1} \end{array} \right)}{D\left(\begin{array}{c} 1, t, \ldots, t^n \\ t_1, \ldots, t_{n+1} \end{array} \right)}. \tag{2.1}$$

The first result shows that the divided differences can be computed recursively.

Theorem 2.2. *Let a sufficiently often differentiable function $f \in C[a, b]$ and points $t_1 \leq \cdots \leq t_{n+1}$ in $[a, b]$ be given. If $t_1 < t_{n+1}$, then*

$$f[t_1, \ldots, t_{n+1}] = \frac{f[t_2, \ldots, t_{n+1}] - f[t_1, \ldots, t_n]}{t_{n+1} - t_1} \tag{2.2}$$

and if $t_1 = t_{n+1}$, then

$$f[t_1, \ldots, t_{n+1}] = \frac{1}{n!} f^{(n)}(t_1). \tag{2.3}$$

Proof. It follows from definition of the divided differences that

$$f[t_1, \ldots, t_{n+1}] = 0 \qquad \text{for } f(t) = t^i, \quad i = 0, \ldots, n-1, \tag{2.4}$$

and that

$$f[t_1, \ldots, t_{n+1}] = 1 \qquad \text{for } f(t) = t^n. \tag{2.5}$$

Moreover,

$$w[t_1, \ldots, t_{n+1}] = 0 \qquad \text{for } w(t) = (t - t_1) \cdots (t - t_{n+1}). \tag{2.6}$$

This implies that

$$0 = w[t_1, \ldots, t_{n+1}] = t^{n+1} [t_1, \ldots, t_{n+1}] - \sum_{i=1}^{n+1} t_i$$

and therefore,

$$f[t_1, \ldots, t_{n+1}] = \sum_{i=1}^{n+1} t_i \qquad \text{for } f(t) = t^{n+1}. \tag{2.7}$$

Now, let $u_1 < \cdots < u_p$ be points such that

$$\{u_1, \ldots, u_p\} = \{t_1, \ldots, t_{n+1}\}$$

and the point u_j appears exactly r_j-times in the set $\{t_1, \ldots, t_{n+1}\}$, $j = 1, \ldots, p$. We first consider the case when $t_1 < t_{n+1}$. In this case we define the linear functional

$$L(f) = f[t_1, \ldots, t_{n+1}] - \frac{f[t_2, \ldots, t_{n+1}] - f[t_1, \ldots, t_n]}{t_{n+1} - t_1}.$$

It follows from (2.1) that

$$f[t_1, \ldots, t_{n+1}] = \sum_{i=1}^{p} \sum_{j=1}^{r_j} a_{ij} \, f^{(j-1)}(u_i)$$

and therefore that

$$L(f) = \sum_{i=1}^{p} \sum_{j=1}^{r_j} b_{ij} \, f^{(j-1)}(u_i). \tag{2.8}$$

By (2.4) we get that

$$L(t^i) = 0, \qquad i = 0, \ldots, n-1.$$

Moreover, by (2.6) we get that

$$L(t^{n+1}) = 1 - \frac{(t_2 + \cdots + t_{n+1}) - (t_1 + \cdots + t_n)}{t_{n+1} - t_1} = 0.$$

We consider the system of linear equations for the real numbers b_{ij},

$$L(t^i) = 0, \qquad i = 0, \ldots, n.$$

Since by (2.8) the corresponding determinant

$$D \begin{pmatrix} 1, t, \ldots, t^n \\ t_1, \ldots, t_{n+1} \end{pmatrix} > 0,$$

all real numbers b_{ij} are zero and therefore $L = 0$. This shows (2.2).

We now consider the case when $t_1 = t_{n+1}$. By (2.1) we get

$$f[t_1, \ldots, t_{n+1}] = \frac{f^{(n)}(t_1) \prod_{j=1}^{n-1} j!}{\prod_{j=1}^{n} j!} = \frac{1}{n!} f^{(n)}(t_1).$$

This shows (2.3) and proves Theorem 2.2.

Theorem 2.2 shows that the divided differences can be easily computed according to the following scheme:

$$f[t_1]$$
$$\qquad f[t_1, t_2]$$
$$f[t_2]$$
$$\qquad\qquad\qquad\qquad f[t_1, \ldots, t_n]$$
$$\vdots \qquad\quad \vdots \qquad \cdots \qquad\qquad\qquad\qquad f[t_1, \ldots, t_{n+1}]$$
$$\qquad\qquad\qquad\qquad f[t_2, \ldots, t_{n+1}]$$
$$f[t_n]$$
$$\qquad f[t_n, t_{n+1}]$$
$$f[t_{n+1}]$$

Remark 2.3. It follows from the proof of Theorem 2.2 that

$$f[t_1, \ldots, t_{n+1}] = \begin{cases} 0, & \text{if } f(t) = t^i, \quad i = 1, \ldots, n-1 \\ 1, & \text{if } f(t) = t^n \\ \sum_{i=1}^{n+1} t_i, & \text{if } f(t) = t^{n+1} \end{cases} \tag{2.9}$$

We now give a formula for computing divided differences of the product of two functions.

Theorem 2.4. *Let sufficiently often differentiable functions $f_1, f_2 \in C[a, b]$ and points $t_1 \leq \cdots \leq t_{n+1}$ in $[a, b]$ be given. Then*

$$(f_1 \, f_2)[t_1, \ldots, t_{n+1}] = \sum_{i=1}^{n+1} f_1[t_1, \ldots, t_i] \, f_2[t_i, \ldots, t_{n+1}]. \tag{2.10}$$

Proof. We first consider the case when $t_1 = t_{n+1}$. It follows from Theorem 2.2 and the well-known Leibniz rule that

$$\begin{aligned} (f_1 \, f_2)[t_1, \ldots, t_{n+1}] &= \frac{1}{n!} (f_1 \, f_2)^{(n)}(t_1) \\ &= \frac{1}{n!} \sum_{i=1}^{n} \binom{n}{i} f_1^{(i)}(t_1) \, f_2^{(n-i)}(t_1) \\ &= \sum_{i=1}^{n+1} \frac{1}{(i-1)!} f_1^{(i-1)}(t_1) \frac{1}{(n-i+1)!} f_2^{(n-i+1)}(t_1) \\ &= \sum_{i=1}^{n+1} f_1[t_1, \ldots, t_i] \, f_2[t_i, \ldots, t_{n+1}]. \end{aligned}$$

Now, we consider the case when $t_1 < t_{n+1}$. In this case we proceed by induction on n. For $n = 0$ we have $(f_1 f_2)[t_1] = (f_1 f_2)(t_1) = f_1(t_1) f_2(t_1) = f_1[t_1] f_2[t_1]$. We assume that (2.10) holds for $n-1$. Then it follows from the induction hypothesis and Theorem 2.2 that

$$(f_1 f_2)[t_1, \ldots, t_{n+1}]$$

$$= \frac{(f_1 f_2)[t_2, \ldots, t_{n+1}] - (f_1 f_2)[t_1, \ldots, t_n]}{t_{n+1} - t_1}$$

$$= \frac{\sum_{i=2}^{n+1} f_1[t_2, \ldots, t_i] f_2[t_i, \ldots, t_{n+1}] - \sum_{i=1}^{n} f_1[t_1, \ldots, t_i] f_2[t_i, \ldots, t_n]}{t_{n+1} - t_1}.$$

We now consider the last numerator.

$$\sum_{i=2}^{n+1} f_1[t_2, \ldots, t_i] f_2[t_i, \ldots, t_{n+1}] - \sum_{i=1}^{n} f_1[t_1, \ldots, t_i] f_2[t_i, \ldots, t_n]$$

$$= \sum_{i=2}^{n+1} ((t_i - t_1) f_1[t_1, \ldots, t_i] + f_1[t_1, \ldots, t_{i-1}]) f_2[t_i, \ldots, t_{n+1}] -$$

$$- \sum_{i=2}^{n+1} f_1[t_1, \ldots, t_{i-1}] (f_2[t_i, \ldots, t_{n+1}] - (t_{n+1} - t_{i-1}) f_2[t_{i-1}, \ldots, t_{n+1}])$$

$$= \sum_{i=2}^{n+1} (t_i - t_1) f_1[t_1, \ldots, t_i] f_2[t_i, \ldots, t_{n+1}]$$

$$+ \sum_{i=1}^{n} (t_{n+1} - t_i) f_1[t_1, \ldots, t_i] f_2[t_i, \ldots, t_{n+1}]$$

$$= (t_{n+1} - t_1) \sum_{i=1}^{n+1} f_1[t_1, \ldots, t_i] f_2[t_i, \ldots, t_n].$$

By combining the above equations we obtain (2.10). This proves Theorem 2.4.

We now show that the mapping $(t_1, \ldots, t_{n+1}) \rightarrow f[t_1, \ldots, t_{n+1}]$ is continuous.

Theorem 2.5. *Let a sufficiently often differentiable function $f \in C[a, b]$, points $t_1 \leq \cdots \leq t_{n+1}$ in $[a, b]$ and for all m, points $t_{1,m} \leq \cdots \leq t_{n+1,m}$ in $[a, b]$ be given. If*

$$\lim_{m \to \infty} t_{i,m} = t_i, \qquad i = 1, \ldots, n+1,$$

then

$$\lim_{m \to \infty} f[t_{1,m}, \ldots, t_{n+1,m}] = f[t_1, \ldots, t_{n+1}].$$

Proof. To simplify the notation, we set $g_i(t) = t^{i-1}$, $i = 1, \ldots, n+1$. Suppose that $t_1 \leq \cdots \leq t_j < t_{j+1} = \cdots = t_{j+p} < t_{j+p+1} \leq \cdots \leq t_{n+1}$. We only consider the case $t_{j+p} \leq t_{j+p,m}$ for all m and $\lim_{m \to \infty} t_{j+p,m} = t_{j+p}$. The other cases follow analogously. We argue as in the proof of Theorem 1.8. It follows from Taylor's theorem that for each m there exist points $y_m, y_{i,m} \in [t_{j+p}, t_{j+p,m}]$, $i = 1, \ldots, n$, such that

$$f[t_1, \ldots, t_{j+p-1}, t_{j+p,m}, t_{j+p+1}, \ldots, t_{n+1}] =$$

$$= \frac{\begin{vmatrix} \vdots & & \vdots & \vdots \\ g_1(t_{j+1}) & \cdots & g_n(t_{j+1}) & f(t_{j+1}) \\ \vdots & & \vdots & \vdots \\ g_1^{(p-2)}(t_{j+1}) & \cdots & g_n^{(p-2)}(t_{j+1}) & f^{(p-2)}(t_{j+1}) \\ g_1^{(p-1)}(y_{1,m}) & \cdots & g_n^{(p-1)}(y_{n,m}) & f^{(p-1)}(y_m) \\ \vdots & & \vdots & \vdots \end{vmatrix}}{\begin{vmatrix} \vdots & & \vdots \\ g_1(t_{j+1}) & \cdots & g_{n+1}(t_{j+1}) \\ \vdots & & \vdots \\ g_1^{(p-2)}(t_{j+1}) & \cdots & g_{n+1}^{(p-2)}(t_{j+1}) \\ g_1^{(p-1)}(y_{1,m}) & \cdots & g_{n+1}^{(p-1)}(y_{n,m}) \\ \vdots & & \vdots \end{vmatrix}}.$$

By taking limits we obtain

$$\lim_{m \to \infty} f[t_1, \ldots, t_{j+p-1}, t_{j+p,m}, t_{j+p+1}, \ldots, t_{n+1}] = f[t_1, \ldots, t_{n+1}].$$

This proves Theorem 2.5.

2.2. Newton Form of Interpolating Polynomials

It is shown that polynomials which solve the Hermite interpolation problem can be computed in the Newton form by using divided differences. Moreover, we derive a formula for the corresponding error. Finally, the Lagrange form of interpolating polynomials is discussed.

Since P_n is an extended Chebyshev space (see Example 1.13), it follows from Theorem 1.8 that the Hermite interpolation problem has a unique solution from P_n. The next result shows that such polynomials can be easily computed by using divided differences.

Theorem 2.6. *Let a sufficiently often differentiable function $f \in C[a, b]$ and points $t_1 \leq \cdots \leq t_{n+1}$ in $[a, b]$ be given. Then the unique polynomial $p \in P_n$ which solves the corresponding Hermite interpolation problem can be written as*

$$p(t) = f[t_1] + f[t_1, t_2](t - t_1) + \cdots + f[t_1, \ldots, t_{n+1}](t - t_1) \cdots (t - t_n) \quad (2.11)$$

for all $t \in [a, b]$. This is the so-called Newton-form of p.

Proof. We first consider the case when $t_1 < \cdots < t_{n+1}$. For all $j \in \{0, \ldots, n\}$, let $p_j \in P_j$ be the polynomial which solves the Lagrange interpolation problem for $t_1 < \cdots < t_{j+1}$. We show that for all $j \in \{0, \ldots, n-1\}$,

$$p_{j+1}(t) = p_j(t) + f[t_1, \ldots, t_{j+2}](t - t_1) \cdots (t - t_{j+1}). \tag{2.12}$$

Let the polynomial on the right side of (2.12) be denoted by $\tilde{p}_{j+1}$. Then we obviously have

$$\tilde{p}_{j+1}(t_i) = f(t_i), \qquad i = 1, \ldots, j+1,$$

and therefore,

$$p_{j+1}(t_i) - \tilde{p}_{j+1}(t_i) = 0, \qquad i = 1, \ldots, j+1.$$

By definition of the divided differences,

$$p_{j+1} - \tilde{p}_{j+1} \in P_j$$

and therefore $p_{j+1} = \tilde{p}_{j+1}$. This shows (2.12). By applying (2.12) inductively, the claim of the theorem is proved for the case when $t_1 < \cdots < t_{n+1}$.

We now consider the general case when $t_1 \leq \cdots \leq t_{n+1}$. Let $q \in P_n$ be the polynomial which solves the corresponding Hermite interpolation problem. Then the polynomial p remains the same if we replace f by q in the definition of p. Therefore, we may assume that $f \in P_n$. For all $i \in \{1, \ldots, n+1\}$ we choose a sequence $(t_{i,m})$ such that

$$t_{1,m} < \cdots < t_{n+1,m}$$

and

$$\lim_{m \to \infty} t_{i,m} = t_i, \qquad i = 1, \ldots, n+1.$$

For all m, let $\tilde{q}_m \in P_n$ be the polynomial which solves the Lagrange interpolation problem for $t_{1,m} < \cdots < t_{n+1,m}$. Since $f \in P_n$, we have $\tilde{q}_m = f$ for all m. Therefore, by the claim of the theorem for distinct points we get for all m,

$$\begin{aligned}
f(t) = \tilde{q}_m(t) &= f[t_{1,m}] + f[t_{1,m}, t_{2,m}](t - t_{1,m}) + \cdots + \\
&\quad + f[t_{1,m}, \ldots, t_{n+1,m}](t - t_{1,m}) \cdots (t - t_{n,m}).
\end{aligned}$$

By taking limits, it follows from Theorem 2.5 that

$$f(t) = f[t_1] + f[t_1, t_2](t - t_1) + \cdots + f[t_1, \ldots, t_{n+1}](t - t_1) \cdots (t - t_n) = p(t).$$

Since $f = p$, the polynomial p obviously solves the Hermite interpolation problem for $t_1 \leq \cdots \leq t_{n+1}$. This proves Theorem 2.6.

Definition 2.7. Let a sufficiently often differentiable function $f \in C[a, b]$ and points $t_1, \ldots, t_{n+1}$ in $[a, b]$ be given which are not necessarily in order. In this case we define the divided difference $f[t_1, \ldots, t_{n+1}]$ as follows. Choose any re-ordering $\tilde{t}_1, \ldots, \tilde{t}_{n+1}$ of $t_1, \ldots, t_{n+1}$ such that $\tilde{t}_1 \leq \cdots \leq \tilde{t}_{n+1}$. Then we define $f[t_1, \ldots, t_{n+1}]$ to be $f[\tilde{t}_1, \ldots, \tilde{t}_{n+1}]$.

A close inspection of the proofs shows that the previous results in this section remain true for arbitrary points $t_1, \ldots, t_{n+1}$ in $[a, b]$, if we use the above definition.

The interpolating polynomial p to a given function f yields an approximation of f on $[a, b]$. A formula for computing the interpolation error $f - p$ is given in the next result.

Theorem 2.8. *Let the assumptions of Theorem 2.6 be satisfied. Then the interpolation error is given by*

$$f(t) - p(t) = f[t_1, \ldots, t_{n+1}, t](t - t_1) \cdots (t - t_{n+1}) \qquad (2.13)$$

for all $t \in [a, b]$.

Proof. Let a point $t \in [a, b]$ be given and let $q \in P_{n+1}$ be the polynomial which solves the Hermite interpolation problem for $t_1, \ldots, t_{n+1}, t$ (in the sense of Remark 1.6). Then it follows from Theorem 2.6 and Definition 2.7 that

$$f(t) = q(t) = p(t) + f[t_1, \ldots, t_{n+1}, t](t - t_1) \cdots (t - t_{n+1}).$$

This proves Theorem 2.8.

Now we consider further properties of the interpolation error. We first generalize property (2.3) of divided differences. To do this, we need the following well-known Rolle's Theorem which can be found in standard books on Analysis.

Theorem 2.9. *Let a function $f \in C^1[a, b]$ and points $y_1, y_2 \in [a, b]$ with $y_1 < y_2$ be given. If $f(y_1) = f(y_2) = 0$, then there exists a point $y_0 \in (y_1, y_2)$ such that $f'(y_0) = 0$.*

Theorem 2.10. *Let a function $f \in C^n[a, b]$ and points $t_1 \leq \cdots \leq t_{n+1}$ in $[a, b]$ be given. Then there exists a point $y \in [t_1, t_{n+1}]$ such that*

$$f[t_1, \ldots, t_{n+1}] = \frac{1}{n!} f^{(n)}(y). \qquad (2.14)$$

Proof. We first consider the case when $t_1 < \cdots < t_{n+1}$. Let $p \in P_n$ be the polynomial with

$$p(t_i) = f(t_i), \qquad i = 1, \ldots, n + 1.$$

Then by applying Theorem 2.9 to $f - p$ several times, there exists a point $y \in (t_1, t_{n+1})$ such that $f^{(n)}(y) - p^{(n)}(y) = 0$. It follows from (2.11) that $p^{(n)}(y) = n! f[t_1, \ldots, t_{n+1}]$ and therefore $f[t_1, \ldots, t_{n+1}] = \frac{1}{n!} f^{(n)}(y)$. We now consider the general case when $t_1 \leq \cdots \leq t_{n+1}$. For all $i \in \{1, \ldots, n+1\}$ we choose a sequence $t_{i,m}$ such that $t_{1,m} < \cdots < t_{n+1,m}$ for all m and $\lim_{m \to \infty} t_{i,m} = t_i, i = 1, \ldots, n+1$. Then it follows that for each m there exists a point $y_m \in (t_{1,m}, t_{n+1,m})$ such that

$$f[t_{1,m}, \ldots, t_{n+1,m}] = \frac{1}{n!} f^{(n)}(y_m).$$

By choosing a subsequence, we may assume that there exists a point $y \in [t_1, t_{n+1}]$ with $\lim_{m \to \infty} y_m = y$. Then by taking limits it follows from Theorem 2.5 that

$$f[t_1, \ldots, t_{n+1}] = \frac{1}{n!} f^{(n)}(y).$$

This proves Theorem 2.10.

The next result is an immediate consequence of Theorem 2.8 and Theorem 2.10.

Corollary 2.11. *Let the assumptions of Theorem 2.6 be satisfied. Then for each $t \in [a, b]$, there exists a point $y_t \in [\min\{t_1, t\}, \max\{t_{n+1}, t\}]$ such that*

$$f(t) - p(t) = \frac{1}{(n+1)!} f^{(n+1)}(y_t)(t - t_1) \cdots (t - t_{n+1}). \tag{2.15}$$

Finally, we describe a further representation of interpolating polynomials which is sometimes used in Constructive Analysis.

Definition 2.12. Let points $t_1 < \cdots < t_{n+1}$ in $[a, b]$ be given. For all $j \in \{1, \ldots, n+1\}$, we define the polynomial $l_j \in P_n$ by

$$l_j(t) = \prod_{\substack{i=1 \\ i \neq j}}^{n+1} \frac{(t - t_i)}{(t_j - t_i)} \tag{2.16}$$

for all $t \in [a, b]$.

Theorem 2.13. *Let a function $f \in C[a, b]$ and points $t_1 < \cdots < t_{n+1}$ in $[a, b]$ be given. Then the unique polynomial $p \in P_n$ which solves the corresponding Lagrange interpolation problem is given by*

$$p(t) = \sum_{j=1}^{n+1} f(t_j) l_j(t) \tag{2.17}$$

for all $t \in [a, b]$. This is the so-called Lagrange form *of p.*

Proof. Obviously for all $i \in \{1, \ldots, n+1\}$ and for all $j \in \{1, \ldots, n+1\}$ we have

$$l_j(t_i) = \begin{cases} 1, & \text{if } i = j \\ 0, & \text{if } i \neq j \end{cases}.$$

Then it follows that

$$p(t_i) = f(t_i), \quad i = 1, \ldots, n+1.$$

This proves Theorem 2.13.

2.3. Nearly Optimal Interpolation Points

We discuss the problem of how to choose the interpolation points such that the resulting error in polynomial interpolation is small. Equidistant interpolation points do not have this property. On the other hand, Lagrange interpolation at the zeros of Chebyshev polynomials — called Chebyshev points — yields a nearly optimal error.

Let the space $C[a, b]$ be endowed with the so–called *uniform norm* or L_∞- *norm*

$$\|h\|_\infty = \sup_{t\in[a,b]} |h(t)|, \qquad h \in C[a, b]. \tag{2.18}$$

It follows from Corollary 2.11 that the error $\|f-p\|_\infty$ for polynomial interpolation at arbitrary points $t_1 < \cdots < t_{n+1}$ in $[a, b]$ can be estimated by

$$\|f - p\|_\infty \leq \frac{1}{(n + 1)!} \|f^{(n+1)}\|_\infty \left\| \prod_{j=1}^{n+1}(\cdot - t_j)\right\|_\infty. \tag{2.19}$$

(The last expression denotes the uniform norm of the function $t \to \prod_{j=1}^{n+1}(t - t_j)$ on $[a, b]$.) This inequality shows that an adequate strategy is to choose those interpolation points which minimize $\| \prod_{j=1}^{n+1}(\cdot - t_j)\|_\infty$.

To simplify the considerations we assume that $[a, b] = [-1, 1]$. The general case can be obtained by transforming the interval $[-1, 1]$ into $[a, b]$ with aid of the mapping $t \to \frac{a+b}{2} - \frac{a-b}{2}t$. In order to choose good interpolation points, the following type of polynomials plays an important role.

The function $T_n : [-1, 1] \to \mathbf{R}$, defined by

$$T_n(t) = \cos(n \arccos t) \tag{2.20}$$

for all $t \in [-1, 1]$, is called the *Chebyshev polynomial of degree n*.

We need the following properties of these polynomials.

Lemma 2.14. *(i)   The function T_n is a polynomial of degree n with $\|T_n\|_\infty = 1$.*
(ii) $T_0(t) = 1$, $T_1(t) = t$ and

$$T_{n+1}(t) = 2tT_n(t) - T_{n-1}(t), \quad t \in [-1, 1], \qquad n = 1, 2, \dots \tag{2.21}$$

(iii) The zeros of T_{n+1} are

$$t_i = \cos\left\{ \frac{2(n + 1 - i) + 1}{2(n + 1)} \pi \right\}, \qquad i = 1, \dots, n + 1. \tag{2.22}$$

(iv)

$$T_{n+1}(u_i) = (-1)^{n+2-i}\|T_{n+1}\|_\infty, \tag{2.23}$$

where

$$u_i = \cos\left\{ \frac{n + 2 - i}{n + 1} \pi \right\}, \qquad i = 1, \dots, n + 2. \tag{2.24}$$

Proof. (i). We expand $\cos(nw)$ in powers of $\cos w$. Then by setting $t = \cos w$, it is easy to verify that $T_n \in P_n$. Obviously, we have for all $t \in [-1, 1]$,

$$|T_n(t)| = |\cos(n \arccos t)| \leq 1$$

and $T_n(t) = 1$. This shows that $\|T_n\|_\infty = 1$.

(ii). It follows from the definition of T_n that $T_0(t) = 1$ and $T_1(t) = t$. Since $\cos\{(n+1)w\} + \cos\{(n-1)w\} = 2\cos w(\cos(nw))$, the recurrence relation in (ii) also follows from the definition of T_n.

The statements (iii) and (iv) can be verified directly. This proves Lemma 2.14. $\blacksquare$

The next result shows that the zeros of T_{n+1}, called *Chebyshev points*, minimize the right side of (2.19).

Theorem 2.15. *The zeros*

$$t_i = \cos\left\{\frac{2(n+1-i)+1}{2(n+1)} \pi\right\}, \qquad i = 1, \ldots, n+1, \qquad (2.25)$$

of T_{n+1} minimize the norm $\| \prod_{i=1}^{n+1}(\cdot - t_i)\|_\infty$ on $[-1, 1]$.

Proof. The result is proved by using a theorem from Section 3 below. By Theorem 3.23 the polynomial

$$p(t) = t^{n+1} - \frac{1}{2^n} T_{n+1}(t)$$

is the best uniform approximation of $f(t) = t^{n+1}$ from P_n on $[-1, 1]$. Since

$$f(t) - p(t) = \frac{1}{2^n} T_{n+1}(t) = (t - t_1) \cdots (t - t_{n+1}),$$

where $t_1, \ldots, t_{n+1}$ are the zeros of T_{n+1}, it follows that the norm $\| \prod_{j=1}^{n+1}(\cdot - t_j)\|_\infty$ on $[-1, 1]$ is minimized for this choice of points. This proves Theorem 2.15. $\blacksquare$

The error estimate (2.19) together with Theorem 2.15 justifies the use of Chebyshev points as interpolation points. Moreover, we will see that interpolation at Chebyshev points provides a nearly optimal error.

We first need some basic facts on projections in normed linear spaces.

Definition 2.16. Let G be a finite–dimensional subspace of a real normed linear space E. A linear mapping $L : E \to G$ is called *bounded* if

$$\sup\{\|L(f)\| : f \in E, \|f\| \leq 1\} < \infty.$$

The *norm* of L is defined by

$$\|L\| = \sup\{\|L(f)\| : f \in E, \|f\| \leq 1\}. \qquad (2.26)$$

The value $\|L\|$ is sometimes called the *Lebesgue constant* of L. A bounded linear mapping $L : E \to G$ is called a *projection* if

$$L(g) = g \qquad \text{for all } g \in G.$$

The next result shows that the norm $\|L\|$ can be used to compare the error $\|f - L(f)\|$ with the smallest possible error $\inf_{g \in G} \|f - g\|$.

Theorem 2.17. *Let G be a finite-dimensional subspace of a real normed linear space E, and let $L : E \to G$ be a projection. Then for every $f \in E$,*

$$\|f - L(f)\| \leq (1 + \|L\|) \inf_{g \in G} \|f - g\|. \tag{2.27}$$

Proof. Let $f \in E$ be given. It follows from Theorem 3.3 below that there exists a function $g_f \in G$ such that

$$\|f - g_f\| = \inf_{g \in G} \|f - g\|.$$

Since L is a projection, we have

$$f - L(f) = (f - g_f) - (L(f) - g_f) = (f - g_f) - L(f - g_f).$$

This implies that

$$
\begin{aligned}
\|f - L(f)\| &\leq \|f - g_f\| + \|L(f - g_f)\| \leq \|f - g_f\| + \|L\|\|f - g_f\| \\
&= (1 + \|L\|) \inf_{g \in G} \|f - g\|.
\end{aligned}
$$

This proves Theorem 2.17.

Example 2.18. Let points $t_1 < \cdots < t_{n+1}$ in $[a, b]$ be given. For each $f \in C[a, b]$ we denote by $L_n(f)$ the unique polynomial from P_n which solves the corresponding Lagrange interpolation problem. This defines a mapping $L_n : C[a, b] \to P_n$ which is called the *Lagrange interpolation operator* (with respect to the points $t_1, \ldots, t_{n+1}$).

By using the uniqueness of interpolating polynomials, it is easy to see that L_n is a linear operator. The next result shows that L_n is bounded. Therefore, L_n is a projection, since obviously

$$L_n(p) = p \qquad \text{for all } p \in P_n.$$

Theorem 2.19. *Let points $t_1 < \cdots < t_{n+1}$ in $[a, b]$ be given and let $L_n : C[a, b] \to P_n$ be the corresponding Lagrange interpolation operator. Then we have*

$$\|L_n\|_\infty = \|\sum_{j=1}^{n+1} |l_j(\cdot)|\|_\infty,$$

where l_j, $j = 1, \ldots, n + 1$, are the polynomials defined in (2.16)

Proof. It follows from Theorem 2.13 that

$$|L_n(f)(t)| = |\sum_{j=1}^{n+1} l_j(t)f(t_j)| \le \sum_{j=1}^{n+1} |l_j(t)||f(t_j)| \le \|\sum_{j=1}^{n+1} |l_j(\cdot)|\|_\infty \|f(t_j)\|_\infty.$$

Therefore, we have

$$\|L_n(f)\|_\infty \le \|\sum_{j=1}^{n+1} |l_j(\cdot)|\|_\infty \|f(t_j)\|_\infty,$$

which implies that

$$\|L_n\|_\infty \le \|\sum_{j=1}^{n+1} |l_j(\cdot)|\|_\infty.$$

On the other hand, we choose a point $t_0 \in [a, b]$ such that

$$\sum_{j=1}^{n+1} |l_j(t_0)| = \|\sum_{j=1}^{n+1} |l_j(\cdot)|\|_\infty.$$

Moreover, we set

$$f(t_j) = \operatorname{sgn} l_j(t_0) \qquad j = 1, \ldots, n+1.$$

Then f can be extended to $[a, b]$ such that $\|f\|_\infty = 1$. Therefore, we have

$$\|L_n(f)\|_\infty \ge |L_n(f)(t_0)| = (\sum_{j=1}^{n+1} |l_j(t_0)|) \|f\|_\infty = \|(\sum_{j=1}^{n+1} |l_j(\cdot)|)\|_\infty \|f\|_\infty,$$

which implies that

$$\|L_n\|_\infty \ge \|\sum_{j=1}^{n+1} |l_j(\cdot)|\|_\infty.$$

This proves Theorem 2.19.

It follows from Theorem 2.17 that for all $f \in [a, b]$,

$$\|f - L_n(f)\|_\infty \le (1 + \|L_n\|_\infty) \inf_{p \in P_n} \|f - p\|. \tag{2.28}$$

Therefore, to obtain small interpolation errors it is natural to choose the interpolation points in such a way that $\|L_n\|_\infty$ is small.

Powell [1967] computed the norm $\|L_n\|_\infty$ for equidistant points and Chebyshev points. The computations show that $\|L_n\|_\infty$ grows extremely with n, when equidistant points are used: $\|L_{10}\|_\infty = 30$ and $\|L_{20}\|_\infty = 10987$. On the other hand, we have

$$\|L_n\|_\infty \le 2.1, \quad \text{if } n \le 10$$
$$\|L_n\|_\infty \le 2.5, \quad \text{if } n \le 20$$

and

$$\|L_n\|_\infty \le 4, \quad \text{if } n \le 100.$$

This together with inequality (2.27) shows that for realistic values of n, interpolation at Chebyshev points provides a nearly optimal error. That is a further reason for using Chebyshev points instead of equidistant points to keep the interpolation error small. Further results on interpolation at Chebyshev points can be found in Rivlin [1969] and Brutman [1978].

Bernstein [1931] conjectured that the norm $\|L_n\|_\infty$ of the Lagrange interpolation operator L_n is minimized if for all $i \in \{1, \ldots, n\}$ the values

$$\max_{t \in [t_i, t_{i+1}]} \sum_{j=1}^{n+1} |l_j(t)| \tag{2.29}$$

are the same, where $t_1 < \cdots < t_{n+1}$ are the interpolation points corresponding to L_n. Later, Erdös [1958] conjectured that the interpolation points for which the values (2.29) coincide are uniquely determined. It took nearly 50 years until independently Kilgore [1978] and de Boor & Pinkus [1978] gave a proof of these conjectures. In particular, they showed that the interpolation points for which $\|L_n\|_\infty$ is minimized are also uniquely determined. However, these points are not known explicitly.

3. Best Uniform Approximation by Chebyshev Spaces

In the previous sections we have seen that interpolation is a relatively simple method to approximate a given function by functions from a finite–dimensional space, in particular by spaces of polynomials. On the other hand, it is natural to investigate those functions from a finite–dimensional space which are best approximations of a function with respect to a given norm.

We begin this section with some elementary facts on best approximation in normed linear spaces. Then we consider best approximation by Chebyshev spaces in the uniform norm. In particular, we give results on characterization, unicity and strong unicity of best uniform approximations. Finally, we describe an algorithm for computing such approximations.

3.1. Best Approximation in Normed Linear Spaces

We give some elementary facts on existence, unicity and continuity concerning best approximation in normed linear spaces.

The problem of best approximation can be stated as follows.

Definition 3.1. Let G be a subset of a real normed linear space E and let an element $f \in E$ be given. The *problem of best approximation* is to determine an element $g_f \in G$ such that

$$\|f - g_f\| = \inf_{g \in G} \|f - g\|. \tag{3.1}$$

Such an element is called a *best approximation* of f from G, and

$$d(f, G) = \inf_{g \in G} \|f - g\|. \tag{3.2}$$

is called the *minimal deviation* of f from G. The set of best approximations of f from G is denoted by $P_G(f)$ and the resulting set-valued mapping $P_G : E \to POW(G)$ is called the *metric projection* onto G, where $POW(G)$ denotes the set of all subsets of G.

Remark 3.2. Standard examples of best approximation problems correspond to $E = C[a, b]$, G is a subspace of polynomials or a subspace of splines and the norm is an L_p−norm, where $1 \leq p \leq \infty$ (see the subsequent sections).
 The basic problems in the theory of best approximation are the existence, unicity and characterization of best approximations, the investigation of continuity properties of the metric projection and the development of efficient methods for computing best approximations. A further important problem is to investigate the approximation power of certain function classes, in particular of polynomials and splines.

We first show that best approximations from finite–dimensional spaces always exist.

Theorem 3.3. *Let G be a finite–dimensional subspace of a real normed linear space E. Then for every $f \in E$, there exists a best approximation from G.*

Proof. Let $f \in E$ be given. Then by definition of the infimum there exists a sequence (g_n) in G such that $\|f - g_n\| \to d(f, G)$. This implies that there exists a constant $K > 0$ such that for all n,

$$\|g_n\| - \|f\| \leq \|f - g_n\| \leq d(f, G) + K \leq \|f\| + K.$$

Therefore, for all n, we have

$$\|g_n\| \leq 2\|f\| + K.$$

Since (g_n) is a bounded sequence and G is finite–dimensional, there exists a subsequence (g_{n_k}) of (g_n) converging to some $g_f \in G$. Then we have

$$\|f - g_f\| = \lim_{k \to \infty} \|f - g_{n_k}\| = d(f, G),$$

i.e. $g_f \in P_G(f)$. This proves Theorem 3.3.

The next result shows that the best approximations from a finite–dimensional subspace of a strictly convex space are always unique.

Definition 3.4. A real normed linear space E is called *strictly convex* if for all $f_1, f_2 \in E$ with $f_1 \neq f_2$ and $\|f_1\| = \|f_2\| = 1$, we have $\|\frac{1}{2}(f_1 + f_2)\| < 1$.

Theorem 3.5. *Let G be a finite–dimensional subspace of a strictly convex space E. Then for every $f \in E$, there exists a unique best approximation from G.*

Proof. Let $f \in E$ be given. Since G is finite–dimensional, by Theorem 3.3 there exists an element $g_f \in P_G(f)$. We will show that $P_G(f) = \{g_f\}$. We first show that $P_G(f)$ is convex. Let $g_1, g_2 \in P_G(f)$ and $0 \leq a \leq 1$ be given. Then we have

$$\begin{aligned}
\|f - (a\, g_1 + (1-a)\, g_2)\| &= \|a\,(f - g_1) + (1-a)(f - g_2)\| \\
&\leq a\,\|f - g_1\| + (1-a)\|f - g_2\| \\
&= a\,d(f, G) + (1-a)\,d(f, G) \\
&= d(f, G).
\end{aligned}$$

Since $a\, g_1 + (1-a)g_2 \in G$, this shows that $a\, g_1 + (1-a)g_2 \in P_G(f)$. Suppose now that $g \in P_G(f)$. Then $\frac{1}{2}\,(g_f + g) \in P_G(f)$ which implies that

$$\|\tfrac{1}{2}\,\{(f - g_f) + (f - g)\}\| = \|f - \tfrac{1}{2}\,(g_f - g)\| = d(f, G).$$

Then, since $\|f - g_f\| = \|f - g\| = d(f, G)$ and E is strictly convex, we have $f - g_f = f - g$, i.e. $g = g_f$. This shows that $P_G(f) = \{g_f\}$ and proves Theorem 3.5.

Remark 3.6. It is well–known that the space $C[a, b]$ endowed with an L_p–norm, $1 < p < \infty$, is strictly convex. On the other hand, this space endowed with the L_1–norm or L_∞–norm is not strictly convex. These results can be found in books on Functional Analysis (e.g. Diestel [1975]).

Therefore, it follows from Theorem 3.5 that best approximations with respect to an L_p–norm, $1 < p < \infty$, are always unique. However, the problem of uniqueness of best approximations with respect to the L_1–norm or L_∞–norm needs further investigations (see the subsequent sections).

We close this section with a continuity result on the metric projection.

Theorem 3.7. *Let G be a finite–dimensional subspace of a real normed linear space E with the property that every function in E has a unique best approximation from G. Then for all $f_1, f_2 \in E$,*

$$|d(f_1, G) - d(f_2, G)| \leq \|f_1 - f_2\|$$

and the metric projection $P_G : E \to G$ is continuous.

Proof. Suppose that P_G is not continuous, i.e. that there exist an element $f \in E$ and a sequence (f_n) in E such that $P_G(f_n)$ does not converge to $P_G(f)$. Since G is finite–dimensional, there exists a subsequence (f_{n_k}) of (f_n) such that $P_G(f_{n_k}) \to g_0$, $g_0 \neq P_G(f)$ and $g_0 \in G$. We now show that the mapping $f \to d(f, G)$ ($f \in E$) is continuous. Let $f_1, f_2 \in E$ be given. Then there exists a $g_2 \in G$ with $\|f_2 - g_2\| = d(f_2, G)$. This implies that

$$d(f_1, G) \leq \|f_1 - g_2\| \leq \|f_1 - f_2\| + \|f_2 - g_2\| = \|f_1 - f_2\| + d(f_2, G).$$

Analogously, we have

$$d(f_2, G) - d(f_1, G) \leq \|f_2 - f_1\|.$$

This shows that

$$|d(f_1, G) - d(f_2, G)| \leq \|f_1 - f_2\|.$$

By continuity it follows that

$$\|f_{n_k} - P_G(f_{n_k})\| = d(f_{n_k}, G) \to d(f, G)$$

and

$$\|f_{n_k} - P_G(f_{n_k})\| \to \|f - g_0\|,$$

which implies that $\|f - g_0\| = d(f, G)$. Therefore, g_0 and $P_G(f)$ are two distinct best approximations of f from G which contradicts the uniqueness of best approximations. This proves Theorem 3.7.

3.2. Characterization of Best Uniform Approximations

In this section we give characterizations of best uniform approximations. In particular, it is shown that best uniform approximations from Chebyshev spaces can be characterized by alternation properties.

We begin with the definition of best uniform approximations.

Definition 3.8. For all functions $h \in C[a, b]$, the *uniform norm* or L_∞*-norm* is defined by

$$\|h\|_\infty = \sup_{t \in [a,b]} |h(t)|. \tag{3.3}$$

Best approximations with respect to this norm are called *best uniform approximations*.

In the subsequent results the following notation is used. The set of *extreme points* of a function $h \in C[a, b]$ is defined by

$$E(h) = \{t \in [a, b] : |h(t)| = \|h\|_\infty\}.$$

We first give a characterization of best uniform approximations due to Kolmogorov [1948].

Theorem 3.9. *Let G be a subspace of $C[a, b]$, $f \in C[a, b]$ and $g_f \in G$. The following statements are equivalent:*
(i) The function g_f is a best uniform approximation of f from G.
(ii) For every function $g \in G$,

$$\min_{t \in E(f - g_f)} (f(t) - g_f(t)) g(t) \leq 0. \tag{3.4}$$

Proof. (ii) $\Rightarrow$ (i). Suppose that (ii) holds and let $y \in G$ be given. By (ii) there exists a point $t \in E(f - g_f)$ such that

$$(f(t) - g_f(t))(g(t) - g_f(t)) \le 0.$$

Then we have

$$\begin{aligned}
\|f - g\|_\infty &\ge |f(t) - g(t)| = |(f(t) - g_f(t)) - (g(t) - g_f(t))| \\
&= |(f(t) - g_f(t))| + |(g(t) - g_f(t))| \ge \|f - g_f\|_\infty
\end{aligned}$$

which shows that (i) holds.

(i) $\Rightarrow$ (ii). Suppose that (ii) fails, i.e. there exists a function $g_1 \in G$ such that for all $t \in E(f - g_f)$,

$$(f(t) - g_f(t))g_1(t) > 0.$$

Since $E(f - g_f)$ is compact, there exists a real number $c > 0$ such that for all $t \in E(f - g_f)$,

$$(f(t) - g_f(t))g_1(t) > c.$$

Moreover, there exists an open neighborhood U of $E(f - g_f)$ such that for all $t \in U$,

$$(f(t) - g_f(t))g_1(t) > \frac{c}{2} \tag{3.5}$$

and

$$|f(t) - g_f(t)| \ge \frac{1}{2}\|f - g_f\|_\infty. \tag{3.6}$$

Since $[a, b] \setminus U$ is compact, there exists a real number $d > 0$ such that for all $t \in [a, b] \setminus U$,

$$|f(t) - g_f(t)| < \|f - g_f\|_\infty - d. \tag{3.7}$$

By multiplying g_1 with an appropriate positive factor, we may assume that

$$\|g_1\|_\infty \le \min\{d, \frac{1}{2}\|f - g_f\|_\infty\}. \tag{3.8}$$

We set $g_2 = g_f + g_1$. Then by (3.7) and (3.8) we have for all $t \in [a, b] \setminus U$,

$$\begin{aligned}
|f(t) - g_2(t)| &= |(f(t) - g_f(t)) - g_1(t)| \\
&\le |f(t) - g_f(t)| + |g_1(t)| \\
&\le \|f - g_f\|_\infty - d + \|g_1\|_\infty < \|f - g_f\|_\infty.
\end{aligned}$$

Moreover, by (3.5), (3.6), and (3.8) we have for all $t \in U$,

$$\begin{aligned}
|f(t) - g_2(t)| &= |(f(t) - g_f(t)) - g_1(t)| \\
&\le |f(t) - g_f(t)| - |g_1(t)| \le \|f - g_f\|_\infty.
\end{aligned}$$

This shows that

$$\|f - g_2\|_\infty < \|f - g_f\|_\infty$$

and therefore, g_f is not best uniform approximation of f. This proves Theorem 3.9.

Condition (3.4) is called the *Kolmogorov criterion* and represents a general criterion for best uniform approximations from arbitrary subspaces of $C[a, b]$. If we consider the special class of Chebyshev spaces, then by using Theorem 3.9, we obtain a characterization of best uniform approximations in terms of alternating extreme points of the error (Theorem 3.12) which is the basis for the algorithm described in Section 3.4. We first need a unicity result of Haar [1918]. (A complete characterization of global unicity is given in Theorem 3.18 of Section 3.3)

Theorem 3.10. *Let G be a Chebyshev subspace of $C[a, b]$. Then for every function $f \in C[a, b]$, there exists a unique best uniform approximation from G.*

Proof. Suppose that G is a Chebyshev subspace of $C[a, b]$, and let $f \in C[a, b] \setminus G$ be given. By Theorem 3.3 there exists a best uniform approximation $g_f \in G$. We first show that $E(f - g_f)$ contains at least $n + 1$ distinct points. Suppose to the contrary that $E(f - g_f) = \{t_1, \ldots, t_p\}$ with $p \leq n$. We choose points $t_{p+1}, \ldots, t_n$ in $[a, b]$ such that the points $t_1, \ldots, t_n$ are distinct. Since G is a Chebyshev space, by Theorem 1.4 there exists a function $g \in G$ such that

$$g(t_i) = f(t_i) - g_f(t_i), \qquad i = 1, \ldots, n.$$

Then for all $t \in E(f - g_f)$, we have

$$(f(t) - g_f(t))g(t) > 0.$$

Therefore, by Theorem 3.9 the function g_f is not a best approximation of f, which is a contradiction. Suppose now that $g_1, g_2 \in P_G(f)$. Then it follows that $g = \frac{1}{2}(g_1 + g_2) \in P_G(f)$. Thus, there exist $n + 1$ distinct points $t_1, \ldots, t_{n+1}$ in $[a, b]$ and signs $\sigma_1, \ldots, \sigma_{n+1} \in \{-1, 1\}$ such that

$$f(t_i) - g(t_i) = \sigma_i \, d(f, G), \qquad i = 1, \ldots, n + 1.$$

Since

$$|f(t_i) - g_j(t_i)| \leq d(f, G), \qquad i = 1, \ldots, n + 1, \quad j = 1, 2,$$

it follows that

$$f(t_i) - g_j(t_i) = \sigma_i \, d(f, G), \qquad i = 1, \ldots, n + 1, \quad j = 1, 2.$$

This implies that

$$g_1(t_i) - g_2(t_i) = 0, \qquad i = 1, \ldots, n + 1.$$

It follows from Theorem 1.14 that $g_1 = g_2$. This proves Theorem 3.10.

The notion of alternating extreme points plays an important role in the description of best approximations.

Definition 3.11. We call points $t_1 < \cdots < t_p$ in $[a, b]$ *alternating extreme points* of a function $h \in C[a, b]$ if there exists a sign $\sigma \in \{-1, 1\}$ such that

$$\sigma(-1)^i h(t_i) = \|h\|_\infty, \qquad i = 1, \ldots, p. \tag{3.9}$$

The next result, due to Chebyshev [1899], shows that best uniform approximations from Chebyshev spaces are characterized by alternation properties of the error.

Theorem 3.12. *Let G be an n-dimensional Chebyshev subspace of $C[a, b]$, $f \in C[a, b]$ and $g_f \in G$. The following statements are equivalent:*
(i)   The function g_f is a best uniform approximation of f from G.
(ii)   The error $f - g_f$ has at least $n + 1$ alternating extreme points in $[a, b]$.

Proof. Let G be an n-dimensional Chebyshev subspace of $C[a, b]$.
    (ii) $\Rightarrow$ (i). Suppose that (ii) holds, i.e. there exist points

$$a \leq t_1 < \cdots < t_{n+1} \leq b$$

and a sign $\sigma \in \{-1, 1\}$ such that

$$\sigma(-1)^i \left(f(t_i) - g_f(t_i)\right) = \|f - g_f\|_\infty, \qquad i = 1, \ldots, n + 1. \tag{3.10}$$

Moreover, suppose that (i) fails, i.e. there exists a function $g \in G$ such that

$$\|f - g\|_\infty < \|f - g_f\|_\infty.$$

Then it follows from (3.10) that

$$\begin{aligned}
\sigma(-1)^i \left(f(t_i) - g(t_i)\right) &\leq \|f - g\|_\infty < \|f - g_f\|_\infty \\
&= \sigma(-1)^i \left(f(t_i) - g_f(t_i)\right), \qquad i = 1, \ldots, n + 1,
\end{aligned}$$

which implies that

$$\sigma(-1)^i \left(g_f(t_i) - g(t_i)\right) < 0, \qquad i = 1, \ldots, n + 1.$$

Therefore, the function $g_f - g$ in G has at least n distinct zeros which contradicts Theorem 1.14.
    (i) $\Rightarrow$ (ii). Suppose that (i) holds, but (ii) fails. We choose a maximal number of alternating extreme points $t_1 < \cdots < t_r$ of $f - g_f$. Then it follows that $r \leq n$. We assume that $f(t_1) - g_f(t_1) = \|f - g_f\|_\infty$. (The other case follows analogously). Thus, there exist points

$$a = u_0 < u_1 < \cdots < u_{r-1} < u_r = b$$

and a real number $c > 0$ such that

$$u_i \in [t_i, t_{i+1}), \qquad i = 1, \ldots, r - 1,$$

and

$$(-1)^{i+1} \left(f(t) - g_f(t) \right) \le \| f - g_f \|_\infty - c, \quad t \in [u_i, u_{i+1}], \quad i = 0, \dots, r-1.$$

Since $r \le n$, by Corollary 1.15 there exists a nontrivial function $g \in G$ such that

$$(-1)^i g(t) \ge 0, \qquad t \in [u_i, u_{i+1}], \quad i = 0, \dots, r-1.$$

By multiplying g with an appropriate positive factor, we may assume that $\|g\|_\infty \le c$. Then we have for all $i \in \{0, \dots, r-1\}$ and for all $t \in [u_i, u_{i+1}]$,

$$
\begin{aligned}
-\| f - g_f \|_\infty &\le (-1)^{i+1} \left(f(t) - g_f(t) \right) \\
&\le (-1)^{i+1} \left(f(t) - g_f(t) \right) - (-1)^{i+1} g(t) \\
&\le \| f - g_f \|_\infty - c + \|g\|_\infty \\
&\le \| f - g_f \|_\infty .
\end{aligned}
$$

This implies that

$$\| f - (g_f + g) \|_\infty \le \| f - g_f \|_\infty .$$

Therefore, the functions g_f and $g_f + g$ are best uniform approximations of f, which contradicts Theorem 3.10. This proves Theorem 3.12.

3.3. Global Unicity and Strong Unicity of Best Uniform Approximations

It is shown that the Chebyshev spaces are exactly those spaces for which best uniform approximations are always unique and even strongly unique. For proving this, we need a characterization of strongly unique best uniform approximations. Moreover, we give formulas for computing the strong unicity constant. Finally, we show that the functions from the closure of the unicity set can be characterized by a Kolmogorov type criterion which is not true for the unicity set itself.

We first introduce the notion of strongly unique best approximations which goes back to Newman & Shapiro [1963] and plays an important role in the computation of best approximations.

Definition 3.13. Let G be a subspace of $C[a, b]$. A function $g_f \in G$ is called a *strongly unique best uniform approximation* of $f \in C[a, b]$ if there exists a constant $K_f > 0$ such that for all $g \in G$,

$$\| f - g \|_\infty \ge \| f - g_f \|_\infty + K_f \|g - g_f\|_\infty . \tag{3.11}$$

The *strong unicity constant* $K(f)$ of f is defined to be the maximum of all such numbers K_f.

We now discuss the relationship between unique and strongly unique best uniform approximations.

Remark 3.14. Obviously, every strongly unique best approximation is a unique best approximation. However, the converse is not true in general, as the following example shows.

We consider the subspace $G = \mathrm{span}\{g_1\}$ of $C[1-,1]$, where $g_1(t) = t$. Let the function $f_0 \in C[-1,1]$ be defined by $f_0(t) = 1 - |t|$. Then e.g. the functions g_1 and 0 are best uniform approximations of f_0 from G. However, if we modify f_0 in an arbitrary small neighborhood of $t = 0$ such that for the resulting function f we have $f \in C[-1,1]$, $\|f\|_\infty = 1$ and $f'(0) = 0$, then $g_f = 0$ is a unique, but not strongly unique best uniform approximation of f from G. But, since f_0 and f coincide except on a small neighborhood of $t = 0$ and

$$\|f - g_1\|_\infty \approx \|f - g_f\|_\infty,$$

in practice we would consider both g_f and g_1 as best uniform approximations of f from G.

Moreover, this example shows that although a function $g_f \in G$ is a unique best uniform approximation of a given function f, we cannot conclude that, if for some $g_1 \in G$, $\|f - g_1\|_\infty - \|f - g_f\|_\infty$ is small, then also $\|g_1 - g_f\|_\infty$ is small. This, however, is true if $g_f \in G$ is a strongly unique best approximation of f, since then for all $g \in G$,

$$\|g - g_f\|_\infty \leq \frac{1}{K_f}(\|f - g\|_\infty - \|f - g_f\|_\infty). \tag{3.12}$$

The above observations show that in practice only strongly unique best approximations can be considered as unique best approximations.

Moreover, it was proved by Nürnberger & Singer [1982] (see also Nürnberger [1985c]) that for a given finite–dimensional subspace G of $C[a,b]$ the set

$$\{f \in C[a,b] : f \text{ has a strongly unique best uniform approximation from } G\}$$

is a dense F_σ–subset of

$$\{f \in C[a,b] : f \text{ has a unique best uniform approximation from } G\}.$$

In addition, these sets even coincide, if we replace $[a,b]$ by a finite set.

We now examine the relationship between strong unicity and Lipschitz-continuity of the metric projection.

Definition 3.15. Let G be a subset of $C[a,b]$ and let $f \in C[a,b]$ have a unique best uniform approximation $g_f \in G$. The metric projection $P_G : C[a,b] \to POW(G)$ is called *Lipschitz-continuous* at f if there exists a constant $L_f > 0$ such that for all $\tilde{f} \in C[a,b]$ and all $g_{\tilde{f}} \in P_G(\tilde{f})$,

$$\|g_f - g_{\tilde{f}}\|_\infty \leq L_f \|f - \tilde{f}\|_\infty. \tag{3.13}$$

The next result is due to Cheney [1966].

Theorem 3.16. *Let G be a subset of $C[a,b]$ and $f \in C[a,b]$. If f has a strongly unique best uniform approximation from G, then $P_G : C[a,b] \to POW(G)$ is Lipschitz-continuous at f.*

Proof. Let $f \in C[a,b]$ have a strongly unique best uniform approximation $g_f \in G$, i.e. there exists a $K_f > 0$ such that for all $g \in G$,

$$\|f - g\|_\infty \geq \|f - g_f\|_\infty + K_f \|g - g_f\|_\infty.$$

Then for all $\tilde{f} \in C[a,b]$ and all $g_{\tilde{f}} \in P_G(\tilde{f})$ we have

$$
\begin{aligned}
K_f \|g_f - g_{\tilde{f}}\|_\infty &\leq \|f - g_{\tilde{f}}\|_\infty - \|f - g_f\|_\infty \\
&\leq \|f - \tilde{f}\|_\infty + \|\tilde{f} - g_{\tilde{f}}\|_\infty - \|f - g_f\|_\infty \\
&\leq \|f - \tilde{f}\|_\infty + \|\tilde{f} - g_f\|_\infty - \|f - g_f\|_\infty \\
&\leq \|f - \tilde{f}\|_\infty + \|f - \tilde{f}\|_\infty = 2\|f - \tilde{f}\|_\infty.
\end{aligned}
$$

Then $L_f = \dfrac{2}{K_f}$ is the desired constant. This proves Theorem 3.16.

For the case when G is a finite–dimensional subspace of $C[a,b]$, Bartelt & Schmidt [1984] proved the following equivalence: $f \in C[a,b]$ has a strongly unique best uniform approximation from G if and only if f has a unique best uniform approximation from G and P_G is Lipschitz-continuous at f.

The next result due to Wulbert [1971] (see also Bartelt & McLaughlin [1973]) shows that strongly unique best uniform approximations can be characterized in terms of *strong Kolmogorov criteria*. The reader should compare this characterization with Theorem 3.9.

Theorem 3.17. *Let G be a finite–dimensional subspace of $C[a,b]$, $f \in C[a,b]\backslash G$ and $g_f \in G$. The following statements are equivalent:*

(i)   *The function g_f is a strongly unique best uniform approximation of f from G.*

(ii)   *For every nontrival function $g \in G$,*

$$\min_{t \in E(f - g_f)} (f(t) - g_f(t)) g(t) < 0. \tag{3.14}$$

(iii) *There exists a constant $K_f > 0$ such that for every function $g \in G$,*

$$\min_{t \in E(f - g_f)} (f(t) - g_f(t)) g(t) \leq -K_f \|f - g_f\|_\infty \|g\|_\infty. \tag{3.15}$$

Proof. (iii) $\Rightarrow$ (i). Suppose that (iii) holds and let a function $g \in G$ be given. By (iii) there exists a point $t \in E(f - g_f)$ such that

$$(f(t) - g_f(t))(g(t) - g_f(t)) \leq -K_f \|f - g_f\|_\infty \|g - g_f\|_\infty.$$

This implies that

$$
\begin{aligned}
\|f - g\|_\infty &= |f(t) - g(t)| = |(f(t) - g_f(t)) - (g(t) - g_f(t))| \\
&= |f(t) - g_f(t)| + |g(t) - g_f(t)| \\
&\geq \|f - g_f\|_\infty + K_f \frac{\|f - g_f\|_\infty}{|f(t) - g_f(t)|} \|g - g_f\|_\infty \\
&= \|f - g_f\|_\infty + K_f \|g - g_f\|_\infty.
\end{aligned}
$$

(i) $\Rightarrow$ (iii). Suppose that (iii) fails, i.e. there exists a function $g_1 \in G$ such that for all $t \in E(f - g_f)$,

$$
(f(t) - g_f(t)) g_1(t) > -K_f \|f - g_f\|_\infty \|g_1\|_\infty.
$$

Since $E(f - g_f)$ is compact, there exists an open neighborhood U of $E(f - g_f)$ such that for all $t \in U$,

$$
(f(t) - g_f(t)) g_1(t) > -K_f \|f - g_f\|_\infty \|g_1\|_\infty.
$$

and

$$
|f(t) - g_f(t)| \geq \frac{1}{2} \|f - g_f\|_\infty. \tag{3.16}
$$

Moreover, we may choose U sufficiently small such that for all $t \in U$ with $(f(t) - g_f(t)) g_1(t) < 0$,

$$
|g_1(t)| < K_f \|g_1\|_\infty. \tag{3.17}
$$

Since $[a, b] \setminus U$ is compact, there exists a real number $c > 0$ such that for all $t \in [a, b] \setminus U$,

$$
|f(t) - g_f(t)| \leq \|f - g_f\|_\infty - c. \tag{3.18}
$$

By multiplying g_1 with an appropriate positive factor, we may assume that

$$
\|g_1\|_\infty \leq \min\{c, \frac{1}{2} \|f - g_f\|_\infty\}. \tag{3.19}
$$

We set $g_2 = g_f + g_1$. Then by (3.18) and (3.19) we have for all $t \in [a, b] \setminus U$,

$$
\begin{aligned}
|f(t) - g_2(t)| &= |(f(t) - g_f(t)) - g_1(t)| \\
&\leq \|f - g_f\|_\infty - c + \|g_1\|_\infty \leq \|f - g_f\|_\infty.
\end{aligned}
$$

Moreover, by (3.16) and (3.17) we have for all $t \in U$ with $(f(t) - g_f(t)) g_1(t) < 0$,

$$
\begin{aligned}
|f(t) - g_2(t)| &= |(f(t) - g_f(t)) - g_1(t)| \\
&= |f(t) - g_f(t)| + |g_1(t)| < \|f - g_f\|_\infty + K_f \|g_1\|_\infty \\
&= \|f - g_f\|_\infty + K_f \|g_2 - g_f\|_\infty.
\end{aligned}
$$

Finally, by (3.16) and (3.19) we have for all $t \in U$ with $(f(t) - g_f(t)) g_1(t) \geq 0$,

$$
\begin{aligned}
|f(t) - g_2(t)| &= |(f(t) - g_f(t)) - g_1(t)| \\
&= |f(t) - g_f(t)| - |g_1(t)| \leq \|f - g_f\|_\infty.
\end{aligned}
$$

This shows that

$$\|f - g_2\|_\infty < \|f - g_f\|_\infty + K_f \|g_2 - g_f\|_\infty,$$

i.e. (i) fails.

(ii) $\Rightarrow$ (iii). Suppose that (ii) holds. Let $F : \{g \in G : \|g\|_\infty = 1\} \to \mathbf{R}$ be the mapping, defined by

$$F(g) = \min_{t \in E(f - g_f)} \frac{(f(t) - g_f(t))}{\|f - g_f\|_\infty} g(t).$$

Since G is finite–dimensional, the set $\{g \in G : \|g\|_\infty = 1\}$ is compact. Therefore, since by (ii) we have $F(g) < 0$ for all $g \in \{g \in G : \|g\|_\infty = 1\}$, there exists a constant $K_f > 0$ such that $F(\frac{g}{\|g\|_\infty}) \leq -K_f$ for all $g \in G$, which shows that (iii) holds. Since the implication (iii) $\Rightarrow$ (ii) is obvious, the proof of Theorem 3.17 is complete.

The next result is a characterization of those finite–dimensional spaces for which best uniform approximations are always unique, respectively, strongly unique. Haar [1918] proved the equivalence (i) $\Leftrightarrow$ (iii) and Newman & Shapiro [1963] proved the implication (iii) $\Rightarrow$ (ii). An extension of this result to optimization was given by Nürnberger [1985b].

Theorem 3.18. *For a finite–dimensional subspace G of $C[a,b]$, the following statements are equivalent:*
(i)   *For every function $f \in C[a,b]$, there exists a unique best uniform approximation from G.*
(ii) *For every function $f \in C[a,b]$, there exists a strongly unique best uniform approximation from G.*
(iii) *G is a Chebyshev subspace.*

Proof. The implication (ii) $\Rightarrow$ (i) is obvious.

(iii) $\Rightarrow$ (ii). Suppose that (iii) holds and let $f \in C[a,b]$ be given. By Theorem 3.10 there exists a unique best uniform approximation $g_f \in G$ of f. Moreover, by Theorem 3.12 there exist points

$$a \leq t_1 < \cdots < t_{n+1} \leq b$$

and a sign $\sigma \in \{-1, 1\}$ such that

$$\sigma(-1)^i (f(t_i) - g_f(t_i)) = \|f - g_f\|_\infty, \quad i = 1, \ldots, n+1.$$

We now assume that (ii) fails. Then by Theorem 3.17 there exists a nontrivial function $g_1 \in G$ such that

$$
\begin{aligned}
0 &\leq (f(t_i) - g_f(t_i)) g_1(t_i) \\
&= \sigma(-1)^i \|f - g_f\|_\infty g_1(t_i), \quad i = 1, \ldots, n+1.
\end{aligned}
\tag{3.20}
$$

Since by Theorem 1.14, the function $g_1 \in G$ has at most $n - 1$ distinct zeros, there exists an integer $j \in \{1, \ldots, n+1\}$ such that $g_1(t_j) \neq 0$. Moreover, by Theorem 1.4 there exists a function $g_2 \in G$ such that

$$\sigma(-1)^i g_2(t_i) = 1, \qquad i = 1, \ldots, n+1, \quad i \neq j. \tag{3.21}$$

Since $\sigma(-1)^j g_1(t_j) > 0$, there exists a sufficiently small $\varepsilon > 0$ such that

$$\sigma(-1)^j \left(g_1(t_j) + \varepsilon g_2(t_j) \right) > 0.$$

Moreover, it follows from (3.20) and (3.21) that

$$\sigma(-1)^i \left(g_1(t_i) + \varepsilon g_2(t_i) \right) > 0, \qquad i = 1, \ldots, n+1, \quad i \neq j.$$

Therefore, the function $g_1 + \varepsilon g_2$ in G has at least n distinct zeros, which contradicts Theorem 1.14.

(i) $\Rightarrow$ (iii). Suppose that (iii) fails. Let $\{g_1, \ldots, g_n\}$ be a basis of G. Then there exist points $t_1 < \cdots < t_n$ such that

$$D \left(\begin{matrix} g_1, \ldots, g_n \\ t_1, \ldots, t_n \end{matrix} \right) = 0. \tag{3.22}$$

Therefore, the rows of the corresponding matrix are linearly dependent. This implies that there exist real numbers $a_1, \ldots, a_n$ such that

$$\sum_{i=1}^{n} |a_i| \neq 0$$

and

$$\sum_{i=1}^{n} a_i g(t_i) = 0 \quad \text{for all } g \in G. \tag{3.23}$$

Moreover, it follows from (3.22) that there exists a function $g_0 \in G$ with $\|g_0\|_\infty = 1$ such that

$$g_0(t_i) = 0, \qquad i = 1, \ldots, n.$$

We now set for all $i \in \{1, \ldots, n\}$, $h(t_i) = 1$, if $a_i = 0$, and $h(t_i) = \operatorname{sgn} a_i$, if $a_i \neq 0$. Then h can be extended to $[a, b]$ such that $h \in C[a, b]$ and $\|h\|_\infty = 1$. Let the function $f \in C[a, b]$ be defined by

$$f(t) = h(t)(1 - |g_0(t)|) \qquad \text{for all } t \in [a, b].$$

Since $\|f\|_\infty = 1$, it follows that $d(f, G) \leq 1$. Suppose that $d(f, G) < 1$. Then there exists a function $g \in G$ such that

$$|f(t_i) - g(t_i)| < 1, \quad i = 1, \ldots, n.$$

Let $i \in \{1, \ldots, n\}$ with $a_i \neq 0$ be given. Then $|(\operatorname{sgn} a_i) - g(t_i)| < 1$ which implies that $(\text{sgn}\, a_i) g(t_i) < 0$. Thus it follows that

$$\sum_{i=1}^{n} a_i g(t_i) > 0,$$

which contradicts (3.23). This shows that $0 \in P_G(f)$. Moreover, for all $t \in [a, b]$,

$$
\begin{aligned}
|f(t) - g_0(t)| &\leq |h(t)(1 - |g_0(t)|)| + |g_0(t)| \\
&\leq 1 - |g_0(t)| + g_0(t) = 1.
\end{aligned}
$$

This shows that $\|f - g_0\|_\infty = 1$ which implies that $g_0 \in P_G(f)$. Therefore, statement (i) is not true. This proves Theorem 3.18.

It was proved by McLaughlin & Somers [1975] that a finite–dimensional sub-space G of $C[a, b]$ is a Chebyshev subspace if and only if unique and strongly unique best uniform approximations from G coincide.

In connection with the above Theorem 3.18 and also Theorem 1.4, we mention that roughly speaking, by a result of Mairhuber [1956], for $n \geq 2$ there do not exist n–dimensional Chebyshev spaces of functions of more than one variable (for details see e.g. Singer [1970]).

In the following we derive formulas for computing the strong unicity constant (see Definition 3.13). We note that strong unicity constants can be used to obtain error estimations in the computation of best uniform approxmations (see Remark 3.29 in this chapter and also Remark 4.19 in Chapter II.)

We begin with a general formula for strong unicity constants, due to Bartelt & Schmidt [1981].

In order to state the result, we use the following notation. For a given function $h \in C[a, b]$, we define the *sign function* of f by

$$
\text{sgn}\, f(t) = \begin{cases}
-1, & \text{if } f(t) < 0 \\
0, & \text{if } f(t) = 0 \\
1, & \text{if } f(t) > 0
\end{cases}
$$

for all $t \in [a, b]$.

Lemma 3.19. *Let G be a finite–dimensional subspace of $C[a, b]$, and let $g_f \in G$ be a strongly unique best uniform approximation of $f \in C[a, b] \backslash G$. Then we have*

$$
\begin{aligned}
K(f) &= \min_{\substack{g \in G \\ \|g\|_\infty = 1}} \max_{t \in E(f - g_f)} \text{sgn}(f(t) - g_f(t))\, g(t) \\
&= \min\{1/\|g\|_\infty : g \in G,\ \text{sgn}(f(t) - g_f(t))\, g(t) \leq 1 \\
&\qquad\qquad\qquad \text{for all } t \in E(f - g_f)\}.
\end{aligned}
$$

Proof. We denote the first (respectively second) minimum by m_1 (respectively m_2). By the proof of Theorem 3.17 the constants $K_f > 0$ which appear in condition (i) and (iii) of Theorem 3.17 are the same. Since condition (3.15) is equivalent to

$$\max_{t \in E(f-g_f)} (f(t) - g_f(t)) g(t) \geq K_f \, \|f - g_f\|_\infty \, \|g\|_\infty$$

for all $g \in G$, it follows that $K(f) = m_1$. We will show that $m_1 = m_2$. Let a nontrivial function $g \in G$ with

$$\text{sgn}(f(t) - g_f(t)) g(t) \leq 1$$

for all $t \in E(f - g_f)$ be given. We set $g_1 = g/\|g\|_\infty$. Then we have $\|g_1\|_\infty = 1$ and

$$m_1 \leq \max_{t \in E(f-g_f)} (f(t) - g_f(t)) g_1(t) \leq 1/\|g_1\|_\infty.$$

This shows that $m_1 \leq m_2$.

On the other hand, let $g \in G$ with $\|g\|_\infty = 1$ be given. We set

$$M = \max_{t \in E(f-g_f)} (f(t) - g_f(t)) g(t)$$

and $g_2 = (1/M) g$. Then we have

$$(f(t) - g_f(t)) g(t) \leq 1$$

for all $t \in E(f - g_f)$ and therefore, $M = 1/\|g_2\|_\infty \geq m_2$. This shows that $m_1 \geq m_2$ and proves Lemma 3.19.

The next result, due to Cline [1973], shows that for Chebyshev spaces G, strong unicity constants can be computed by the norm of interpolating functions from G. A more general formula was given by Schmidt [1980].

Theorem 3.20. *Let G be an n−dimensional Chebyshev subspace of $C[a,b]$, and let $g_f \in G$ be a strongly unique best uniform approximation of $f \in C[a,b] \setminus G$ such that $E(f - g_f) = \{t_1, \ldots, t_{n+1}\}$. For each $j \in \{1, \ldots, n+1\}$, let $g_j \in G$ be the uniquely determined function such that*

$$g_j(t_i) = \text{sgn}(f(t_i) - g_f(t_i)), \qquad i = 1, \ldots, n+1, \quad i \neq j.$$

Then we have

$$K(f) = \min\{1/\|g_j\|_\infty : j = 1, \ldots, n+1\}.$$

Proof. For simplicity we set $E = E(f - g_f)$ and $\sigma = \text{sgn}(f - g_f)$. By Lemma 3.19, there exists a $g_0 \in G$ such that $\sigma(t) g_0(t) \leq 1$ for all $t \in E$ and $K(f) = 1/\|g_0\|_\infty$. We choose an extreme point t_0 of g_0. Moreover, we set $A = \{t \in E : g_0(t) = \sigma(t)\}$ and have to show that A contains exactly n points. To do this we first show that there does not exist a $g \in G$ such that

$$\sigma(t) g(t) < 0, \qquad t \in A, \tag{3.24}$$

and

$$\operatorname{sgn} g_0(t_0)\, g(t_0) > 0. \tag{3.25}$$

Suppose to the contrary that such a function $g \in G$ exists. For all $\varepsilon > 0$ we set $g_\varepsilon = g_0 + \varepsilon g$. Then by (3.25) we have

$$\|g_\varepsilon\|_\infty \geq |g_\varepsilon(t_0)| = |g_0(t_0)| + \varepsilon |g(t_0)| > \|g_0\|_\infty. \tag{3.26}$$

By (3.24) there exists a neighborhood U of A such that for all $t \in U$,

$$\sigma(t)\, g(t) < 0, \qquad t \in U.$$

Then for all $t \in E \cap U$,

$$\sigma(t)\, g_\varepsilon(t) \leq 1 + \varepsilon\, \sigma(t)\, g(t) < 1.$$

Moreover, there exists a real number $c > 0$ such that

$$\sigma(t)\, g_0(t) \leq 1 - c, \qquad t \in E \setminus U.$$

Then for all $t \in E \setminus U$,

$$\sigma(t)\, g_\varepsilon(t) \leq 1 - c + \varepsilon\, \|g\|_\infty \leq 1,$$

if ε is sufficiently small. Therefore, by (3.26) and Lemma 3.19 $K(f) \leq 1/\|g_\varepsilon\|_\infty < 1/\|g_0\|_\infty = K(f)$, which is a contradiction. This shows that (3.24) and (3.25) hold.

We now show that A contains at least n points. Suppose to the contrary that $A = \{u_1, \ldots, u_m\}$, where $m < n$. If necessary, we add further points $u_{m+1}, \ldots, u_{n-1}$. Since G is a n–dimensional Chebyshev space, there exists a $\tilde{g} \in G$ with

$$\tilde{g}(u_i) = -\sigma(u_i), \qquad i = 1, \ldots, n-1,$$

and

$$\tilde{g}(t_0) = \operatorname{sgn} g_0(t_0),$$

which contradicts (3.24) and (3.25).

If A contains $n+1$ points, then by the assumption $A = E$ and by the definition of A,

$$(f(t) - g_f(t))\, g_0(t) > 0, \qquad t \in E.$$

It follows from Theorem 3.9 that g_f is not a best uniform approximation of f, which is a contradiction. This proves Theorem 3.20.

We briefly discuss an open problem concerning strong unicity constants. Given a function $f \in C[a, b]$, we denote by $K_n(f)$ the strong unicity constant of f w.r.t. the space P_n. Poreda [1976] posed the problem to characterize those functions $f \in C[a, b]$ for which $\liminf_{n \to \infty} K_n(f) = 0$. It was proved by several authors that certain classes of functions have this property (see Poreda [1976], Henry

& Roulier [1978], Schmidt [1978], Blatt [1982], [1984] and Bartelt & Schmidt [1981]). Henry & Roulier [1978] even conjectured that $\liminf_{n\to\infty} K_n(f) = 0$ for all nonpolynomial functions $f \in C[a, b]$. However, these problems are unsolved at present.

We note that if as in Theorem 3.20 the set $E(f - g_f)$ contains exactly $n + 1$ points, then by Theorem 3.12 these points are alternating extreme points of $f - g_f$.

In the following we will show that functions f with $f^{(n+1)}(t) \neq 0$ for all $t \in (a, b)$ have best uniform approximations g_f from P_n with this property.

To do this, we use the following result of Bernstein [1926].

Lemma 3.21. *Let a function $f \in C^{n+1}[a, b]$ with $f^{(n+1)}(t) \neq 0$ for all $t \in (a, b)$ be given. Then the space* $\mathrm{span}\{P_n \cup \{f\}\}$ *is an* $(n + 2)$*-dimensional Chebyshev subspace of $C[a, b]$.*

Proof. We set $H = \mathrm{span}\{P_n \cup \{f\}\}$. Suppose now that H is not a Chebyshev space. Then by Theorem 1.14 there exists a nontrivial function $h \in H$ with at least $n + 2$ distinct zeros. Moreover, it follows from Theorem 2.9 that

$$h' \in \mathrm{span}\{P_{n-1} \cup \{f'\}\}$$

has at least $n + 1$ zeros. We consider further derivatives of h' and finally get that $h^{(n+1)} \in \mathrm{span}\{f^{(n+1)}\}$ has at least one zero which is a contradiction. This proves Lemma 3.21.

Theorem 3.22. *Let a function $f \in C^{n+1}[a, b]$ with $f^{(n+1)}(t) \neq 0$ for all $t \in (a, b)$ be given, and let $g_f \in P_n$ be the best uniform approximation of f. Then the error $f - g_f$ has exactly $n + 2$ alternating extreme points*

$$a = t_1 < t_2 < \cdots < t_{n+1} < t_{n+2} = b$$

such that $f'(a) - g_f'(a) \neq 0$ and $f'(b) - g_f'(b) \neq 0$.

Proof. By Theorem 3.12 the error $f - g_f$ has at least $n + 2$ alternating extreme points. Then it follows that $f - g_f$ has exactly $n + 2$ alternating extreme points; otherwise $f - g_f$ has at least $n + 2$ zeros which by Theorem 1.14 is a contradiction to Lemma 3.21. Suppose now that $f'(a) - g_f'(a) = 0$. (The other case follows analogously.) Then $f' - g_f'$ has $n + 1$ zeros at $t_1, \ldots, t_{n+1}$. As in the proof of Lemma 3.21 it follows that $f^{(n+1)} = f^{(n+1)} - g_f^{(n+1)}$ has at least one zero, which is a contradiction. This proves Theorem 3.22.

For example, the functions $f_1(t) = e^t$ ($t \in [a, b]$), $f_2(t) = \sqrt{t}$ ($t \in [0, b]$), $f_3(t) = \cos t$ ($t \in [0, \pi]$) and $f_4(t) = \sin t$ ($t \in [-\frac{\pi}{2}, \frac{\pi}{2}]$) satisfy the assumption of Lemma 3.21 and Theorem 3.22 for all n.

Only in very rare cases are best approximations known explicitly. An exceptional situation is the approximation of the function $f(t) = t^{n+1}$ by polynomials of degree n. The next result is due to Chebyshev [1899]. Further results of this type can be found e.g. in Meinardus [1967].

Theorem 3.23. *The best uniform approximation of $f(t) = t^{n+1}$ from P_n on $[-1, 1]$ is the polynomial*

$$t^{n+1} - \frac{1}{2^n} T_{n+1}(t),$$

where T_{n+1} is the Chebyshev polynomial of degree $n + 1$.

Proof. We set

$$g_f(t) = t^{n+1} - \frac{1}{2^n} T_{n+1}(t)$$

for all $t \in [-1, 1]$. By Lemma 2.14 (ii), the coefficient of t^{n+1} of the polynomial T_{n+1} is 2^n. Therefore, we have $g_f \in P_n$. Moreover, it follows from Lemma 2.14 that

$$
\begin{aligned}
(-1)^{n+2-i} \left(f(u_i) - g_f(u_i) \right) &= (-1)^{n+2-i} \frac{1}{2^n} T_{n+1}(u_i) \\
&= \left\| \frac{1}{2^n} T_{n+1} \right\|_\infty = \| f - g_f \|_\infty, \quad i = 1, \ldots, n+2,
\end{aligned}
$$

where $u_1, \ldots, u_{n+2}$ are the points in statement (iv) of Lemma 2.14. Then by Theorem 3.12 the polynomial $g_f \in P_n$ is a best uniform approximation of f. This proves Theorem 3.23.

Theorem 3.9 and Theorem 3.17 show that best uniform approximations and strongly unique best uniform approximations can be characterized by Kolmogorov criteria. However, this is not true for unique best uniform approximations, because by Theorem 4.6 in Chapter II, unique spline approximations depend on certain flatness properties of the error. On the other hand, Berens & Nürnberger [1987] proved that for finite–dimensional subspaces G of $C[a, b]$, the functions from the closure of

$$U(G) = \{ f \in C[a, b] : f \text{ has a unique best uniform approximation from } G \}$$

can actually be characterized by a Kolmogorov type criterion.

Theorem 3.24. *Let G be a finite–dimensional subspace of $C[a, b]$, and $f \in C[a, b] \setminus G$. The following statements are equivalent:*
(i) $f \in \overline{U(G)}$.
(ii) f has a best uniform approximation $g_f \in G$ such that for every nontrivial function $g \in G$ and every neighborhood U of $E(f - g_f)$,

$$\inf_{t \in U} \left(f(t) - g_f(t) \right) g(t) < 0. \tag{3.27}$$

Proof. (ii) $\Rightarrow$ (i). Suppose that (ii) holds. We may assume that $g_f = 0$. Let (U_m) be an open neighborhood basis of $E(f)$ such that $\overline{U}_m \subset U_{m-1}$ for all m. We define for sufficiently large m,

$$f_m(t) = f(t), \qquad t \in [a, b] \setminus U_{m-1}$$

and

$$f_m(t) = \|f\|_\infty \operatorname{sgn} f(t), \qquad t \in \overline{U}_m.$$

Then there exists a continuous extension of f_m to $[a, b]$ such that $\|f_m\|_\infty = \|f\|_\infty$. By construction, the sequence (f_m) converges to f. Now, let a sufficiently large m and a nontrivial $g \in G$ be given. Since $\overline{U}_m$ is a neighborhood of $E(f)$ and $\overline{U}_m \subset E(f_m)$, it follows from (ii) that

$$\min_{t \in E(f_m)} f(t) \, g(t) < 0.$$

Since $f_m \to f$, this implies that

$$\min_{t \in E(f_m)} f_m(t) \, g(t) < 0.$$

Then it follows from Theorem 3.17 that the zero function is a unique best uniform approximation of f_m, i.e. (i) holds.

(i) $\Rightarrow$ (ii). Suppose that (i) holds, i.e. there exists a sequence (f_m) in $C[a, b]$ converging to f such that for all m, f_m has a unique best uniform approximation $g_m \in G$. Since G is finite–dimensional, by going to a subsequence we may assume that (g_m) converges to a best uniform approximation $g_f \in G$ of f. Suppose now that (ii) fails. Then there exists a nontrivial $g_0 \in G$ and an open neighborhood U of $E(f - g_f)$ such that

$$(f(t) - g_f(t)) \, g_0(t) \geq 0, \qquad t \in U. \tag{3.28}$$

Since $[a, b] \setminus U$ is compact, there exists a real number $c > 0$ such that

$$|f(t) - g_f(t)| \leq \|f - g_f\|_\infty - c, \qquad t \in [a, b] \setminus U. \tag{3.29}$$

Moreover, we may assume that U is chosen sufficiently small such that

$$|f(t) - g_f(t)| \geq \frac{1}{2} \|f - g_f\|_\infty, \qquad t \in U. \tag{3.30}$$

Since $f_m \to f$ and $g_m \to g_f$, it follows from (3.29) that for sufficiently large m,

$$|f_m(t) - g_m(t)| < \|f_m - g_m\|_\infty - \frac{c}{2}, \qquad t \in [a, b] \setminus U. \tag{3.31}$$

Moreover, it follows from (3.28) and (3.30) that

$$(f_m(t) - g_m(t)) \, g_0(t) \geq 0, \qquad t \in U. \tag{3.32}$$

By multiplying g_0 with an appropriate positive factor, we may assume that

$$\|g_0\|_\infty < \frac{c}{2} \tag{3.33}$$

and

$$|f_m(t) - g_m(t)| \geq |g_0(t)|, \qquad t \in U. \tag{3.34}$$

Then it follows from (3.31) and (3.33) that for all $t \in [a,b] \setminus U$,

$$|f_m(t) - g_m(t) - g_0(t)| \leq \|f_m - g_m\|_\infty - \frac{c}{2} + \|g_0\|_\infty < \|f_m - g_m\|_\infty.$$

Moreover, it follows from (3.32) and (3.34) that for all $t \in U$,

$$|f_m(t) - g_m(t) - g_0(t)| \leq |f_m(t) - g_m(t)| - |g_0(t)| \leq \|f_m - g_m\|_\infty.$$

This shows that

$$\|f_m - (g_m - g_0)\|_\infty \leq \|f_m - g_m\|_\infty$$

which is a contradiction, since g_m is the unique best uniform approximation of f_m. This proves Theorem 3.24.

Finally, we note that the above Theorems 3.9, 3.17, 3.18, 3.20 (respectively 3.24) remain true if we replace the interval $[a,b]$ by an arbitrary compact Hausdorff space (respectively metric space) — with the same proofs.

3.4. Algorithm

One of the fundamental problems in Approximation Theory is the development of efficient algorithms for computing best approximations. In this section we describe an iterative method for best uniform approximations from Chebyshev spaces which goes back to Remez [1934a]–[1934c].

Let $G = \text{span}\{h_1, \ldots, h_n\}$ be an n–dimensional Chebyshev subspace of $C[a,b]$ and a function $f \in C[a,b] \setminus G$ be given.

The idea of Remez [1934a]–[1934c] is to iteratively compute a sequence of functions (g_p) in G converging to a function $g_f \in G$ with the property that there exist points $a \leq t_1 < \cdots < t_{n+1} \leq b$ and a sign $\sigma \in \{-1, 1\}$ such that

$$\sigma(-1)^i (f(t_i) - g_f(t_i)) = \|f - g_f\|_\infty, \qquad i = 1, \ldots, n+1.$$

It follows from Theorem 3.12 that g_f is a best uniform approximation of f from G.

In the following we describe this approach.

Description of the Algorithm

In the first step of the algorithm we choose a set of points

$$M_1 = \{t_{1,1}, \ldots, t_{n+1,1}\}$$

such that

$$a \leq t_{1,1} < \cdots < t_{n+1,1} \leq b.$$

Then we determine the unique function $g_1 \in G$ and the unique real number $\lambda_1 \in \mathbf{R}$ satisfying

$$(-1)^i (f(t_{i,1}) - g_1(t_{i,1})) = \lambda_1, \qquad i = 1, \ldots, n+1. \tag{3.35}$$

As shown below, this system of linear equations has a unique solution $a_{1,1}, \ldots,$ $a_{n,1}$, λ_1, where $g_1 = \sum_{i=1}^{n} a_{i,1} h_i$. For $p \geq 1$ we proceed by induction as follows. We determine a point $t_p \in [a, b]$ such that

$$|\lambda_p| < |f(t_p) - g_p(t_p)| \approx \|f - g_p\|_\infty \qquad (3.36)$$

and replace a point from

$$M_p = \{t_{1,p}, \ldots, t_{n+1,p}\},$$

where

$$a \leq t_{1,p} < \cdots < t_{n+1,p} \leq b,$$

such that a new set

$$M_{p+1} = \{t_{1,p+1}, \ldots, t_{n+1,p+1}\},$$

where

$$a \leq t_{1,p+1} < \cdots < t_{n+1,p+1} \leq b,$$

is obtained. In the following we describe which point from M_p will be replaced by t_p.

Rule of Exchange

We set $t_{0,p} = -\infty$ and $t_{n+2,p} = \infty$. Then there exists an index $j \in \{0, \ldots, n+1\}$ such that $t_{j,p} < t_p < t_{j+1,p}$.

Case 1. $j \in \{1, \ldots, n\}$
If $\mathrm{sgn}(f(t_{j,p}) - g_p(t_{j,p})) = \mathrm{sgn}(f(t_p) - g_p(t_p))$ (respectively $\mathrm{sgn}(f(t_{j+1,p}) - g_p(t_{j+1,p}))$ $= \mathrm{sgn}(f(t_p) - g_p(t_p)))$, then we replace $t_{j,p}$ (respectively $t_{j+1,p}$) by t_p.

Case 2. $j = 0$
If $\mathrm{sgn}(f(t_{1,p}) - g_p(t_{1,p})) = \mathrm{sgn}(f(t_p) - g_p(t_p))$, then we replace $t_{1,p}$ by t_p. Otherwise, we replace $t_{n+1,p}$ by t_p.

Case 3. $j = n + 1$
If $\mathrm{sgn}(f(t_{n+1,p}) - g_p(t_{n+1,p})) = \mathrm{sgn}(f(t_p) - g_p(t_p))$, then we replace $t_{n+1,p}$ by t_p. Otherwise, we replace $t_{1,p}$ by t_p.

By replacing a point from M_p by t_p according to the above rule, we obtain a new set

$$M_{p+1} = \{t_{1,p+1}, \ldots, t_{n+1,p+1}\}.$$

Then we determine the unique function $g_{p+1} \in G$ and the unique real number $\lambda_{p+1} \in \mathbf{R}$ satisfying

$$(-1)^i \left(f(t_{i,p+1}) - g_{p+1}(t_{i,p+1}) \right) = \lambda_{p+1}, \qquad i = 1, \ldots, n+1, \qquad (3.37)$$

and proceed by induction.

In this way we obtain a sequence (g_p) in G and a sequence (λ_p) in $\mathbf{R}$. We will show that (g_p) converges to the unique best uniform approximation g_f of f from G, and that $(|\lambda_p|)$ converges monotonely increasing to the minimal deviation $d(f, G)$.

We first consider the system of linear equations

$$(-1)^i \left(f(t_{i,p}) - g_p(t_{i,p}) \right) = \lambda_p, \qquad i = 1, \ldots, n+1. \tag{3.38}$$

If we set $g_p = \sum_{j=1}^{n} a_{j,p} \, h_j$, then (3.38) is equivalent to

$$\sum_{j=1}^{n} a_{j,p} \, h_j(t_{i,p}) + (-1)^i \lambda_p = f(t_{i,p}), \qquad i = 1, \ldots, n+1. \tag{3.39}$$

This is a system of $n+1$ linear equations with $n+1$ unknowns $a_{1,p}, \ldots, a_{n,p}$ and λ_p. We denote the corresponding determinant by

$$D(M_p) = \begin{vmatrix} h_1(t_{1,p}) & \cdots & h_n(t_{1,p}) & (-1)^1 \\ \vdots & & \vdots & \vdots \\ h_1(t_{n+1,p}) & \cdots & h_n(t_{n+1,p}) & (-1)^{n+1} \end{vmatrix}.$$

The first result shows that the system (3.39) has a unique solution, and that the sequence $(|\lambda_p|)$ is monotonely increasing and bounded above by $d(f, G)$.

Lemma 3.25. *For all $p = 1, 2, \ldots$ the following properties hold:*
(i) $D(M_p) \neq 0$.
(ii) There exists a sign $\sigma_p \in \{-1, 1\}$ such that

$$\text{sgn} \left(f(t_{i,p+1}) - g_p(t_{i,p+1}) \right) = \sigma_p \, \text{sgn} \left(f(t_{i,p}) - g_p(t_{i,p}) \right), \quad i = 1, \ldots, n+1.$$

(iii) $|\lambda_p| < |\lambda_{p+1}| \leq d(f, G)$.

Proof. (i). Expanding $D(M_p)$ along the last column, we get $D(M_p) \neq 0$, since G is a Chebyshev space.
   (ii). This property follows from the rule of exchange.
   (iii). For all $p = 1, 2, \ldots$ we set

$$D_{i,p} = D \begin{pmatrix} h_1, & \cdots\cdots & , h_n \\ t_{1,p}, \ldots, t_{i-1,p}, t_{i+1,p}, \ldots, t_{n+1,p} \end{pmatrix}. \tag{3.40}$$

Since G is a Chebyshev space, it follows that

$$|D(M_p)| = \sum_{i=1}^{n+1} |D_{i,p}|. \tag{3.41}$$

Moreover, we have

$$\sum_{i=1}^{n+1} (-1)^i D_{i,p} \, g(t_{i,p}) = 0 \tag{3.42}$$

for all $g \in G$. This property follows from the fact that

$$\begin{vmatrix} h_1(t_{1,p}) & \cdots & h_n(t_{1,p}) & g(t_{1,p}) \\ \vdots & & \vdots & \vdots \\ h_1(t_{n+1,p}) & \cdots & h_n(t_{n+1,p}) & g(t_{n+1,p}) \end{vmatrix} = 0$$

for all $g \in G$. Then it follows from (ii), (3.39), and (3.42) that there exists a sign $\sigma \in \{-1, 1\}$ such that

$$\lambda_{p+1} = \sigma \frac{1}{D(M_{p+1})} \sum_{i=1}^{n+1} D_{i,p+1} (-1)^i f(t_{i,p+1})$$

$$= \sigma \frac{1}{D(M_{p+1})} \sum_{i=1}^{n+1} D_{i,p+1} (-1)^i (f(t_{i,p+1}) - g_p(t_{i,p+1}))$$

$$= \sigma \frac{1}{D(M_{p+1})} \sum_{i=1}^{n+1} D_{i,p+1} \sigma_p \, \text{sgn}(\lambda_p) \, |f(t_{i,p+1}) - g_p(t_{i,p+1})|.$$

Then it follows from (3.41) that

$$|\lambda_{p+1}| = \frac{1}{|D(M_{p+1})|} \sum_{i=1}^{n+1} |D_{i,p+1}| \, |f(t_{i,p+1}) - g_p(t_{i,p+1})|$$

$$= |\lambda_p| + \frac{1}{|D(M_{p+1})|} \sum_{i=1}^{n+1} |D_{i,p+1}| (|f(t_{i,p+1}) - g_p(t_{i,p+1})| - |\lambda_p|).$$

Moreover, it follows from (3.36) that

$$|\lambda_{p+1}| > |\lambda_p|.$$

Moreover, by using (3.41), we get that

$$|\lambda_{p+1}| = |\sigma \frac{1}{D(M_{p+1})} \sum_{i=1}^{n+1} D_{i,p+1} (-1)^i (f(t_{i,p+1}) - g_f(t_{i,p+1}))|$$

$$\leq \frac{1}{|D(M_{p+1})|} \sum_{i=1}^{n+1} |D_{i,p+1}| \, \|f - g_f\|_\infty = \|f - g_f\|_\infty = d(f, G),$$

where $g_f \in G$ is the best uniform approximation of f. This proves Lemma 3.25.

The next result, due to Remez [1934a]–[1934c], shows that the sequence (g_p) constructed in the algorithm converges to the best uniform approximation of f from G and that $(|\lambda_p|)$ converges monotonely increasing to $d(f, G)$.

Theorem 3.26. *The following statements hold:*
(i) $\lim_{p \to \infty} |\lambda_p| = d(f, G)$.
(ii) $\lim_{p \to \infty} \|g_p - g_f\|_\infty = 0$,
   where $g_f \in G$ is the best uniform approximation of f.

Proof. We will show that there exists a constant $L > 0$ such that for all p,

$$|\lambda_{p+1}| - |\lambda_p| \geq L (\|f - g_p\|_\infty - |\lambda_p|). \tag{3.43}$$

Suppose for the moment that (3.43) holds. By Lemma 3.25 (iii), the sequence $(|\lambda_p|)$ is monotone increasing and bounded by $d(f, G)$. Therefore, this sequence converges. Moreover, it follows from (3.43) that for all p,

$$(d(f, G) - |\lambda_p|) \leq \frac{1}{L} (|\lambda_{p+1}| - |\lambda_p|)$$

and, since $\lim_{p\to\infty}(|\lambda_{p+1}| - |\lambda_p|) = 0$ we get $\lim_{p\to\infty}|\lambda_p| = d(f,G)$. This proves (i).
Furthermore, it follows from (3.43) that for all p,

$$d(f,G) \le \|f - g_p\|_\infty \le |\lambda_p| + \frac{1}{L}(|\lambda_{p+1}| - |\lambda_p|) \le d(f,G) + \frac{1}{L}(|\lambda_{p+1}| - |\lambda_p|).$$

This implies that

$$\lim_{p\to\infty}\|f - g_p\|_\infty = d(f,G).$$

By Theorem 3.18 there exists a constant $K_f > 0$ such that for all p,

$$\|f - g_p\|_\infty - \|f - g_f\|_\infty \ge K_f\|g_p - g_f\|_\infty$$

which implies that

$$\|f - g_p\|_\infty - d(f,G) \ge K_f\|g_p - g_f\|_\infty.$$

This proves (ii).
Thus, it remains to prove (3.43). We first show that there exists a constant
$M > 0$ such that for all p and all $i \in \{1, \ldots, n\}$,

$$|t_{i+1,p} - t_{i,p}| \ge M. \tag{3.44}$$

Suppose that (3.44) fails. Then by going to a subsequence, we may assume that
there exist points

$$a \le t_1 \le \cdots \le t_{n+1} \le b,$$

where at least two of these points are equal, such that

$$\lim_{p\to\infty} t_{i,p} = t_i, \qquad i = 1, \ldots, n+1. \tag{3.45}$$

Then by Theorem 1.4 there exists a unique function $g \in G$ such that

$$(-1)^i(f(t_i) - g(t_i)) = \lambda = 0, \qquad i = 1, \ldots, n+1. \tag{3.46}$$

Moreover, by using continuity arguments, it follows from (3.38),(3.45) and (3.46)
that there exists a sufficiently large p such that $|\lambda_p| < |\lambda_1|$ which is a contradiction
to property (iii) in Lemma 3.25. This proves (3.44).
We now show that there exists a constant $L > 0$ such that for all p and all
$i \in \{1, \ldots, n+1\}$,

$$\frac{|D_{i,p}|}{|D(M_p)|} \ge L. \tag{3.47}$$

Let $i \in \{1, \ldots, n\}$ be given. Then by Theorem 1.4, for all p there exists a unique
function $\tilde{g}_p \in G$ such that

$$\tilde{g}_p(t_{j,p}) = f(t_{j,p}), \qquad j \ne i, \quad j \in 1, \ldots, n+1.$$

By using continuity arguments, it follows from (3.44) that there exists a constant
$L_i > 0$ such that for all p,

$$\|f - \tilde{g}_p\|_\infty \le L_i.$$

Moreover, it follows from the proof of Lemma 3.25 that

$$|\lambda_1| < |\lambda_p| = \frac{1}{|D(M_p)|} \sum_{j=1}^{n+1} |D_{j,p}| |f(t_{j,p}) - \tilde{g}_p(t_{j,p})|$$

$$= \frac{|D_{i,p}|}{|D(M_p)|} |f(t_{i,p}) - \tilde{g}_p(t_{i,p})| \leq \frac{|D_{i,p}|}{|D(M_p)|} L_i.$$

Therefore, we have

$$\frac{|D_{i,p}|}{|D(M_p)|} \geq \frac{|\lambda_1|}{L_i}.$$

This shows that

$$L = \min_{i=1,\dots,n+1} \frac{|\lambda_1|}{L_i} > 0$$

is the desired constant in (3.47). Finally, it follows from (3.47) and the proof of Lemma 3.25 that

$$|\lambda_{p+1}| - |\lambda_p| = \frac{1}{|D(M_{p+1})|} \sum_{i=1}^{n+1} |D_{i,p+1}| (|f(t_{i,p+1}) - g_p(t_{i,p+1})| - |\lambda_p|)$$

$$\geq L (\|f - g_p\|_\infty - |\lambda_p|).$$

This proves (3.43) and Theorem 3.26.

Remark 3.27. (Simultaneous exchange method). In order to obtain faster convergence of the Remez algorithm, one should exchange not only one "almost extreme point" of $f - g_p$ in the p-th step of the algorithm, but $n + 1$ such points according to the above exchange rule such that property (ii) in Lemma 3.25 holds.

It can be shown that under certain assumptions the simultaneous exchange method is equivalent to the Newton method and therefore, the Remez algorithm converges quadratically, since the Newton method has this property. This theorem due to Veidinger [1960] is given in the next theorem. (For a detailed proof, see e.g. Meinardus [1967].)

Theorem 3.28. *Let* $G = \mathrm{span}\{h_1, \dots, h_n\}$ *be an* n-*dimensional subspace of* $C[a, b]$ *such that* $h_1, \dots, h_n \in C^2[a, b]$, *and let* $g_f \in G$ *be the best uniform approximation of* $f \in C[a, b]$ *such that* $E(f - g_f) = \{t_1, \dots, t_{n+1}\}$. *We assume that for all* $i \in \{1, \dots, n+1\}$,

$$f'(t_i) - g_f'(t_i) \neq 0, \quad \text{if } t_i \in \{a, b\} \tag{3.48}$$

and

$$f''(t_i) - g_f''(t_i) \neq 0, \quad \text{if } t_i \notin \{a, b\}. \tag{3.49}$$

Then the simultaneous exchange method converges quadratically, i.e. there exists a constant $C > 0$ *such that for all* p,

$$\|g_f - g_{p+1}\|_\infty \leq C \|g_f - g_p\|_\infty^2.$$

Theorem 3.22 shows that for $G = P_{n-1}$ and for functions f with $f^{(n)}(t) \neq 0$, for all $t \in (a, b)$ the assumptions $E(f-g_f) = \{t_1, \ldots, t_{n+1}\}$ and (3.48) of Theorem 3.28 are satisfied.

Remark 3.29. (Error estimation). It is easy to verify that for all $p = 1, 2, \ldots$ the following estimations hold:

(i) $\|f - g_p\|_\infty - \|f - g_f\|_\infty \leq \|f - g_p\|_\infty - |\lambda_p|$.

(ii) $\|g_p - g_f\|_\infty \leq \frac{1}{K(f)} (\|f - g_p\|_\infty - |\lambda_p|)$,

where $g_f \in G$ is the best uniform approximation of f and $K(f) > 0$ is the strong unicity constant of f.

Since the values $\|f - g_p\|_\infty$ and $|\lambda_p|$ are computed during the algorithm (and also the strong unicity constant $K(f)$ can be computed (compare Theorem 3.20)), we stop the algorithm, if $\|f - g_p\|_\infty - |\lambda_p| \leq \varepsilon$, where ε is the desired accuracy.

We note that numerical examples are given at the end of Section 4.3 in Chapter II.

3.5. Approximation Power of Polynomials

In order to decide which set of functions should be used to approximate a given function efficiently, it is important to investigate the approximation power of certain function classes. Here we briefly discuss how well functions can be approximated by the class of polynomials. We will see that analytic functions can be approximated efficiently by polynomials. However, this is not the case for nonsmooth functions. Since this subject is treated in detail in most books on Approximation Theory, we only give some typical results without proof.

Before giving estimates on the optimal error in best uniform approximation by polynomials, we need some notation.

The *modulus of continuity* of a function $h \in C[a, b]$ is defined by

$$\omega(h; \delta) = \sup\{|h(t_1) - h(t_2)| : t_1, t_2 \in [a, b], |t_1 - t_2| \leq \delta\},$$

where $\delta > 0$ is a positive real number.

Note, that for every function $h \in C^1[a, b]$ we have

$$\omega(h; \delta) \leq \delta \|h'\|_\infty.$$

The minimal deviation of a function $f \in C[a, b]$ from a space of polynomials P_m in the the uniform norm is denoted by

$$d_\infty(f, P_m) = \inf_{p \in P_m} \|f - p\|_\infty.$$

In the following we give some results on the approximation power of polynomials. The proofs of these theorems can be found in many books on Approximation Theory (listed in the references).

The first theorem on functions which are differentiable up to a certain order is due to Jackson [1930].

Theorem 3.30. *Let intergers $m \geq 1$ and $j \in \{0, \ldots, m-1\}$ be given. Then there exists a constant $K > 0$ (depending only on j) such that for every function $f \in C^j[a, b]$, we have*

$$d_\infty(f, P_m) \leq K \left(\frac{b-a}{m} \right)^j \omega \left(f^{(j)}; \frac{b-a}{2(m-j)} \right). \tag{3.50}$$

In particular, if $f \in C^{j+1}[a, b]$, then

$$\omega \left(f^{(j)}; \frac{b-a}{2(m-j)} \right) \leq \frac{b-a}{2(m-j)} \| f^{(j+1)} \|_\infty. \tag{3.51}$$

Next, we give a result on infinitely often differentiable functions which is also due to Jackson [1930].

Theorem 3.31. *If $f \in C^\infty[a, b]$, then for all $p \in (0, \infty)$,*

$$\lim_{m \to \infty} m^p \, d_\infty(f, P_m) = 0. \tag{3.52}$$

The following result, due to Bernstein [1926], deals with analytic functions $f : \mathbf{C} \to \mathbf{C}$ which are real–valued on $[a, b]$. Here $\mathbf{C}$ denotes the set of complex numbers.

Theorem 3.32. *If $f : \mathbf{C} \to \mathbf{C}$ is an analytic function such that $f(t) \in \mathbf{R}$ for all $t \in [a, b]$, then*

$$\lim_{m \to \infty} \sqrt[m]{d_\infty(f, P_m)} = 0. \tag{3.53}$$

Remark 3.33. It follows from (3.52) and (3.53) that analytic functions can be approximated efficiently by polynomials.

However, the estimate (3.50) in Theorem 3.30 shows that, if $f \in C^r[a, b] \setminus C^{r+1}[a, b]$ and r is small, then $d_\infty(f, P_m)$ may converge slowly to zero as m tends to infinity. A typical example is the function $f(t) = \sqrt{t}$ on $[0, 2]$. It follows from (3.50) that

$$d_\infty(f, P_m) \leq K \omega \left(f; \frac{1}{m} \right) = K \frac{1}{\sqrt{m}}.$$

In fact, numerical results actually show that in this case the decrease of the error $d_\infty(f, P_m)$ is extremely slow when m increases.

On the other hand, the estimate (3.50) shows that the error $d_\infty(f, P_m)$ is small if the value $\frac{b-a}{m}$ is small. Therefore, to obtain better approximations we can divide the given interval $[a, b]$ into several subintervals and work with polynomials on these smaller intervals. This is the main motivation for the use of piecewise polynomials or splines as approximating functions. (Compare Section 4.4 in Chapter II on the approximation power of splines.)

4. Best L_1–Approximation by Chebyshev Spaces

Best approximation can be considered with respect to various norms. The choice of the norm depends on the given minimization problem. Up to now we have investigated best uniform approximation. In this section we consider best approximation by Chebyshev spaces in the L_1–norm. In particular, we give results on characterization and unicity of best L_1–approximations. Moreover, it is shown that for certain functions best L_1–approximations from a given Chebyshev space can be computed by solving an interpolation problem.

4.1. Global Unicity of Best L_1–Approximations

While the existence of best approximations is guaranteed for every finite-dimensional space, best approximations need not be unique in general. However, it will be shown that best L_1–approximations from Chebyshev spaces are always unique. To prove this, we apply a characterization of best L_1–approximations.

We begin with the definition of best L_1–approximations.

Definition 4.1. For all functions $h \in C[a, b]$ the L_1–*norm* is defined by

$$\|h\|_1 = \int_a^b |h(t)|\, dt. \tag{4.1}$$

Best approximations with respect to this norm are called *best L_1–approximations* (compare Definition 3.1).

Let G be a subspace of $C[a, b]$ and $f \in C[a, b]$. Then by definition the problem of best L_1–approximation is to minimize the area between f and g for $g \in G$.
For every function $h \in C[a, b]$ we call

$$Z(h) = \{t \in [a, b] : h(t) = 0\} \tag{4.2}$$

the *zero set of* h.
We first give a characterization of best L_1–approximations due to Kripke & Rivlin [1965].

Theorem 4.2. *Let G be a subspace of $C[a, b]$, $f \in C[a, b]$ and $g_f \in G$. The following statements are equivalent:*
(i)   The function g_f is a best L_1–approximation of f from G.
(ii) For every function $g \in G$,

$$\int_a^b g(t)\, \mathrm{sgn}(f(t) - g_f(t))\, dt \le \int_{Z(f-g_f)} |g(t)|\, dt. \tag{4.3}$$

Proof. (ii) $\Rightarrow$ (i). Suppose that (ii) holds and let $g \in G$ be given. For simplicity we set $Z = Z(f - g_f)$. Since $g - g_f \in G$, it follows from (ii) that

$$\int_a^b (g(t) - g_f(t))\, \mathrm{sgn}(f(t) - g_f(t))\, dt \le \int_Z |g(t) - g_f(t)|\, dt.$$

and therefore,

$$
\|f - g\|_1 - \|f - g_f\|_1
$$

$$
= \int_a^b |f(t) - g(t)|\, dt - \int_a^b |f(t) - g_f(t)|\, dt
$$

$$
= \int_Z |f(t) - g(t)|\, dt + \int_{[a,b]\setminus Z} |f(t) - g(t)|\, dt - \int_a^b |f(t) - g_f(t)|\, dt
$$

$$
= \int_Z \{(f(t) - g_f(t)) - (g(t) - g_f(t))\}\, \mathrm{sgn}\{(f(t) - g_f(t)) - (g(t) - g_f(t))\}\, dt
$$

$$
+ \int_{[a,b]\setminus Z} |f(t) - g(t)|\, dt
$$

$$
- \int_a^b \{(f(t) - g(t)) + (g(t) - g_f(t))\}\, \mathrm{sgn}(f(t) - g_f(t))\, dt
$$

$$
= \int_Z |g(t) - g_f(t)|\, dt + \int_{[a,b]\setminus Z} |f(t) - g(t)|\, dt
$$

$$
- \int_a^b (g(t) - g_f(t))\, \mathrm{sgn}(f(t) - g_f(t))\, dt
$$

$$
- \int_a^b (f(t) - g(t))\, \mathrm{sgn}(f(t) - g_f(t))\, dt \geq 0.
$$

Therefore, the function $g_f \in G$ is a best L_1–approximation of f.

(i) $\Rightarrow$ (ii). Suppose that (ii) fails, i.e. there exists a $g \in G$ such that

$$
\int_a^b g(t)\, \mathrm{sgn}(f(t) - g_f(t))\, dt > \int_Z |g(t)|\, dt.
$$

By multiplying g with a positive factor, we may assume that $\|g\|_\infty = 1$. For all $\varepsilon > 0$ we set

$$
A(\varepsilon) = \{t \in [a, b] : |f(t) - g_f(t)| \leq \varepsilon\}.
$$

Then it follows that

$$
\|f - (g_f + \varepsilon g)\|_1 - \|f - g_f\|_1
$$

$$
= \int_a^b |f(t) - g_f(t) - \varepsilon g(t)|\, dt - \int_a^b |f(t) - g_f(t)|\, dt
$$

$$
= \int_Z |\varepsilon g(t)|\, dt + \int_{[a,b]\setminus Z} |f(t) - g_f(t) - \varepsilon g(t)|\, dt - \int_a^b |f(t) - g_f(t)|\, dt
$$

$$
= \int_Z |\varepsilon g(t)|\, dt
$$

$$
+ \int_{[a,b]\setminus A(\varepsilon)} (f(t) - g_f(t) - \varepsilon g(t))\, \mathrm{sgn}(f(t) - g_f(t))\, dt
$$

$$
+ \int_{A(\varepsilon)\setminus Z} (f(t) - g_f(t) - \varepsilon g(t))\, \mathrm{sgn}(f(t) - g_f(t) - \varepsilon g(t))\, dt
$$

$$
- \int_{[a,b]\setminus A(\varepsilon)} (f(t) - g_f(t))\, \mathrm{sgn}(f(t) - g_f(t))\, dt
$$

$$-\int_{A(\varepsilon)\backslash Z}(f(t)-g_f(t))\,\mathrm{sgn}(f(t)-g_f(t))\,dt$$

$$= \int_Z |\varepsilon\,g(t)|\,dt - \int_{[a,b]\backslash A(\varepsilon)}\varepsilon\,g(t)\,\mathrm{sgn}(f(t)-g_f(t))\,dt$$

$$+\int_{A(\varepsilon)\backslash Z}(f(t)-g_f(t)-\varepsilon\,g(t))\,\mathrm{sgn}(f(t)-g_f(t)-\varepsilon\,g(t))\,dt$$

$$-\int_{A(\varepsilon)\backslash Z}(f(t)-g_f(t))\,\mathrm{sgn}(f(t)-g_f(t))\,dt$$

$$= \int_Z |\varepsilon\,g(t)|\,dt - \int_{[a,b]\backslash Z}\varepsilon\,g(t)\,\mathrm{sgn}(f(t)-g_f(t))\,dt$$

$$-\int_{A(\varepsilon)\backslash Z}\varepsilon\,g(t)\,\mathrm{sgn}(f(t)-g_f(t))\,dt$$

$$+\int_{A(\varepsilon)\backslash Z}(f(t)-g_f(t)-\varepsilon\,g(t))\,\mathrm{sgn}(f(t)-g_f(t)-\varepsilon\,g(t))\,dt$$

$$-\int_{A(\varepsilon)\backslash Z}(f(t)-g_f(t))\,\mathrm{sgn}(f(t)-g_f(t))\,dt$$

$$= \varepsilon\Big(\int_Z |g(t)|\,dt - \int_a^b g(t)\,\mathrm{sgn}(f(t)-g_f(t))\,dt\Big)$$

$$+\int_{A(\varepsilon)\backslash Z}(f(t)-g_f(t)-\varepsilon\,g(t))\cdot$$

$$\cdot\{\mathrm{sgn}(f(t)-g_f(t)-\varepsilon\,g(t))-\mathrm{sgn}(f(t)-g_f(t))\}\,dt$$

$$\leq \varepsilon\Big(\int_Z |g(t)|\,dt - \int_a^b g(t)\,\mathrm{sgn}(f(t)-g_f(t))\,dt\Big) + \varepsilon\int_{A(\varepsilon)\backslash Z}2\,dt.$$

Since the first expression in the last sum is strictly negative and $\int_{A(\varepsilon)\backslash Z}2\,dt$ tends to zero for $\varepsilon\to 0$, for sufficiently small $\varepsilon>0$, we have

$$\|f-(g_f+\varepsilon\,g)\|_1 - \|f-g_f\|_1 < 0,$$

i.e. $g_f\in G$ not a best L_1-approximation of f. This proves Theorem 4.2.

In the following we give a global unicity result for Chebyshev spaces. We first need a lemma.

Lemma 4.3. *Let G be a subspace of $C[a,b]$ and $f\in C[a,b]$. If $g_1,g_2\in G$ are best L_1-approximations of f, then*

$$(f(t)-g_1(t))\,(f(t)-g_2(t))\geq 0, \qquad t\in[a,b]. \tag{4.4}$$

Proof. Since g_1 and g_2 are best L_1-approximations of f from G, the function $\frac{1}{2}(g_1+g_2)\in G$ is also a best L_1-approximation of f, i.e.

$$\int_a^b |f(t)-g_1(t)|\,dt = \int_a^b |f(t)-g_2(t)|\,dt = \int_a^b |f(t)-\tfrac{1}{2}(g_1+g_2)|\,dt.$$

This implies that

$$\int_a^b \left(|f(t) - g_1(t)| + |f(t) - g_2(t)| - |(f(t) - g_1(t)) + (f(t) - g_2(t))| \right) dt = 0.$$

Therefore, we have for all $t \in [a, b]$,

$$|f(t) - g_1(t)| + |f(t) - g_2(t)| - |(f(t) - g_1(t)) + (f(t) - g_2(t))| = 0,$$

which implies

$$(f(t) - g_1(t))(f(t) - g_2(t)) \geq 0.$$

This proves Lemma 4.3.

The next result, due to Jackson [1930], shows that best L_1-approximations from Chebyshev spaces are always unique.

Theorem 4.4. *Let G be a Chebyshev subspace of $C[a, b]$. Then for every function $f \in C[a, b]$, there exists a unique best L_1-approximation from G.*

Proof. Suppose that there exists a function $\tilde{f} \in C[a, b]$ with two different best L_1-approximations $g_1 \in G$ and $g_2 \in G$. We set $f = \tilde{f} - g_1$ and $g_0 = g_2 - g_1$. Then $0 \in G$ and $g_0 \in G$ are best L_1-approximations of f. It follows from Lemma 4.3 that

$$f(t)(f(t) - g_0(t)) \geq 0, \qquad t \in [a, b].$$

This implies that

$$\begin{aligned}
|f(t) - \tfrac{1}{2} g_0(t)| &= |\tfrac{1}{2}(f(t) - g_0(t)) + \tfrac{1}{2} f(t)| \\
&= \tfrac{1}{2}|f(t) - g_0(t)| + \tfrac{1}{2}|f(t)|, \qquad t \in [a, b].
\end{aligned}$$

Therefore, if $f(t) - \tfrac{1}{2} g_0(t) = 0$, then

$$\tfrac{1}{2}|f(t) - g_0(t)| + \tfrac{1}{2}|f(t)| = 0$$

which implies that $g_0(t) = 0$. This shows that

$$Z(f - \tfrac{1}{2} g_0) \subset Z(g_0). \tag{4.5}$$

Since G is a Chebyshev space, the function $g_0 \in G$ has only finitely many zeros. Therefore, by (4.5) the function $f - \tfrac{1}{2} g_0$ has only finitely many zeros.
Claim. There exists a nontrivial $g \in G$ such that

$$(f(t) - \tfrac{1}{2} g_0(t)) g(t) \geq 0, \qquad t \in [a, b]. \tag{4.6}$$

Suppose for the moment that the claim is true. Then we have

$$\int_a^b g(t) \, \text{sgn}(f(t) - \tfrac{1}{2} g_0(t)) \, dt > 0 = \int_{Z(f - \frac{1}{2} g_0)} |g(t)| \, dt.$$

Therefore, it follows from Theorem 4.2 that $\frac{1}{2} g_0 \in G$ is not a best L_1-approxima- tion of f. This is a contradiction, since the fact that 0 and g_0 are best L_1-appro- ximations of f implies that $\frac{1}{2} g_0$ is a best L_1-approximation of f. Therefore, it remains to prove the claim. Let the dimension of G be n. Since G is a Chebyshev space, it follows from Theorem 1.14 that the function $g_0 \in G$ has at most $n - 1$ distinct zeros. Then by (4.5) the function $f - \frac{1}{2} g_0$ has at most $n - 1$ sign changes. This implies that there exist points

$$a = t_0 \leq t_1 < \cdots < t_{m-1} \leq t_m = b,$$

where $m \in \{1, \ldots, n\}$, and a sign $\sigma \in \{-1, 1\}$ such that

$$\sigma (-1)^i \left(f(t) - \tfrac{1}{2} g_0(t) \right) \geq 0, \qquad t \in [t_{i-1}, t_i], \quad i = 1, \ldots, m.$$

By Corollary 1.15 there exists a nontrivial $g \in G$ such that

$$\sigma (-1)^i g(t) \geq 0, \qquad t \in [t_{i-1}, t_i], \quad i = 1, \ldots, m.$$

Therefore, g is the desired function. This proves Theorem 4.4.

Remark 4.5. In contrast to best uniform approximation (see Theorem 3.18), the converse of Theorem 4.4 is not true, since for example best L_1-approxima- tions from spline spaces are always unique (see Theorem 6.1 in Chapter II), but spline spaces are not Chebyshev spaces.

4.2. Interpolation at Canonical points

It is shown that for functions from the so-called convexity cone, best L_1-appro- ximations from a given Chebyshev space can be computed by solving a Lagrange interpolation problem.

The points which correspond to such interpolation problems are of the fol- lowing type.

Definition 4.6. Let G be an n-dimensional subspace of $C[a, b]$. We call points $t_1 < \cdots < t_r$ in (a, b), where $1 \leq r \leq n$, *canonical points* for G if

$$\sum_{i=0}^{r} (-1)^i \int_{t_i}^{t_{i+1}} g(t)\, dt = 0 \tag{4.7}$$

for all $g \in G$, where $t_0 = a$ and $t_{r+1} = b$

We first give a result on the existence and uniqueness of canonical points due to Micchelli [1977].

Theorem 4.7. *For every n-dimensional Chebyshev subspace of $C[a, b]$, there exists a unique set of n canonical points.*

Proof. Let $G = \text{span}\{g_1, \ldots, g_n\}$ be an n-dimensional Chebyshev subspace of $C[a, b]$. Then by Zalik [1975] there exists a function $g_f \in C[a, b]$ such that $G_f = \text{span}\{G, f\}$ is an $(n + 1)$-dimensional Chebyshev subspace of $C[a, b]$. By Theorem 4.4 the function f has a unique best L_1-approximation $g_f \in G$. It follows from Theorem 4.2 that for all $g \in G$,

$$\int_a^b g(t) \, \text{sgn}(f(t) - g_f(t)) \, dt \leq \int_{Z(f - g_f)} |g(t)| \, dt. \tag{4.8}$$

Since G_f is a Chebyshev space, it follows from Theorem 1.14 that the function $f - g_f$ has at most n zeros. Thus, there exist a sign $\sigma \in \{-1, 1\}$ and points

$$a = t_0 \leq t_1 < \cdots < t_r \leq t_{r+1} = b,$$

where $r \in \{0, \ldots, n\}$, such that

$$\text{sgn}(f(t) - g_f(t)) = \sigma \, (-1)^i, \qquad t \in (t_i, t_{i+1}), \quad i = 0, \ldots, r. \tag{4.9}$$

Then it follows from (4.8) and (4.9) that for all $g \in G$,

$$\sum_{i=0}^r (-1)^i \int_{t_i}^{t_{i+1}} g(t) \, dt = 0. \tag{4.10}$$

We will show that $r = n$. Suppose to the contrary that $r < n$. Then by Corollary 1.15 there exists a nontrivial $g_0 \in G$ such that

$$(-1)^i g_0(t) \geq 0, \qquad t \in [t_i, t_{i+1}], \quad i = 0, \ldots, r.$$

Then it follows that

$$\sum_{i=0}^r (-1)^i \int_{t_i}^{t_{i+1}} g_0(t) \, dt > 0,$$

which contradicts (4.10).

Finally, we show that $\{t_1, \ldots, t_n\}$ is the only set of canonical points for G. Suppose to the contrary that there exists a further set $\{u_1, \ldots, u_n\}$ of canonical points for G which is distinct from $\{t_1, \ldots, t_n\}$. We set $t_0 = u_0 = a$ and $t_{n+1} = u_{n+1} = b$ and define

$$s_1(t) = (-1)^i, \qquad t \in [t_i, t_{i+1}), \quad i = 0, \ldots, n,$$

and

$$s_2(t) = (-1)^i, \qquad t \in [u_i, u_{i+1}), \quad i = 0, \ldots, n.$$

Moreover, we set

$$t_j = \min\{t_i : t_i \neq u_i, \ i = 1, \ldots, n\}.$$

We may assume that $t_j < u_j$. Then it follows that

$$s_1(t) - s_2(t) = 0, \quad t \in [t_0, t_j), \tag{4.11}$$

$$(-1)^j \, (s_1(t_j) - s_2(t_j)) > 0 \tag{4.12}$$

and

$$(-1)^i \, (s_1(t) - s_2(t)) \geq 0 \qquad t \in [t_i, t_{i+1}], \quad i = j, \ldots, n. \tag{4.13}$$

We now define $g_1 \in G$ by

$$g_1(t) = \overline{D}\left(\begin{array}{cccc} g_1, & \cdots\cdots & , g_n \\ t_1, \ldots, t_{j-1}, t_{j+1}, \ldots, t_n, t \end{array}\right), \qquad t \in [a, b].$$

(For the definition of the above determinant see (1.27).) Since G is a Chebyshev space, by replacing g_1 by $-g_1$, if necessary, we get

$$(-1)^i g_1(t) \geq 0, \qquad t \in [t_i, t_{i+1}], \quad i = j, \ldots, n \qquad (4.14)$$

and

$$(-1)^j g_1(t_j) > 0. \qquad (4.15)$$

It follows from (4.11)–(4.15) that

$$g_1(t)\,(s_1(t) - s_2(t)) \geq 0, \qquad t \in [t_i, t_{i+1}], \quad i = 0, \ldots, n,$$

and

$$g_1(t_j)\,(s_1(t_j) - s_2(t_j)) > 0.$$

This implies that

$$\sum_{i=0}^{n}(-1)^i \int_{t_i}^{t_{i+1}} g_1(t)\,dt - \sum_{i=0}^{n}(-1)^i \int_{u_i}^{u_{i+1}} g_1(t)\,dt = \int_a^b (s_1(t_j) - s_2(t))\,dt > 0.$$

But this is a contradiction, since $\{t_1, \ldots, t_n\}$ and $\{u_1, \ldots, u_n\}$ are sets of canonical points for G. This proves Theorem 4.7.

We remark that Hobby & Rice [1965] showed that for every arbitrary n–dimensional subspace of $C[a, b]$ there exists a set of r canonical points, where $1 \leq r \leq n$.

In the following we investigate the relationship between best L_1–approximation and interpolation at canonical ponts. The convexity cone plays a central role in this context.

Definition 4.8. Let G be a Chebyshev subspace of $C[a, b]$. The set

$$C(G) = \{f \in C[a, b] : \ \text{span}(G \cup \{f\}) \text{ is a} \atop \hspace{3.5em} \text{Chebyshev subspace of } C[a, b]\} \qquad (4.16)$$

is called the *convexity cone* of G.

It follows from Lemma 3.21 that functions $f \in C^{n+1}[a, b]$ with $f^{(n+1)}(t) \neq 0$ for all $t \in (a, b)$ belong to the convexity cone $C(P_n)$.

The following theorem, due to Micchelli [1977], shows that for a function from the convexity cone, its best L_1–approximations from a given Chebyshev space can be computed by interpolation at canonical points. A similar result was proved by Bernstein [1926] for spaces of polynomials.

Theorem 4.9. *Let G be an n-dimensional Chebyshev subspace of $C[a,b]$ and $t_1 < \cdots < t_n$ in (a,b) be the canonical points for G. Then for every function $f \in C(G)$ its unique best L_1-approximation g_f from G is uniquely determined by*

$$g_f(t_i) = f(t_i), \qquad i = 1, \ldots, n. \tag{4.17}$$

Proof. Let $G = \operatorname{span}\{g_1, \ldots, g_n\}$ be an n-dimensional Chebyshev subspace of $C[a,b]$ and $f \in C(G)$. By Theorem 4.7 there exists a set $\{t_1, \ldots, t_n\}$ of canonical points for G. Since G is a Chebyshev space, we have

$$D \left(\begin{matrix} g_1, \ldots, g_n \\ t_1, \ldots, t_n \end{matrix} \right) \neq 0.$$

This implies that there exists a unique function $g_f \in G$ such that

$$g_f(t_i) = f(t_i), \qquad i = 1, \ldots, n. \tag{4.18}$$

We will show that g_f is the best L_1-approximation of f from G. It follows from (4.18) and Cramer's rule that

$$f(t) - g_f(t) = \frac{\overline{D} \left(\begin{matrix} g_1, \ldots, g_n, f \\ t_1, \ldots, t_n, t \end{matrix} \right)}{\overline{D} \left(\begin{matrix} g_1, \ldots, g_n \\ t_1, \ldots, t_n \end{matrix} \right)}$$

for all $t \in [a,b]$. Then since G is Chebyshev and $f \in C(G)$, there exists a sign $\sigma \in \{-1, 1\}$ such that

$$\sigma (-1)^i \left(f(t) - g_f(t) \right) \geq 0, \qquad t \in [t_i, t_{i+1}], \quad i = 0, \ldots, n.$$

Then, since $\{t_1, \ldots, t_n\}$ is a set of canonical points for G, for all $g \in G$,

$$
\begin{aligned}
\|f - g_f\|_1 &= \int_a^b |f(t) - g_f(t)| \, dt \\
&= \sum_{i=0}^n \sigma (-1)^i \int_{t_i}^{t_{i+1}} \left(f(t) - g_f(t) \right) dt \\
&= \sum_{i=0}^n \sigma (-1)^i \int_{t_i}^{t_{i+1}} \left(f(t) - g(t) \right) dt \\
&\leq \int_a^b |f(t) - g(t)| \, dt = \|f - g\|_1.
\end{aligned}
$$

Then it follows from Theorem 4.4 that g_f is the unique best L_1-approximation from G. This proves Theorem 4.9.

By a result of Bernstein [1926], the canonical points of spaces of polynomials are explicitly known.

Theorem 4.10. *The canonical points for P_n on $[-1,1]$ are the extreme points of the Chebyshev polynomial T_{n+2} in $(-1,1)$:*

$$t_i = \cos\{\tfrac{n+2-i}{n+2}\pi\}, \qquad i = 1,\dots,n+1. \tag{4.19}$$

Proof. Let $t_1,\dots,t_{n+1}$ be the points as defined in (4.19) and $t_0 = -1$, $t_{n+2} = 1$. Since the set $\{T_0,\dots,T_n\}$ of Chebyshev polynomials is a basis of P_n, it suffices to show that

$$\sum_{i=0}^{n+1}(-1)^i \int_{t_i}^{t_{i+1}} T_j(t)\,dt = 0, \qquad j = 0,\dots,n. \tag{4.20}$$

We set

$$\tilde{\sigma}(t) = \begin{cases} (-1)^i, & \text{if } t \in (t_i, t_{i+1}), \\ 0, & \text{if } t \in \{t_0,\dots,t_{n+2}\} \end{cases} \qquad i = 0,\dots,n+1$$

and

$$\sigma(w) = \tilde{\sigma}(\cos w), \qquad w \in [0,\pi].$$

We extend the function σ to $(-\infty,\infty)$ by defining

$$\sigma(w) = -\sigma(-w), \qquad w \in [-\pi, 0],$$

and by letting $\sigma : (-\infty,\infty) \to \mathbf{R}$ be a 2π-periodic function. By definition of the points $t_0,\dots,t_{n+2}$, we have

$$\sigma(w + \tfrac{\pi}{n+2}) = -\sigma(w). \tag{4.21}$$

Then by setting $t = \cos w$, it follows that for all $j \in \{0,\dots,n\}$,

$$\begin{aligned}
\sum_{i=0}^{n+1}(-1)^i \int_{t_i}^{t_{i+1}} T_j(t)\,dt &= \int_{-1}^{1} \tilde{\sigma}\, T_j(t)\,dt \\
&= \int_{0}^{\pi} \sigma(w)\cos(jw)\sin w\,dw \\
&= \tfrac{1}{2} \int_{0}^{\pi} \sigma(w)\{\sin((j+1)w) - \sin((j-1)w)\}\,dw \\
&= \tfrac{1}{4} \int_{-\pi}^{\pi} \sigma(w)\{\sin((j+1)w) - \sin((j-1)w)\}\,dw.
\end{aligned}$$

The last equality sign holds since σ is an odd function. Therefore, it suffices to show that

$$\int_{-\pi}^{\pi} \sigma(w)\sin(kw)\,dw = 0, \qquad k = 0,\dots,n+1.$$

Let $k \in \{0,\dots,n+1\}$ be given. Then it follows from (4.21) that

$$\begin{aligned}
\int_{-\pi}^{\pi} \sigma(w)\sin(kw)\,dw &= \int_{-\pi}^{\pi} \sigma(w + \tfrac{\pi}{n+2})\sin\{k(w + \tfrac{\pi}{n+2})\}\,dw \\
&= \cos(\tfrac{k\pi}{n+2}) \int_{-\pi}^{\pi} \sigma(w)\sin(kw)\,dw + \sin(\tfrac{k\pi}{n+2}) \int_{-\pi}^{\pi} \sigma(w)\cos(kw)\,dw \\
&= \cos(\tfrac{k\pi}{n+2}) \int_{-\pi}^{\pi} \sigma(w)\sin(kw)\,dw.
\end{aligned}$$

The last equality sign holds since $w \to \sigma(w) \sin(kw)$ is an odd function and therefore

$$\int_{-\pi}^{\pi} \sigma(w) \cos(kw) \, dw = 0.$$

Finally, since $\cos(\frac{k\pi}{n+2}) \neq 1$, we get

$$\int_{-\pi}^{\pi} \sigma(w) \sin(kw) \, dw = 0.$$

This shows (4.20) and proves Theorem 4.10.

5. Best One–Sided L_1–Approximation by Chebyshev Spaces and Quadrature Formulas

The problem of best one–sided L_1–approximation by functions from Chebyshev spaces is investigated. We give results on characterization and unicity of best one–sided L_1–approximations. It is shown that there is a close relationship between one–sided L_1–approximation and quadrature formulas. In particular, we prove the existence and uniqueness of Gauss quadrature formulas for Chebyshev spaces. Moreover, it is shown that for certain functions, best one–sided L_1–approximations from a Chebyshev space can be computed by Hermite interpolation at points of the corresponding Gauss quadrature formula.

5.1. Unicity of Best One–Sided L_1–Approximations

It is shown that in contrast to best uniform approximation and best L_1–approximation, there exist continuous functions with more than one best one–sided L_1–approximation from a given Chebyshev space. On the other hand, global unicity holds for best one–sided L_1–approximations from extended Chebyshev spaces, if the functions to be approximated are differentiable.

We begin with the definition of best one–sided L_1–approximations.

Definition 5.1. Let G be a subspace of $C[a, b]$ and $f \in C[a, b]$. A function $g_f \in G$ with $g_f \leq f$ (i.e. $g_f(t) \leq f(t)$ for all $t \in [a, b]$) is called a *best one–sided L_1–approximation* of f from G, if

$$\|f - g_f\|_1 = \inf_{\substack{g \in G \\ g \leq f}} \|f - g\|_1. \tag{5.1}$$

The general Existence Theorem 3.3 cannot be applied to the special approximation problem (5.1). But the following result on the existence of best one-sided L_1–approximations from finite-dimensional spaces can be shown.

Theorem 5.2. *Let G be a finite-dimensional subspace of $C[a, b]$ with a strictly positive function. Then for every $f \in C[a, b]$, there exists a best one–sided L_1–approximation from G.*

Proof. Let $f \in C[a, b]$ be given. Then there exists a sequence (g_n) in G such that $g_n \leq f$ for all n and $\|f - g_n\|_1 \to d(f, G)$, where $d(f, G) = \inf\{\|f - g\|_1 : g \in G, g \leq f\}$. This implies that there exists a constant $K > 0$ such that for all n,

$$\|g_n\|_1 \leq \|f - g_n\|_1 + \|f\|_1 \leq d(f, G) + K + \|f\|_1.$$

Since (g_n) is a bounded sequence and G is finite–dimensional, we may assume that (g_n) converges to some $g_f \in G$ in the L_1–norm. Moreover, since G is finite–dimensional, the sequence (g_n) also converges to g_f in the L_∞-norm. Therefore, since $g_n \leq f$ for all n, by taking limits we get that $g_f \leq f$. Moreover, we have

$$\|f - g_f\|_1 = \lim_{n \to \infty} \|f - g_n\|_1 = d(f, G),$$

i.e. g_f is a best one–sided L_1–approximation of f from G. This proves Theorem 5.2.

We now investigate the uniqueness of the best approximations. Bojanic & DeVore [1966] showed that in contrast to the results in the Sections 3 and 4, best one–sided L_1–approximations from Chebyshev spaces are not always unique.

Theorem 5.3. *Let $n \geq 2$ and G be an n–dimensional Chebyshev subspace of $C[a, b]$. Then there exists a function in $C[a, b]$ which has more than one best one–sided L_1–approximation from G.*

Proof. Since G is an n–dimensional Chebyshev space with $n \geq 2$, it follows from Theorem 5.9 below that there exists a Gauss quadrature formula of G. This implies that there exist points $a \leq t_1 < \cdots < t_r < b$ and weights $a_1, \ldots, a_r > 0$ with $r \leq n - 1$ such that

$$\int_a^b g(t)\,dt = \sum_{i=1}^r a_i\, g(t_i) \qquad \text{for all } g \in G. \tag{5.2}$$

We now choose additional distinct points $t_{r+1}, \ldots, t_n$. Since G is a Chebyshev space, by Theorem 1.4 there exists a function $g_1 \in G$ such that

$$g_1(t_i) = 0, \qquad i = 1, \ldots, r, \tag{5.3}$$

and

$$g_1(t_i) = 1, \qquad i = r + 1, \ldots, n. \tag{5.4}$$

We will show that $f = |g_1|$ has two distinct best one–sided L_1–approximations. Let $g \in G$ with $g \leq f$ be given. Then it follows from (5.2), (5.3) and $g \leq |g_1|$ that

$$\int_a^b g(t)\,dt = \sum_{i=1}^r a_i\, g(t_i) \leq \sum_{i=1}^r a_i\, g_1(t_i) = \int_a^b g_1(t)\,dt.$$

Obviously, this implies that g_1 is a best one–sided L_1–approximation of f from G. Moreover, since

$$\int_a^b g_1(t)\,dt = \sum_{i=1}^r a_i\, g_1(t_i) = 0 = \int_a^b g_2(t)\,dt,$$

where $g_2 = 0$, the function g_2 is a further best one–sided L_1–approximation of f from G. This proves Theorem 5.3.

We remark that independently Nürnberger [1985a], [1985b] and Pinkus & Totik [1984] proved nonunicity as in Theorem 5.3 for arbitrary n–dimensional subspaces G of $C[a, b]$ which contain a strictly positive function, if $n \geq 2$. The first author gave a characterization of unicity spaces by using optimization techniques.

In the following we will show that in contrast to Theorem 5.3, differentiable functions always have unique best one–sided L_1–approximations from extended Chebyshev spaces.

We need the following sufficient condition for global unicity due to Strauß [1982].

For every function $h \in C^1[a, b]$ we set

$$Z_1(h) = \{t \in [a, b]: \; h(t) = h'(t) = 0, \quad \text{if } t \in (a, b),$$
$$\text{or } h(t) = 0, \qquad \text{if } t \in \{a, b\}\}. \tag{5.5}$$

Theorem 5.4. *Let G be a subspace of $C^1[a, b]$ which contains a strictly positive function. We consider the following statements:*
(i) *For every function $f \in C^1[a, b]$, there exists a unique best one–sided L_1–approximation from G.*
(ii) *For every nontrivial function $g_0 \in G$, there exists a nontrivial function $g \in G$ such that*

$$g(t) \geq 0, \qquad t \in [a, b], \tag{5.6}$$

and

$$Z_1(g_0) \subset Z_1(g). \tag{5.7}$$

Then (ii) $\Rightarrow$ (i).

Proof. Suppose that (ii) holds, but (i) fails, i.e. there exists a function $f \in C^1[a, b]$ such that $0 \in G$ and $g_0 \in G$, $g_0 \neq 0$, are best one–sided L_1–approximations of f from G. We will show that this leads to a contradiction. It follows from (ii) that there exists a nontrivial $g \in G$ such that (5.6) and (5.7) hold. Moreover, by assumption there exists a function $g_1 \in G$ with $g_1 > 0$. Then by (5.6) and (5.7) there exists a sufficiently small $\varepsilon_1 > 0$ and a neighborhood U of $Z_1(g_0)$ such that the function $g_2 = g - \varepsilon_1 g_1$ satisfies

$$g_2(t) \leq |g_0(t)|, \qquad t \in U, \tag{5.8}$$

and

$$\int_a^b g_2(t) \, dt > 0. \tag{5.9}$$

By using (5.8), we will show that there exists a sufficiently small $\varepsilon > 0$ such that

$$\tfrac{1}{2} g_0 + \varepsilon \, g_2 \leq f. \tag{5.10}$$

Moreover, it follows from (5.9) that

$$\int_a^b (f(t) - \tfrac{1}{2} g_0(t))\, dt > \int_a^b (f(t) - \tfrac{1}{2} g_0(t) - \varepsilon\, g_2(t))\, dt.$$

This shows that $\tfrac{1}{2} g_0$ is not a best one–sided L_1–approximation of f from G. On the other hand, since 0 and g_0 are best approximations of f, the function $\tfrac{1}{2} g_o \in G$ is also a best approximation of f. This is the desired contradiction. Thus, it remains to prove (5.10). We first show that

$$Z_1(f - \tfrac{1}{2} g_0) \subset Z_1(g_0). \qquad (5.11)$$

Let $t \in (a, b)$ be given such that $f(t) - \tfrac{1}{2} g_0(t) = f'(t) - \tfrac{1}{2} g_0'(t) = 0$. Then we have $\tfrac{1}{2}(f(t) - g_0(t)) + \tfrac{1}{2} f(t) = f(t) - \tfrac{1}{2} g_0(t) = 0$ and, since $f \geq 0$ and $f - g_0 \geq 0$, we get $f(t) - g_0(t) = f(t) = 0$ and $f'(t) - g_0'(t) = f'(t) = 0$. It follows that $g_0(t) = 0$ and $g_0'(t) = 0$. This proves (5.11). We now show that

$$\tfrac{1}{2} g_2(t) \leq f(t) - \tfrac{1}{2} g_0(t), \qquad t \in U. \qquad (5.12)$$

Let $t \in U$ be given. If $g_2(t) \leq 0$, then, since $f \geq \tfrac{1}{2} g_0$ we get $\tfrac{1}{2} g_2(t) \leq 0 \leq f(t) - \tfrac{1}{2} g_0(t)$. If $g_2(t) > 0$ and $g_0(t) \leq 0$, then by (5.8) and $f \geq 0$ we get $\tfrac{1}{2} g_2(t) \leq -\tfrac{1}{2} g_0(t) \leq f(t) - \tfrac{1}{2} g_0(t)$. If $g_2(t) > 0$ and $g_0(t) > 0$, then by (5.8) and $f \geq g_0$ we get $\tfrac{1}{2} g_2(t) \leq \tfrac{1}{2} g_0(t) \leq f(t) - \tfrac{1}{2} g_0(t)$. This proves (5.12).

It follows from (5.11) that U is a neighborhood of $Z_1(f - \tfrac{1}{2} g_0)$. This implies, since $f - \tfrac{1}{2} g_0 \geq 0$, that there exists a constant $c > 0$ such that

$$f(t) - \tfrac{1}{2} g_0(t) \geq c, \qquad t \in [a, b] \setminus U.$$

Then there exists a sufficiently small $\varepsilon > 0$ such that

$$\varepsilon\, g_2(t) \leq c \leq f(t) - \tfrac{1}{2} g_0(t), \qquad t \in [a, b] \setminus U.$$

Moreover, it follows from (5.12) that

$$\varepsilon\, g_2(t) \leq f(t) - \tfrac{1}{2} g_0(t), \qquad t \in U.$$

This shows (5.10) and proves Theorem 5.4.

By using Theorem 5.4 we can now prove the following global unicity result for differentiable functions due to DeVore [1968].

Theorem 5.5. *Let G be an n–dimensional extended Chebyshev subspace of $C^{n-1}[a, b]$. Then for every function $f \in C^1[a, b]$, there exists a unique best one–sided L_1–approximation from G.*

Proof. We will show that condition (ii) in Theorem 5.4 is satisfied. Let a non-trivial $g_0 \in G$ be given. We set

$$Z_1(g_0) = \{t_1, \ldots, t_r\},$$

where $a \leq t_1 < \cdots < t_r \leq b$. Moreover, let $i \in \{1,\ldots,r\}$ be given. If $t_i \in \{a,b\}$, then we associate with t_i a sequence $(t_{i,m})$ converging to t_i such that $t_{i,m} \neq t_i$ for all m. If $t_i \in (a,b)$, then we associate with t_i two sequences $(u_{i,m})$ and $(v_{i,m})$ converging to t_i such that

$$u_{i,m} < t_i < v_{i,m} \qquad \text{for all } m.$$

We denote the resulting sequences by $(w_{j,m})$, $j = 1,\ldots,q$, such that

$$a = w_{0,m} < w_{1,m} < \cdots < w_{q,m} < w_{q+1,m} = b$$

for sufficiently large m. Since by Theorem 1.17 the sum of multiplicities of the zeros $t_1,\ldots,t_r \in Z_1(g_0)$ is at most $n-1$, it follows from the definition of the sequences $(w_{j,m})$, $j = 1,\ldots,q$, that $q \leq n-1$. Then by Corollary 1.15 for each m there exists a nontrivial $g_m \in G$ such that

$$(-1)^i g_m(t) \geq 0, \qquad t \in [w_{i,m}, w_{i+1,m}], \quad i = 0,\ldots,q. \qquad (5.13)$$

By multiplying each g_m with $1/\|g_m\|_\infty$, we may assume that $\|g_m\|_\infty = 1$. Since G is finite–dimensional, there exists a subsequence of (g_m) converging to a nontrivial function $g \in G$. By taking limits, it follows from (5.13) and the definition of the sequences $(w_{j,m})$ that $g(t) \geq 0$ for all $t \in [a,b]$ and $Z_1(g_0) \subset Z_1(g)$. This shows that condition (ii) in Theorem 5.4 is satisfied and proves Theorem 5.5.

5.2. Gauss Quadrature Formulas for Chebyshev Spaces

We give a characterization of best one–sided L_1–approximations from a given finite–dimensional space by using quadrature formulas which are exact for all functions from this space. Moreover, it is shown that for a given Chebyshev space there exists a unique Gauss quadrature formula, and that for differentiable functions from the convexity cone, best one–sided L_1–approximations from an extended Chebyshev space can be computed by Hermite interpolation at the points of the quadrature formula.

Best uniform approximations and best L_1–approximations can be characterized by Kolmogorov type conditions (see Theorem 3.9 and Theorem 4.2). We begin with a well-known result which characterizes best one–sided L_1–approximations by the existence of certain quadrature formulas.

Theorem 5.6. *Let G be an n–dimensional subspace of $C[a,b]$ which contains a strictly positive function, $f \in C[a,b]$ and $g_f \in G$ with $g_f \leq f$. The following statements are equivalent:*

(i)   The function g_f is a best one–sided L_1–approximation of f from G.

(ii) There exist an integer $r \in \{1,\ldots,n\}$, points $a \leq t_1 < \cdots < t_r \leq b$ and real numbers $a_1,\ldots,a_r > 0$ such that

$$f(t_i) - g_f(t_i) = 0, \qquad i = 1,\ldots,r, \qquad (5.14)$$

and

$$\int_a^b g(t)\,dt = \sum_{i=1}^r a_i\, g(t_i) \qquad \text{for all } g \in G. \qquad (5.15)$$

Proof. (ii) $\Rightarrow$ (i). Suppose that (ii) holds and let $g \in G$ with $g \leq f$ be given. Then it follows that

$$\int_a^b g(t)\,dt \;=\; \sum_{i=1}^r a_i\, g(t_i) \leq \sum_{i=1}^r a_i\, f(t_i)$$

$$=\; \sum_{i=1}^r a_i\, g_f(t_i) = \int_a^b g_f(t)\,dt.$$

Obviously, this implies that g_f is a best one–sided L_1–approximation of f from G.

(i) $\Rightarrow$ (ii). Let $\{g_1, \ldots, g_n\}$ be a basis of G. Suppose that (ii) fails. Then there do not exist points $t_1, \ldots, t_r$ in $Z(f - g_f)$ and real numbers $a_1, \ldots, a_r \geq 0$ such that

$$\left(\int_a^b g_1(t)\,dt, \ldots, \int_a^b g_n(t)\,dt\right) = \sum_{i=1}^r a_i\, (g_1(t_i), \ldots, g_n(t_i)).$$

Then by the theorem of Carathéodory and the lemma of Farkas (see e.g. Hettich & Zencke [1982, p.54-55]), there exist real numbers $b_1, \ldots, b_n$ such that

$$\sum_{i=1}^n b_i \int_a^b g_i(t)\,dt > 0 \tag{5.16}$$

and

$$\sum_{i=1}^n b_i\, g_i(t) \leq 0, \qquad t \in Z(f - g_f). \tag{5.17}$$

We set $g = \sum_{i=1}^n b_i\, g_i$. By assumption there exists a strictly positive function $h \in G$. Then by (5.16) and (5.17) there exists a sufficiently small $\varepsilon > 0$ such that for $g_\varepsilon = g - \varepsilon\, h$ we have

$$\int_a^b g_\varepsilon(t)\,dt > 0 \tag{5.18}$$

and

$$g_\varepsilon(t) < 0, \qquad t \in Z(f - g_f). \tag{5.19}$$

Since $Z(f - g_f)$ is compact, there exists a real number $c > 0$ such that

$$g_\varepsilon(t) < -c, \qquad t \in Z(f - g_f).$$

Moreover, there exists an open neighborhood U of $Z(f - g_f)$ such that

$$g_\varepsilon(t) < -\frac{c}{2}, \qquad t \in U. \tag{5.20}$$

Since $[a, b] \setminus U$ is compact, there exists a real number $d > 0$ such that

$$g_f(t) \leq f(t) - d, \qquad t \in [a, b] \setminus U. \tag{5.21}$$

By multiplying g_ε with an appropriate positive factor, we may assume that $\|g_\varepsilon\|_\infty \leq d$. We set $g_0 = g_f + g_\varepsilon$. Then by (5.18) we have

$$\int_a^b g_0(t)\, dt = \int_a^b (g_f(t) + g_\varepsilon(t))\, dt > \int_a^b g_f(t)\, dt. \qquad (5.22)$$

Moreover, by (5.20) we have

$$g_0(t) = g_f(t) + g_\varepsilon(t) < f(t), \qquad t \in U,$$

and by (5.21) we have

$$g_0(t) = g_f(t) + g_\varepsilon(t) \leq f(t) - d + d = f(t), \qquad t \in [a,b] \setminus U.$$

Therefore, it follows from (5.22) and $g_0 \leq f$ that g_f is not a best one–sided L_1–approximation from G. This proves Theorem 5.6.

A fundamental problem in Numerical Analysis is to compute the integral $\int_a^b f(t)\,dt$ of functions $f \in C[a,b]$ approximatively by so–called *quadrature formulas* of the form

$$Q(f) = \sum_{i=1}^r a_i f(t_i),$$

where $a \leq t_1 < \cdots < t_r \leq b$ and $a_1, \ldots, a_r$ are real coefficients. It follows from Theorem 5.2 and Theorem 5.6 that for every n–dimensional subspace G of $C[a,b]$, there exists a quadrature formula with at most n points and positive coefficients which is exact for all functions from G.

In the following we will give results on the existence and uniqueness of such quadrature formulas with approximately $n/2$ points for n–dimensional Chebyshev spaces.

For a set $T = \{t_1, \ldots, t_r\}$ with $a \leq t_1 < \cdots < t_r \leq b$, the *index* of T, denoted by $I(T)$, is the number of points in T counting each point of $T \cap \{a, b\}$ as $1/2$.

We now introduce the notion of Gauss quadrature formulas.

Definition 5.7. Let G be an n–dimensional subspace of $C[a,b]$. A quadrature formula

$$Q(f) = \sum_{i=1}^r a_i\, f(t_i) \qquad (5.23)$$

with points $a \leq t_1 < \cdots < t_r < b$ and weights $a_1, \ldots, a_r > 0$ is called a *Gauss quadrature formula* for G if

$$I(\{t_1, \ldots, t_r\}) = n/2 \qquad (5.24)$$

and

$$\int_a^b g(t)\, dt = Q(g) \qquad \text{for all } g \in G. \qquad (5.25)$$

(Note that we require that the quadrature formula Q be exact for all $g \in G$.)

Remark 5.8. In the definition of Gauss quadrature formulas we require that the point t_r satisfies $t_r < b$. In the following we will only consider this case, although analogous results hold for the case when $t_r = b$.

Krein [1951] proved the existence and uniqueness of Gauss quadrature formulas for Chebyshev spaces.

Theorem 5.9. *Let G be a Chebyshev subspace of $C[a,b]$. Then there exists a unique Gauss quadrature formula for G.*

Proof. Let G be an n–dimensional Chebyshev subspace of $C[a,b]$. It follows from Zalik [1975] that there exists a function $f \in C[a,b]$ such that span$\{G,f\}$ is an $(n+1)$–dimensional Chebyshev subspace of $C[a,b]$. By Theorem 5.2 there exists a best one–sided L_1–approximation $g_f \in G$ of f. Then it follows from Theorem 5.6 that there exists an integer $r \in \{1,\dots,n\}$, points $a \le t_1 < \cdots < t_r \le b$ and real numbers $a_1,\dots,a_r > 0$ such that (5.14) and (5.15) hold. Since span$\{G,f\}$ is a Chebyshev space and $f - g_f \ge 0$, it follows from (5.14) and Theorem 1.19 that $I(\{t_1,\dots,t_r\}) \le \frac{n}{2}$. If the strict inequality sign were to hold, then by the proof of Theorem 5.5 there exists a nontrivial $g \in G$ such that $g \ge 0$ and $g(t_i) = 0$, $i = 1,\dots,r$. By using (5.15) we get

$$0 < \int_a^b g(t)\,dt = \sum_{i=1}^r a_i\,g(t_i) = 0,$$

which is a contradiction. This shows that $I(\{t_1,\dots,t_r\}) = \frac{n}{2}$. We will now show that there exists a Gauss quadrature formula of G with points in $[a,b)$. To do this, we choose a best one–sided L_1–approximation $g_{-f} \in G$ of $-f$. As above there exist quadrature formulas

$$Q_1(f) = \sum_{i=1}^r a_i\,f(t_i) \tag{5.26}$$

with

$$f(t_i) - g_f(t_i) = 0, \qquad i = 1,\dots,r, \tag{5.27}$$

and

$$Q_2(f) = \sum_{i=1}^r b_i\,f(u_i) \tag{5.28}$$

with

$$-f(u_i) - g_{-f}(u_i) = 0, \qquad i = 1,\dots,r. \tag{5.29}$$

We will show that

$$\{t_1,\dots,t_r\} \cap \{u_1,\dots,u_r\} = \emptyset. \tag{5.30}$$

Suppose to the contrary that this is not true. Since $I(\{t_1,\dots,t_r\}) = I(\{u_1,\dots,u_r\}) = \frac{n}{2}$, it follows that $\{t_1,\dots,t_r\} \cup \{u_1,\dots,u_r\}$ contains at most n points. Then it follows that

$$\{t_1,\dots,t_r\} = \{u_1,\dots,u_r\}. \tag{5.31}$$

Indeed, if there exists an index $j \in \{1, \ldots, r\}$ such that $t_j \neq u_j$, then by Theorem 1.4 there exists a function $g \in G$ satisfying

$$\left. \begin{array}{l} g(t_i) = g(u_i) = 0, \quad i = 1, \ldots, r, \quad i \neq j, \\ g(t_j) = 0, \quad g(u_j) = 1. \end{array} \right\} \tag{5.32}$$

Then it follows from (5.26) and (5.28) that

$$0 = Q_1(g) = \int_a^b g(t)\, dt = Q_2(g) = 1,$$

which is a contradiction. This shows that (5.31) holds. Then it follows from (5.27) and (5.29) that

$$g_f(t_i) + g_{-f}(t_i) = 0, \quad i = 1, \ldots, r.$$

Since $g_f + g_{-f} \leq 0$ and $I(\{t_1, \ldots, t_r\}) = \frac{n}{2}$, Theorem 1.19 implies that $g_f + g_{-f} = 0$. Since $f - g_f \geq 0$ and $-f - g_{-f} \geq 0$, it follows that $f - g_f = 0$ which is a contradiction. This shows that (5.30) holds and proves the existence of a Gauss quadrature formula with points in $[a, b)$. We now suppose that there exist two different such quadrature formulas Q_1 and Q_2 as in (5.26) and (5.28). Since $I(\{t_1, \ldots, t_r\}) = \frac{n}{2}$, $I(\{u_1, \ldots, u_r\}) = \frac{n}{2}$, $t_r < b$ and $u_r < b$, it follows from Theorem 1.4 that there exists a function $g \in G$ which satisfies (5.32). This implies that

$$0 = Q_1(g) = \int_a^b g(t)\, dt = Q_2(g) = 1,$$

which is a contradiction. This proves Theorem 5.9.

It was shown in the previous section that for functions from the convexity cone its best L_1-approximations from a given Chebyshev space can be computed by Lagrange interpolation at canonical points (Theorem 4.9). A similar result, due to DeVore [1968], holds for best one–sided L_1-approximation where Hermite interpolation at the points of Gauss quadrature formulas is involved.

Theorem 5.10. *Let G be an n–dimensional extended Chebyshev subspace of $C^{n-1}[a, b]$ and $a \leq t_1 < \cdots < t_r < b$ be the points of the Gauss quadrature formula for G. Then for every function $f \in C(G) \cap C^1[a, b]$, the best one–sided L_1-approximation g_f from G is uniquely determined by*

$$g_f(t_i) = f(t_i), \quad i = 1, \ldots, r, \tag{5.33}$$

and

$$g'_f(t_i) = f'(t_i), \quad \text{if } t_i \neq a, \quad i = 1, \ldots, r. \tag{5.34}$$

Proof. Let a function $f \in C(G) \cap C^1[a, b]$ be given. By Theorem 1.8 there exists a unique function $g_f \in G$ which satisfies (5.33) and (5.34). We will show that g_f is a best one–sided L_1-approximation of f from G. Let $\{g_1, \ldots, g_n\}$ be a basis of G. We count each point $t_i \in (a, b)$, $i = 1, \ldots, r$, twice and denote the resulting points by $u_1 \leq \cdots \leq u_n$. Note, that we have $I(\{t_1, \ldots, t_r\}) = \frac{n}{2}$. Then it follows from (5.33) and (5.34) and Cramer's rule that

$$f(t) - g_f(t) = \frac{D\left(\begin{array}{c} g_1, \ldots, g_n, f \\ u_1, \ldots, u_n, t \end{array}\right)}{D\left(\begin{array}{c} g_1, \ldots, g_n \\ u_1, \ldots, u_n \end{array}\right)}$$

for all $t \in [a, b]$. Since by assumption the spaces G and $\operatorname{span}(G \cup \{f\})$ are Chebyshev, it follows from the proof of Theorem 1.8, (i) $\Rightarrow$ (ii) that the above determinats are non-negative which implies that $f - g_f \geq 0$. Moreover, it follows from (5.33) that for all $g \in G$ with $g \leq f$,

$$\int_a^b g(t)\,dt \;=\; \sum_{i=1}^r a_i\, g(t_i) \leq \sum_{i=1}^r a_i\, f(t_i)$$

$$=\; \sum_{i=1}^r a_i\, g_f(t_i) = \int_a^b g_f(t)\,dt.$$

This implies that g_f is a best one-sided L_1-approximation of f which by Theorem 5.5 is uniquely determined. This proves Theorem 5.10.

By a result of C.F. Gauss, the points of Gauss quadrature formulas for spaces of polynomials are the zeros of the Legendre polynomials defined as follows.

Definition 5.11. The function $q_n : [-1, 1] \to \mathbf{R}$, defined by

$$q_n(t) = \frac{1}{2^n n!} \{(t^2 - 1)^n\}^{(n)} \tag{5.35}$$

for all $t \in [-1, 1]$, is called the *Legendre polynomial of degree n*.

The following facts can be found in standard books on Numerical Analysis:
(i)   The function q_n is a polynomial of degree n.
(ii) $q_0(t) = 1$, $q_1(t) = -t$ and $(n+1)\, q_{n+1}(t) = (2n+1)\, q_n(t) - n\, q_{n-1}(t)$, $t \in [-1, 1]$, $n = 1, 2, \ldots$
(iii) $\int_{-1}^1 q_i(t)\, q_j(t) = \begin{cases} 0, & \text{if } i \neq j \\ \frac{2}{2i+1}, & \text{if } i = j \end{cases}$, $i, j = 0, 1, \ldots$
(iv) For all $n = 1, 2, \ldots$, the function q_n has exactly n zeros in $(-1, 1)$ which are symmetrical to the point $t = 0$.

Theorem 5.12. *The points of the Gauss quadrature formula for P_{2n+1} on $[-1, 1]$ are the zeros of the Legendre polynomial of degree $n + 1$.*

Proof. Let $-1 < t_1 < \cdots < t_{n+1} < 1$ be the zeros of the Legendre polynomial q_{n+1}. It follows from Theorem 1.4 that for each $i \in \{1, \ldots, n+1\}$ there exists a unique polynomial $l_i \in P_n$ such that for all $j \in \{1, \ldots, n+1\}$,

$$l_i(t_j) = \begin{cases} 1, & \text{if } i = j \\ 0, & \text{if } i \neq j \end{cases} .$$

We set

$$Q(f) = \sum_{i=1}^{n+1} a_i \, f(t_i),$$

where

$$a_i = \int_{-1}^{1} l_i(t) \, dt, \qquad i = 1, \ldots, n+1.$$

In the following we will show that

$$\int_{-1}^{1} p(t) \, dt = Q(p), \qquad p \in P_{2n+1} \tag{5.36}$$

and

$$a_i > 0, \qquad i = 1, \ldots, n+1. \tag{5.37}$$

Then it follows from Theorem 5.9 that Q is the unique Gauss quadrature formula for P_{2n+1}. By Theorem 2.13 every polynomial $p \in P_n$ can be written as

$$p(t) = \sum_{i=1}^{n+1} p(t_i) \, l_i(t), \qquad t \in [a, b].$$

This implies that for all $p \in P_n$,

$$\int_{-1}^{1} p(t) \, dt = \sum_{i=1}^{n+1} a_i \, p(t_i) = Q(p). \tag{5.38}$$

Now, let a polynomial $p \in P_{2n+1}$ be given. Then there exist polynomials p_1 and p_2 in P_n such that $p = p_1 \, q_{n+1} + p_2$. It follows from (iii) that

$$\int_{-1}^{1} p_1(t) \, q_{n+1}(t) \, dt = 0,$$

and therefore by (5.38) we have

$$\begin{aligned}
\int_{-1}^{1} p(t) \, dt &= \int_{-1}^{1} p_1(t) \, q_{n+1}(t) \, dt + \int_{-1}^{1} p_2(t) \, dt \\
&= \sum_{i=1}^{n+1} a_i \, p_1(t_i) \, q_{n+1}(t_i) + \sum_{i=1}^{n+1} a_i \, p_2(t) = Q(p).
\end{aligned}$$

This shows that (5.36) holds. Finally, let an integer $j \in \{1, \ldots, n+1\}$ be given. Since $l_j^2 \in P_{2n}$, it follows from (5.36) that

$$a_j = \sum_{i=1}^{n+1} a_i \, (l_j(t_i))^2 = \int_{-1}^{1} (l_j(t))^2 \, dt > 0.$$

This shows (5.37) and proves Theorem 5.12.

6. Best L_2–Approximation

Best L_2–approximation of functions is part of the general theory of best approximation in Hilbert spaces. We give some elementary results of this theory. In particular, it follows that — in contrast to other norms — best L_2–approximations from arbitrary finite–dimensional spaces are always unique and can be computed by solving a system of linear equations or, if an orthogonal basis is available, in a direct way.

The space

$$L_2[a, b] = \{f : [a, b] \to \mathbf{R} : f \text{ is Lebesgue measurable and } \int_a^b (f(t))^2 \, dt < \infty\},$$

endowed with the scalar product

$$\langle f_1, f_2 \rangle = \int_a^b f_1(t) \, f_2(t) \, dt.$$

for all $f_1, f_2 \in L_2[a, b]$, is a real Hilbert space. Therefore, in the following we consider best approximation in arbitrary Hilbert spaces.

Let a real Hilbert space E with the scalar product $\langle \cdot, \cdot \rangle$ be given. Then E, endowed with the norm

$$\|f\| = (\langle f, f \rangle)^{\frac{1}{2}}$$

for all $f \in E$, is a real normed linear space. By using this norm, the approximation problem is defined as in Definition 3.1.

Since Hilbert spaces are strictly convex, it follows from Theorem 3.5 that best approximations from a finite-dimensional subspace of a Hilbert space are always unique.

Theorem 6.1. *Let G be a finite–dimensional subspace of a real Hilbert space E. Then for every $f \in E$, there exists a unique best approximation from G.*

Proof. Let E be a Hilbert space. It suffices to show that E is strictly convex. Then the result follows from Theorem 3.5. By using the properties of the scalar product a direct computation shows that the parallelogram law

$$\|f_1 + f_2\|^2 + \|f_1 - f_2\|^2 = 2\|f_1\|^2 + 2\|f_2\|^2 \tag{6.1}$$

for all $f_1, f_2 \in E$ holds. Now, let $f_1, f_2 \in E$ with $f_1 \neq f_2$ and $\|f_1\| = \|f_2\| = 1$ be given. Then by (6.1) we have $\|f_1 + f_2\|^2 < 4$, i.e. $\|\frac{1}{2}(f_1 + f_2)\| < 1$. This shows that E is strictly convex and proves Theorem 6.1.

Knowing that best approximations are unique in Hilbert spaces, we now show that they can be characterized by an orthogonality relation.

Theorem 6.2. *Let G be a subspace of a real Hilbert space E, $f \in E$ and $g_f \in G$. The following statements are equivalent:*
(i) The element g_f is a best approximation of f from G.

(ii) For all $g \in G$,

$$\langle f - g_f, g \rangle = 0. \tag{6.2}$$

Proof. (ii) $\Rightarrow$ (i). Suppose that (ii) holds and let $g \in G$ be given. Then it follows from (ii) and the properties of the scalar product that

$$
\begin{aligned}
\|f - g\|^2 - \|f - g_f\|^2 &= \|g\|^2 - 2\langle f, g \rangle - \|g_f\|^2 + 2\langle f, g_f \rangle \\
&= \|g - g_f\|^2 + 2\langle f - g_f, g_f - g \rangle \\
&= \|g - g_f\|^2 \geq 0,
\end{aligned}
$$

which shows that (i) holds.

(i) $\Rightarrow$ (ii). Suppose that (ii) fails, i.e. there exists a function $g_0 \in G$ such that

$$\langle f - g_f, g_0 \rangle \neq 0.$$

This implies that

$$
\left\| f - \left(g_f + \frac{\langle f - g_f, g_0 \rangle}{\langle g_0, g_0 \rangle} g_0 \right) \right\|^2 = \|f - g_f\|^2 - \frac{\langle f - g_f, g_0 \rangle^2}{\langle g_0, g_0 \rangle} < \|f - g_f\|^2,
$$

which implies that g_f is not a best approximation of f. This proves Theorem 6.2.

The following result is a consequence of the characterization Theorem 6.2 and shows that best approximations from a finite–dimensional subspace of a Hilbert space can be computed by solving a system of linear equations.

Corollary 6.3. *Let* $G = \text{span}\{g_1, \ldots, g_n\}$ *be an* n–*dimensional subspace of a real Hilbert space* E, $f \in E$ *and* $g_f = \sum_{i=1}^n a_i g_i \in G$. *The following statements are equivalent:*

(i) *The element* g_f *is a best approximation of* f *from* G.

(ii) *The coefficients* $a_1, \ldots, a_n$ *satisfy the following system of linear equations*

$$\sum_{i=1}^n a_i \langle g_i, g_j \rangle = \langle f, g_j \rangle, \qquad j = 1, \ldots, n. \tag{6.3}$$

Moreover, the corresponding determinant

$$\det(\langle g_i, g_j \rangle)_{i,j=1}^n, \tag{6.4}$$

called the Gram determinant, *is different from zero.*

Proof. Condition (6.2) in Theorem 6.2 is equivalent to

$$\langle f - g_f, g_j \rangle = 0, \qquad j = 1, \ldots, n. \tag{6.5}$$

Since $g_f = \sum_{i=1}^n a_i\, g_i$, condition (6.3) is equivalent to (6.5). Suppose now that the determinant (6.4) is zero. Then there exist real numbers $b_1, \ldots, b_n$ such that

$$\sum_{i=1}^n |b_i| \neq 0 \tag{6.6}$$

and

$$\sum_{i=1}^n b_i\, \langle g_i, g_j \rangle = 0, \qquad j = 1, \ldots, n. \tag{6.7}$$

For each $j \in \{1, \ldots, n\}$, we multiply (6.7) with b_j and form the sum of all equations. Then we get

$$\left\langle \sum_{i=1}^n b_i\, g_i, \sum_{j=1}^n b_j\, g_j \right\rangle = 0$$

which implies $\sum_{i=1}^n b_i\, g_i = 0$. This contradicts the linear independence of $g_1, \ldots, g_n$ and proves Corollary 6.3.

Corollary 6.3 shows that best approximations in Hilbert spaces can be obtained easily, if we are able to solve the linear system (6.2) numerically. However, the Gram matrix (6.4) corresponding to (6.3) can be ill–conditioned and therefore, the numerical solution of (6.2) may be impossible.

A standard example is the following. If $E = L_2[0,1]$ and $G = P_n$ then the elements of the Gram matrix, in this case called the *Hilbert matrix*, are given by

$$\langle t^i, t^j \rangle = \int_0^1 t^i\, t^j\, dt = \frac{1}{i+j+1}, \qquad i, j = 0, \ldots, n. \tag{6.8}$$

It is well known that the Hilbert matrix is highly ill–conditioned.

On the other hand, if $E = L_2[0,1]$ and $G = S_m(x_1, \ldots, x_k)$ is a spline space (see Definition 1.15 in Chapter II) and we use the B–spline basis (see Definition 2.3 and Theorem 2.6 in Chapter II), then the corresponding Gram matrix is relatively well conditioned.

Further details on the numerical solution of the linear system (6.3) can be found in Forsythe & Moler [1967], Lawson & Hanson [1974] and de Boor [1978].

In the following we will describe an alternative approach for computing best approximations which avoids possible difficulties arising from the solution of (6.3).

A basis $\{g_1, \ldots, g_n\}$ of an n–dimensional subspace G of a Hilbert space is called *orthogonal* if

$$\langle g_i, g_j \rangle = 0, \qquad i \neq j, \quad i, j = 1, \ldots, n. \tag{6.9}$$

It is easy to verify that best approximations from spaces with an orthogonal basis can be computed in a direct way.

Corollary 6.4. *Let E be a real Hilbert space, $\{g_1, \ldots, g_n\}$ be an orthogonal basis of an n–dimensional subspace G of E, and $f \in E$. Then the best approximation g_f of f from G is given by*

$$g_f = \sum_{i=1}^n \frac{\langle f, g_i \rangle}{\langle g_i, g_i \rangle}\, g_i. \tag{6.10}$$

Proof. If $g_f = \sum_{i=1}^{n} a_i \, g_i$ is a best approximation of f from G, then by Corollary 6.3 the coefficients $a_1, \ldots, a_n$ satisfy

$$\sum_{i=1}^{n} a_i \, \langle g_i, g_j \rangle = \langle f, g_j \rangle, \qquad j = 1, \ldots, n. \tag{6.11}$$

Since by assumption $\{g_1, \ldots, g_n\}$ is an orthogonal basis of G, it follows from condition (6.11) that

$$a_j \, \langle g_j, g_j \rangle = \langle f, g_j \rangle, \qquad j = 1, \ldots, n.$$

This proves Corollary 6.4.

Example 6.5. (i) It follows from the properties (i) and (iii) after Definition 5.11 that for all n, the set of Legendre polynomials $\{q_0, \ldots, q_n\}$ is an orthogonal basis of the space P_n.

(ii) The set $\{1, \cos t, \ldots, \cos(nt), \sin t, \ldots, \sin(nt)\}$ is an orthogonal basis of the space Q_n of trigonometric polynomials of degree n (see Example 1.3). Therefore, it follows from Corollary 6.4 that for a given function $f \in L_2[-\pi, \pi]$, its best L_2-approximation from Q_n is given by

$$g_f(t) = \frac{1}{2} a_0 + \sum_{j=1}^{n} (a_j \, \cos(jt) + b_j \, \sin(jt)), \ t \in [-\pi, \pi],$$

where

$$a_j = \frac{1}{\pi} \int_{-\pi}^{\pi} f(t) \cos(jt) \, dt, \ j = 0, \ldots, n,$$

and

$$b_j = \frac{1}{\pi} \int_{-\pi}^{\pi} f(t) \sin(jt) \, dt, \ j = 1, \ldots, n,$$

are the so-called *Fourier coefficients* of f.

We note that the results in this section also hold for inner product spaces E (i.e. without assuming the completeness of E).

Chapter II. Splines and Weak Chebyshev Spaces

1. Weak Chebyshev Spaces

The theory of Chebyshev spaces is not applicable to spline spaces. However, spline spaces are prototypes of the more general class of weak Chebyshev spaces. We give some basic theorems concerning these spaces which in the subsequent sections will have important applications to splines. A fundamental result says that weak Chebyshev spaces can be "approximated" by Chebyshev spaces. Therefore, although there are some crucial differences between Chebyshev and weak Chebyshev spaces, certain properties of Chebyshev spaces can be carried over to weak Chebyshev spaces. In particular, for every continous function, there exists a best uniform approximation from a given weak Chebyshev space such that the error has the classical alternation property.

1.1. Basic Properties

In this section we introduce weak Chebyshev spaces and give some fundamental results on these spaces. In particular, it is shown that weak Chebyshev space can be "approximated" by a sequence of Chebyshev spaces.

We begin with the definition of weak Chebyshev spaces which goes back to Karlin & Studden [1966].

Definition 1.1. An n–dimensional subspace G of $C[a, b]$ is called a *weak Chebyshev subspace* if there exists a basis $\{g_1, \ldots, g_n\}$ of G such that

$$D \left(\begin{array}{c} g_1, \ldots, g_n \\ t_1, \ldots, t_n \end{array} \right) \geq 0 \tag{1.1}$$

for all $t_1 < \cdots < t_n$ in $[a, b]$.

In the following we discuss the "approximation" of weak Chebyshev spaces by Chebyshev spaces.

To do this, we need the following result of Polya & Szegö [1972]. The proof given here is due to Meinardus [1987].

Theorem 1.2. *Let functions $g_1, \ldots, g_n$ and $f_1, \ldots, f_n$ in $C(-\infty, \infty)$ be given, and let*

$$\Delta_n = \{(t_1, \ldots, t_n) \in \mathbf{R}^n : t_1 \leq \cdots \leq t_n\}.$$

Then

$$\begin{vmatrix} \int_{-\infty}^{\infty} g_1(t) f_1(t) \, dt & \cdots & \int_{-\infty}^{\infty} g_n(t) f_1(t) \, dt \\ \vdots & & \vdots \\ \int_{-\infty}^{\infty} g_1(t) f_n(t) \, dt & \cdots & \int_{-\infty}^{\infty} g_n(t) f_n(t) \, dt \end{vmatrix}$$

$$= \int_{\Delta_n} \overline{D} \begin{pmatrix} g_1, \ldots, g_n \\ t_1, \ldots, t_n \end{pmatrix} \overline{D} \begin{pmatrix} f_1, \ldots, f_n \\ t_1, \ldots, t_n \end{pmatrix} dt_1 \ldots dt_n, \qquad (1.2)$$

if the integrals in the first determinant exist.

Proof. We denote the expression on the left (respectively right) side of (1.2) by I_1 (respectively I_2). Let S_n be the symmetric group of all permutations of the set $\{1, \ldots, n\}$. (Compare standard books on Linear Algebra.) Then it follows that

$$I_1 = \sum_{\sigma \in S_n} \operatorname{sgn} \sigma \int_{-\infty}^{\infty} f_1(t_1) g_{\sigma(1)}(t_1) \, dt_1 \cdots \int_{-\infty}^{\infty} f_n(t_n) g_{\sigma(n)}(t_n) \, dt_n$$

$$= \int_{\mathbf{R}^n} \sum_{\sigma \in S_n} \operatorname{sgn} \sigma \, f_1(t_1) \cdots f_n(t_n) g_{\sigma(1)}(t_1) \cdots g_{\sigma(n)}(t_n) \, dt_1 \cdots dt_n.$$

Moreover, we set

$$I_3 = \int_{\mathbf{R}^n} \overline{D} \begin{pmatrix} g_1, \ldots, g_n \\ t_1, \ldots, t_n \end{pmatrix} \overline{D} \begin{pmatrix} f_1, \ldots, f_n \\ t_1, \ldots, t_n \end{pmatrix} dt_1 \cdots dt_n$$

and have

$$I_3 = \int_{\mathbf{R}^n} \sum_{\rho \in S_n} \operatorname{sgn} \rho \, f_1(t_{\rho(1)}) \cdots f_n(t_{\rho(n)}) \cdot$$

$$\cdot \sum_{\tau \in S_n} \operatorname{sgn} \tau \, g_1(t_{\tau(1)}) \cdots g_n(t_{\tau(n)}) \, dt_1 \cdots dt_n$$

$$= \sum_{\rho \in S_n} \operatorname{sgn} \rho \int_{\mathbf{R}^n} f_1(t_1) \cdots f_n(t_n) \cdot$$

$$\cdot \sum_{\tau \in S_n} \operatorname{sgn} \tau \, g_{\tau^{-1}\rho(1)}(t_1) \cdots g_{\tau^{-1}\rho(n)}(t_n) \, dt_1 \cdots dt_n$$

$$= \sum_{\rho \in S_n} \operatorname{sgn} \rho \int_{\mathbf{R}^n} f_1(t_1) \cdots f_n(t_n) \operatorname{sgn} \rho \cdot$$

$$\cdot \sum_{\sigma \in S_n} \operatorname{sgn} \sigma \, g_{\sigma(1)}(t_1) \cdots g_{\sigma(n)}(t_n) \, dt_1 \cdots dt_n$$

$$= n! \, I_1,$$

since the cardinality of S_n is $n!$. Finally, since I_3 is invariant under each permutation of S_n, and since the cardinality of S_n is $n!$, we have $I_3 = n! \, I_2$ which implies that $I_1 = I_2$. This proves Theorem 1.2.

The next result due to Jones & Karlovitz [1970] says that weak Chebyshev spaces are characterized by the property that they can be "approximated" by Chebyshev spaces.

Theorem 1.3. *For an n-dimensional subspace $G = \operatorname{span}\{g_1, \ldots, g_n\}$ of $C[a, b]$, the following statements are equivalent:*
(i) G is a weak Chebyshev subspace.
(ii) There exists a sequence $G_m = \operatorname{span}\{g_{1,m}, \ldots, g_{n,m}\}$ of Chebyshev subspace of $C[a, b]$ such that

$$\lim_{m \to \infty} \|g_i - g_{i,m}\|_\infty = 0, \qquad i = 1, \ldots, n. \tag{1.3}$$

Proof. (ii) $\Rightarrow$ (i). Suppose that (ii) holds and let points $a \le t_1 < \cdots < t_n \le b$ be given. Then for all m,

$$\overline{D}\left(\begin{array}{c} g_{1,m}, \cdots, g_{n,m} \\ t_1 \ , \ldots, \ t_n \end{array} \right) > 0,$$

since G_m is a Chebyshev space. By taking limits and using (1.3) we get

$$\overline{D}\left(\begin{array}{c} g_1, \ldots, g_n \\ t_1, \ldots, t_n \end{array} \right) \ge 0.$$

This shows that G is weak Chebyshev.

(i) $\Rightarrow$ (ii). Suppose that (i) holds. For all m, the so called *Gauss-kernel* is defined by

$$K_m(s, t) = \frac{m}{\sqrt{2\pi}} \, e^{-\frac{m^2}{2}(s-t)^2}, \qquad (s, t) \in (-\infty, \infty).$$

Let a bounded function $h \in C(-\infty, \infty)$ be given. For all m, we set

$$h_m(t) = \int_{-\infty}^{\infty} h(t) \, K_m(s, t) \, dt, \qquad t \in (-\infty, \infty). \tag{1.4}$$

We will show that

$$\lim_{m \to \infty} \|h - h_m\|_\infty = 0. \tag{1.5}$$

Suppose for the moment that (1.5) holds. For all $i \in \{1, \ldots, n\}$ we extend $g_i \in G$ to $(-\infty, \infty)$ by defining $g_i(t) = g_i(a)$ for all $t \in (-\infty, a)$ and $g_i(t) = g_i(b)$ for all $t \in (b, \infty)$. Then obviously $\operatorname{span}\{g_1, \ldots, g_n\}$ is a weak Chebyshev subspace of $C(-\infty, \infty)$. We now define for all $i \in \{1, \ldots, n\}$,

$$g_{i,m}(t) = \int_{-\infty}^{\infty} g_i(s) \, K_m(s, t) \, ds, \qquad t \in (-\infty, \infty).$$

Then it follows from (1.5) that

$$\lim_{m \to \infty} \|g_i - g_{i,m}\|_\infty = 0, \qquad i = 1, \ldots, n.$$

We now show that for all m, G_m is a Chebyshev space. Let a positive integer m be given. Since by Example 1.13 in Chapter I, spaces of exponentials are Chebyshev subspaces of $C(-\infty, \infty)$, a simple computation shows that

$$\text{span}\{K_m(\cdot, t_1), \ldots, K_m(\cdot, t_n)\}$$

is a Chebyshev subspace of $C(-\infty, \infty)$. Now, let points $t_1 < \cdots < t_n$ in $[a, b]$ be given. Then by Theorem 1.2

$$\overline{D}\left(\begin{array}{c} g_{1,m}, \ldots, g_{n,m} \\ t_1, \ldots, t_n \end{array}\right)$$

$$= \int_{\Delta_n} \overline{D}\left(\begin{array}{c} g_1, \ldots, g_n \\ s_1, \ldots, s_n \end{array}\right) \overline{D}\left(\begin{array}{c} K_m(\cdot, t_1), \ldots, K_m(\cdot, t_n) \\ s_1, \ldots, s_n \end{array}\right) ds_1 \cdots ds_n > 0.$$

Therefore, it remains to prove (1.5). Since $h \in C(-\infty, \infty)$ is a bounded function, there exists a constant $C > 0$ such that

$$|h(t)| \leq C, \qquad t \in (-\infty, \infty).$$

Moreover, since h is uniformly continuous on $[a, b]$, for each $\varepsilon > 0$ there exists a $\delta > 0$ such that for all $s \in (-\infty, \infty)$ and $t \in [a, b]$

$$|h(t) - h(s)| < \frac{\varepsilon}{2} + \frac{2C(s - t)^2}{\delta^2}.$$

Since for all m and all $t \in [a, b]$,

$$\int_{-\infty}^{\infty} K_m(s, t)\, ds = 1,$$

and

$$\int_{-\infty}^{\infty} (s - t)^2 K_m(s, t)\, ds = \frac{1}{m^2},$$

it follows that for all $m > \frac{2}{\delta}\left(\frac{C}{\varepsilon}\right)^{\frac{1}{2}}$ and all $t \in [a, b]$,

$$|h(t) - h_m(t)| = |\int_{-\infty}^{\infty} (h(t) - h(s)) K_m(s, t)\, ds|$$

$$\leq \int_{-\infty}^{\infty} \left(\frac{\varepsilon}{2} + \frac{2C(s - t)^2}{\delta^2}\right) K_m(s, t)\, ds = \frac{\varepsilon}{2} + \frac{2C}{\delta^2 m^2} < \varepsilon.$$

This proves Theorem 1.3.

Remark 1.4. Let $G = \text{span}\{g_1, \ldots, g_n\}$ be an n–dimensional subspace of $C[a, b]$. The proof of Theorem 1.3 shows that the sequence $G_m = \text{span}\{g_{1,m}, \ldots, g_{n,m}\}$ of Chebyshev subspaces which "approximates" the space G is defined by

$$g_{i,m}(t) = \int_{-\infty}^{\infty} g_i(s) K_m(s, t)\, ds, \qquad t \in [a, b], \quad i = 1, \ldots, n, \qquad (1.6)$$

where

$$K_m(s,t) = \frac{m}{\sqrt{2\pi}} e^{-\frac{m^2}{2}(s-t)^2}, \qquad s,t \in (-\infty, \infty), \tag{1.7}$$

is the *Gauss-kernel*.

Bastien & Dubuc [1976] gave a different approach by using the *Bernstein operator* $B_m : C[a,b] \to P_m$, defined by

$$B_m(f)(t) = \sum_{i=0}^{m} \binom{m}{i} f\left(a + \frac{i(b-a)}{m}\right) \left(\frac{t-a}{b-a}\right)^i \left(\frac{b-t}{b-a}\right)^{m-i} \tag{1.8}$$

for all $t \in [a,b]$ and all $f \in C[a,b]$. Let $G_m = \operatorname{span}\{g_{1,m}, \ldots, g_{n,m}\}$ be defined by

$$g_{i,m} = B_m(g_i), \qquad i = 1, \ldots, n. \tag{1.9}$$

It is well known that $\lim_{m\to\infty} \|f - B_m(f)\|_\infty = 0$ for all $f \in C[a,b]$ (see e.g. Lorentz [1953] and Meinardus [1967]). Moreover, it can be shown by using variation diminishing properties of the Bernstein operator (see Polya & Schoenberg [1958]) that G_m is a Chebyshev space on (a,b) for sufficiently large m.

The next result, due to Jones & Karlovitz [1970], gives conditions wich characterize weak Chebyshev spaces. We first need the notion of sign changes.

Definition 1.5. A function $h \in C[a,b]$ is said to have at least r *sign changes* if there exist points $a \le t_1 < \cdots < t_{r+1} \le b$ such that

$$h(t_i)\, h(t_{i+1}) < 0, \qquad i = 1, \ldots, r.$$

Theorem 1.6. *For an n-dimensional subspace G of $C[a,b]$, the following statements are equivalent:*
(i) G is a weak Chebyshev subspace.
(ii) Every function $g \in G$ has at most $n-1$ sign changes.
(iii) For all points $a = t_0 < t_1 < \cdots < t_{n-1} < t_n = b$, there exists a nontrivial function $g \in G$ such that

$$(-1)^i g(t) \ge 0, \qquad t \in [t_{i-1}, t_i], \quad i = 1, \ldots, n. \tag{1.10}$$

Proof. (i) $\Rightarrow$ (ii). Suppose that (i) holds, but (ii) fails, i.e. there exists a function $g \in G$ and points

$$a \le t_1 < \cdots < t_{n+1} \le b$$

such that

$$g(t_i)\, g(t_{i+1}) < 0, \qquad i = 1, \ldots, n. \tag{1.11}$$

Then by Theorem 1.3 there exists a Chebyshev subspace G_m of $C[a,b]$ and a function $g_m \in G_m$ such that

$$\operatorname{sgn} g_m(t_i) = \operatorname{sgn} g(t_i), \qquad i = 1, \ldots, n+1.$$

Then it follows from (1.11) that g_m has n distinct zeros which is impossible, since G_m is a Chebyshev space.

(i) $\Rightarrow$ (iii). Let points

$$a = t_0 < t_1 < \cdots < t_{n-1} < t_n = b$$

be given and let (G_m) be a sequence of subspaces as in Theorem 1.3. Let m be given. Since G_m is a Chebyshev space, it follows from Theorem 1.14 in Chapter I that there exists a nontrivial function $g_m \in G_m$ such that

$$(-1)^i g_m(t) \geq 0, \qquad t \in [t_{i-1}, t_i], \quad i = 1, \ldots, n. \tag{1.12}$$

By multiplying g_m with an appropriate factor, we may assume that $\|g_m\|_\infty = 1$. For all m, let

$$g_m = \sum_{i=1}^{n} a_{i,m} \, g_{i,m}.$$

Then for all $i \in \{1, \ldots, n\}$, the sequence

$$(a_{i,m}) \text{ is bounded.} \tag{1.13}$$

If not, then by passing to a subsequence and reindexing, if necessary, we may assume that $|a_{1,m}| \geq |a_{i,m}|$, $i = 2, \ldots, n$, and

$$0 < |a_{1,m}| \to \infty.$$

Since $\left| \dfrac{a_{i,m}}{a_{1,m}} \right| \leq 1$, again by passing to a subsequence, we may assume that

$$\frac{a_{i,m}}{a_{1,m}} \to b_i, \qquad i = 2, \ldots, n.$$

Then it follows that

$$\left\| g_1 + \sum_{i=2}^{n} b_i \, g_i \right\|_\infty = \lim_{m \to \infty} \left\| g_{1,m} + \sum_{i=2}^{n} \frac{a_{i,m}}{a_{1,m}} \, g_{i,m} \right\|_\infty$$

$$= \lim_{m \to \infty} \frac{1}{|a_{1,m}|} \|g_m\|_\infty = 0.$$

Therefore, we have

$$-g_1 = \sum_{i=2}^{n} b_i \, g_i$$

which contradicts the linear independence of $g_1, \ldots, g_n$. Now, it follows from (1.13) that by passing to a subsequence we may assume that

$$\lim_{m \to \infty} a_{i,m} = a_i, \qquad i = 1, \ldots, n,$$

which implies that

$$g_m \to g,$$

where $g = \sum_{i=1}^{n} a_i g_i \in G$. Since $\|g_m\|_\infty = 1$ for all m, it follows that $\|g\|_\infty = 1$.
Moreover, by taking limits it follows from (1.12) that

$$(-1)^i g(t) \geq 0, \qquad t \in [t_{i-1}, t_i], \quad i = 1, \ldots, n.$$

(iii) $\Rightarrow$ (i). Suppose that (iii) holds. Let a basis $\{g_1, \ldots, g_n\}$ of G and points $s_1 < \cdots < s_n$ and $t_1 < \cdots < t_n$ in (a, b) be given such that

$$D\left(\begin{matrix} g_1, \ldots, g_n \\ s_1, \ldots, s_n \end{matrix} \right) \neq 0 \tag{1.14}$$

and

$$D\left(\begin{matrix} g_1, \ldots, g_n \\ t_1, \ldots, t_n \end{matrix} \right) \neq 0. \tag{1.15}$$

We will show that the product of these determinants is positive, which implies that (i) holds. It follows from (1.14) that for each $j \in \{1, \ldots, n\}$ there is a function $h_j \in G$ such that

$$h_j(s_i) = (-1)^{j+1} \delta_{ij}, \qquad i = 1, \ldots, n. \tag{1.16}$$

Since for all $j \in \{1, \ldots, n\}$, the function h_j can be written as a unique linear combination of the basis $\{g_1, \ldots, g_n\}$, by using Cramer's rule we get

$$h_j(t) = \frac{\overline{D}\left(\begin{matrix} g_1, & \cdots\cdots & , g_n \\ t, s_1, \ldots, s_{j-1}, s_{j+1}, \ldots, s_n \end{matrix} \right)}{D\left(\begin{matrix} g_1, \ldots, g_n \\ s_1, \ldots, s_n \end{matrix} \right)} \tag{1.17}$$

for all $t \in [a, b]$. Now, let $j \in \{1, \ldots, n\}$ be given. We set

$$\{u_0, \ldots, u_n\} = \{s_0, \ldots, s_{j-1}, s_{j+1}, \ldots, s_{n+1}\},$$

where $s_0 = a$ and $s_{n+1} = b$. Then by (iii) there exists a function $f_j \in G$ such that

$$(-1)^{i+1} f_j(t) \geq 0, \qquad t \in [u_{i-1}, u_i], \quad i = 1, \ldots, n. \tag{1.18}$$

Since f_j can be written as a unique linear combination of the basis $\{h_1, \ldots, h_n\}$, it follows from (1.16) and (1.18) that there exists a real number $a_j > 0$ such that

$$f_j = a_j h_j. \tag{1.19}$$

We may assume that $\{s_1, \ldots, s_n\} \neq \{t_1, \ldots, t_n\}$. Let $j \in \{1, \ldots, n\}$ be given such that $s_j \notin \{t_1, \ldots, t_n\}$. It follows from (1.15) that there exists an integer $k \in \{1, \ldots, n\}$ such that $h_j(t_k) \neq 0$. Then it follows from (1.17) that $t_k \notin \{s_1, \ldots, s_{j-1}, s_{j+1}, \ldots, s_n\}$. We set

$$\{u_1, \ldots, u_n\} = \{s_1, \ldots, s_{j-1}, s_{j+1}, \ldots, s_n\} \cup \{t_k\}$$

such that $u_1 < \cdots < u_n$. Then it follows from (1.17) and (1.19) that

$$D \left(\begin{array}{c} g_1, \ldots, g_n \\ s_1, \ldots, s_n \end{array} \right) D \left(\begin{array}{c} g_1, \ldots, g_n \\ u_1, \ldots, u_n \end{array} \right) > 0.$$

By repeating this process, we finally get

$$D \left(\begin{array}{c} g_1, \ldots, g_n \\ s_1, \ldots, s_n \end{array} \right) D \left(\begin{array}{c} g_1, \ldots, g_n \\ t_1, \ldots, t_n \end{array} \right) > 0.$$

(ii) $\Rightarrow$ (i). Suppose that (ii) holds. Let a basis $\{g_1, \ldots, g_n\}$ of G and points $s_1 < \cdots < s_n$ and $t_1 < \cdots < t_n$ as in the proof of (iii) $\Rightarrow$ (i) be given. Moreover, let $h_1, \ldots, h_n \in G$ be defined as in the proof of (iii) $\Rightarrow$ (i). Let $j \in \{1, \ldots, n\}$ be given. We set

$$\{u_0, \ldots, u_n\} = \{s_0, \ldots, s_{j-1}, s_{j+1}, \ldots, s_{n+1}\},$$

where $s_0 = a$ and $s_{n+1} = b$. We will show that

$$(-1)^{i+1} h_j(t) \geq 0, \qquad t \in [u_{i-1}, u_i], \quad i = 1, \ldots, n. \tag{1.20}$$

Then exactly as in the proof of (iii) $\Rightarrow$ (i) it follows from (1.20) that

$$D \left(\begin{array}{c} g_1, \ldots, g_n \\ s_1, \ldots, s_n \end{array} \right) D \left(\begin{array}{c} g_1, \ldots, g_n \\ t_1, \ldots, t_n \end{array} \right) > 0,$$

which implies that (i) holds. Thus it remains to prove (1.20). Suppose that (1.20) fails, i.e. there exists an integer $k \in \{1, \ldots, n\}$ and a point $v \in (u_{k-1}, u_k)$ such that $(-1)^{i+1} h_j(v) < 0$. By (1.14) there exists a function $h \in G$ such that

$$h(s_i) = \left\{ \begin{array}{ll} (-1)^i, & \text{if } s_i < v \\ (-1)^{i+1}, & \text{if } s_i > v \end{array} \right. \quad , i = 1, \ldots, n.$$

Then it is easy to verify that by (1.16) the function $h_\varepsilon = h_j + \varepsilon h$ has n sign changes at

$$s_1 < \cdots < s_l < v < s_{l+1} < \cdots < s_n,$$

if ε has appropriate sign and $|\varepsilon|$ is sufficiently small, which contradicts (ii). This proves Theorem 1.6.

By using Theorem 1.6 and the proof of Corollary 1.15 in Chapter I, we obtain the following extension of this corollary.

Corollary 1.7. *Let G be an n–dimensional weak Chebyshev subspace of $C[a, b]$. Then for all integers $m \in \{1, \ldots, n\}$ and all points $a = t_0 < t_1 < \cdots < t_{m-1} < t_m = b$, there exists a nontrivial function $g \in G$ such that*

$$(-1)^i g(t) \geq 0, \qquad t \in [t_{i-1}, t_i], \quad i = 1, \ldots, m. \tag{1.21}$$

Finally, we note that there exists a vast literature on weak Chebyshev spaces. Further details concerning spaces of this type can be found e.g. in the books of Karlin & Studden [1966], Karlin [1968], Zielke [1979] and Schumaker [1981].

1.2. Best Uniform Approximation by Weak Chebyshev Spaces

We give results on best uniform approximations and strongly unique best uniform approximations from weak Chebyshev spaces in terms of alternation properties of the error. In particular, weak Chebyshev spaces are characterized by the property that for every continuous function, there exists a best uniform approximation such that the error has the classical alternation property (compare Theorem 3.12 of Chapter I).

We begin with this statement which was proved by Jones & Karlovitz [1970]. Extensions of this result were obtained by Deutsch, Nürnberger & Singer [1980].

Theorem 1.8. *For an n–dimensional subspace G of $C[a, b]$, the following statements are equivalent:*
(i)   *For every function $f \in C[a, b]$, there exists a best uniform approximation $g_f \in G$ such that $f - g_f$ has at least $n + 1$ alternating extreme points in $[a, b]$.*
(ii) *For every function $f \in C[a, b]$ having a unique best uniform approximation $g_f \in G$, the error $f - g_f$ has at least $n + 1$ alternating extreme points in $[a, b]$.*
(iii) *G is a weak Chebyshev subspace.*

Proof. (i) $\Rightarrow$ (ii). This implication is obvious.

(ii) $\Rightarrow$ (iii). Suppose that (ii) holds. We will show that condition (iii) in Theorem 1.6 holds, which implies that G is weak Chebyshev. Let points

$$a = t_0 < t_1 < \cdots < t_{n-1} < t_n = b$$

be given. For all m, let $f_m \in C[a, b]$ be defined by

$$f_m(t) = \begin{cases} 0, & \text{if} \quad t \in \{t_{i-1}, t_i\} \\ (-1)^i, & \text{if} \quad t \in [t_{i-1} + \frac{1}{m}, t_i - \frac{1}{m}] \\ \text{linear}, & \text{elsewhere} \quad \text{on } [t_{i-1}, t_i] \end{cases}$$

$i = 1, \ldots, n$. If f_m has a unique best uniform approximation $g_m \in G$, then it follows from (ii) that $f_m - g_m$ has at least $n + 1$ alternating extreme points. This implies that $g_m \neq 0$. Moreover, we have $\|f_m - g_m\|_\infty \leq \|f_m\|_\infty = 1$. Then it follows that

$$(-1)^i g_m(t) \geq 0, \qquad t \in [t_{i-1} + \frac{1}{m}, t_i - \frac{1}{m}], \quad i = 1, \ldots, n. \qquad (1.22)$$

If f_m does not have a unique best uniform approximation from G, then there exists a best uniform approximation $g_m \in G$ of f_m with $g_m \neq 0$. Analogously as above, it follows that g_m has property (1.22). By multiplying g_m with $1/\|g_m\|_\infty$, we may assume that $\|g_m\|_\infty = 1$. Since G is finite–dimensional, by going to a subsequence, we may assume that (g_m) converges to a function $g \in G$. By taking limits, it follows from (1.22) that

$$(-1)^i g(t) \geq 0, \qquad t \in [t_{i-1}, t_i], \quad i = 1, \ldots, n.$$

(iii) $\Rightarrow$ (i). Suppose that (iii) holds. Let $f \in C[a, b]$ be given. Moreover, let (G_m) be a sequence of spaces as in Theorem 1.3. Let a positive interger m be given. Since G_m is a Chebyshev space, it follows from Theorem 3.12 and Theorem 3.18 in Chapter I that f has a unique best approximation $g_m \in G_m$ such that there exist points $t_{1,m} < \cdots < t_{n+1,m}$ in $[a, b]$ and a sign $\sigma_m \in \{-1, 1\}$ with

$$\sigma_m (-1)^i (f(t_{i,m}) - g_m(t_{i,m})) = \|f - g_m\|_\infty, \qquad i = 1, \ldots, n+1. \tag{1.23}$$

Then it follows that

$$\|g_m\|_\infty - \|f\|_\infty \leq \|f - g_m\|_\infty \leq \|f\|_\infty,$$

which implies that $\|g_m\|_\infty \leq 2\|f\|_\infty$. Since the sequence (g_m) is bounded, it follows from the proof of (i) $\Rightarrow$ (iii) of Theorem 1.6 that by passing to a subsequence we may assume that $g_m \to g_f$ for some $g_f \in G$. Moreover, again passing to a subsequence, we may assume that

$$\lim_{m \to \infty} t_{i,m} = t_i, \qquad i = 1, \ldots, n+1$$

and $\sigma_m = \sigma$ for some $\sigma \in \{-1, 1\}$. Then by taking limits, it follows from (1.23) that

$$\sigma (-1)^i (f(t_i) - g_f(t_i)) = \|f - g_f\|_\infty, \qquad i = 1, \ldots, n+1. \tag{1.24}$$

Moreover, the next Theorem 1.9 shows that by (1.24) the function g_f is a best uniform approximation of f from G. This shows that (i) holds and proves Theorem 1.8.

It follows from Theorem 1.8 that if G is an n–dimensional weak Chebyshev subspace of $C[a, b]$, then for every function $f \in C[a, b]$, there exists a best uniform approximation $g_f \in G$ such that $f - g_f$ has at least $n + 1$ alternating extreme points. The next result shows that in a certain sense the converse is also true.

Theorem 1.9. *Let G be an n–dimensional weak Chebyshev subspace of $C[a, b]$, $f \in C[a, b]$ and $g_f \in G$. If $f - g_f$ has at least $n+1$ alternating extreme points in $[a, b]$, then g_f is a best uniform approximation of f from G.*

Proof. Let $f \in C[a, b]$ and $g_f \in G$ be given. Suppose that there exist points $a \leq t_1 < \cdots < t_{n+1} \leq b$ and a sign $\sigma \in \{-1, 1\}$ such that

$$\sigma (-1)^i (f(t_i) - g_f(t_i)) = \|f - g_f\|_\infty, \qquad i = 1, \ldots, n+1. \tag{1.25}$$

We assume that g_f is not a best uniform approximation of f from G, i.e. there exists a function $g \in G$ such that

$$\|f - g\|_\infty < \|f - g_f\|_\infty.$$

Then it follows from (1.23) that

$$\begin{aligned}\sigma \,(-1)^i \,(f(t_i) - g(t_i)) \;&\leq\; \|f - g\|_\infty < \|f - g_f\|_\infty \\ &\leq\; \sigma \,(-1)^i \,(f(t_i) - g_f(t_i)), \qquad i = 1, \ldots, n + 1,\end{aligned}$$

which implies that

$$\sigma \,(-1)^i \,(g_f(t_i) - g(t_i)) < 0, \qquad i = 1, \ldots, n + 1.$$

Therefore, the function $g_f - g$ in G has at least n sign changes, which by Theorem 1.3 is a contradiction to G being weak Chebyshev. This proves Theorem 1.9.

It follows from Theorem 3.18 in Chapter I that best uniform approximations from a given weak Chebyshev space which is not Chebyshev are not always unique. In the following we give results on strongly unique best uniform approximations of a fixed continuous function.

For this we need the notion of poised sets.

Definition 1.10. Let $\{g_1, \ldots, g_n\}$ be a basis of an n–dimensional subspace G of $C[a, b]$. A subset $\{t_1, \ldots, t_n\}$ of $[a, b]$ is called *poised* with respect to G, if

$$\overline{D} \left(\begin{array}{c} g_1, \ldots, g_n \\ t_1, \ldots, t_n \end{array} \right) \neq 0.$$

For proving results on strong unicity we need the following statement.

Lemma 1.11. *Let G be an n–dimensional subspace of $C[a, b]$, and let points $t_1 < \cdots < t_{n+1}$ in $[a, b]$ be given. The following statements are equivalent:*
(i)   There does not exist a nontrivial function $g \in G$ such that

$$(-1)^i \,g(t_i) \geq 0, \qquad i = 1, \ldots, n + 1.$$

(ii) For all $i \in \{1, \ldots, n + 1\}$, the set

$$\{t_1, \ldots, t_{i-1}, t_{i+1}, \ldots, t_{n+1}\}$$

is poised with respect to G.

Proof. (i) $\Rightarrow$ (ii). Suppose that (ii) fails, i.e. there exists an integer $j \in \{1, \ldots, n + 1\}$ such that the set

$$\{t_1, \ldots, t_{j-1}, t_{j+1}, \ldots, t_{n+1}\}$$

is not poised with respect to G. Then there exists a nontrivial function $g \in G$ such that

$$g(t_i) = 0, \qquad i = 1, \ldots, j - 1, j + 1, \ldots, n + 1.$$

If necessary, we replace g by $-g$ and get

$$(-1)^i g(t_i) \geq 0, \qquad i = 1, \ldots, n+1,$$

which implies that (i) fails.

(ii) $\Rightarrow$ (i). Suppose that (ii) holds, but (i) fails, i.e. there exists a nontrivial function $g \in G$ such that

$$(-1)^i g(t_i) \geq 0, \qquad i = 1, \ldots, n+1.$$

Let $\{g_1, \ldots, g_n\}$ be a basis of G. Then it follows that

$$\begin{aligned}
0 &= D\left(\begin{array}{c} g, \ g_1, \ \cdots, g_n \\ t_1, \ldots\ldots, t_{n+1} \end{array} \right) \\
&= \sum_{i=1}^{n+1} (-1)^{i-1} g(t_i) D\left(\begin{array}{c} g_1, \qquad \cdots \cdots \qquad, g_n \\ t_1, \ldots, t_{i-1}, t_{i+1}, \ldots, t_{n+1} \end{array} \right)
\end{aligned}$$

Since G is weak Chebyshev and (ii) holds, all determinants in the above sum have the same sign. This implies that $g(t_i) = 0$, $i = 1, \ldots, n+1$. Then it follows from (ii) that $g = 0$, which is a contradiction. This proves Lemma 1.11.

By using Lemma 1.11, we now give a result on strongly unique best uniform approximations from weak Chebyshev spaces.

Theorem 1.12. *Let G be an n–dimensional weak Chebyshev subspace of $C[a,b]$, $f \in C[a,b]$ and $g_f \in G$. Consider the following statements:*

(i) *The function g_f is a strongly unique best uniform approximation of f from G.*

(ii) *The error $f - g_f$ has at least $n+1$ alternating extreme points $t_1 < \cdots < t_{n+1}$ in $[a,b]$ such that for all $i \in \{1, \ldots, n+1\}$, the set*

$$\{t_1, \ldots, t_{i-1}, t_{i+1}, \ldots, t_{n+1}\}$$

is poised with respect to G.

Then (ii) $\Rightarrow$ (i). Moreover, (i) $\Rightarrow$ (ii), if $f - g_f$ has exactly $n+1$ extreme points in $[a,b]$.

Proof. (ii) $\Rightarrow$ (i). Suppose that (ii) holds, but (i) fails. It follows from (ii) that there exist points $a \leq t_1 < \cdots < t_{n+1} \leq b$ and a sign $\sigma \in \{-1, 1\}$ such that

$$\sigma(-1)^i (f(t_i) - g_f(t_i)) = \|f - g_f\|_\infty, \qquad i = 1, \ldots, n+1. \qquad (1.26)$$

Since (i) fails, by Theorem 3.17 in Chapter I there exists a nontrivial function $g \in G$ such that

$$(f(t_i) - g_f(t_i)) g(t_i) \geq 0, \qquad i = 1, \ldots, n+1. \qquad (1.27)$$

Then it follows from (1.26) and (1.27) that by replacing g by $-g$, if necessary,

$$(-1)^i g(t_i) \geq 0, i = 1, \ldots, n+1.$$

By condition (ii) and Lemma 1.11 this is a contradiction.

(i) $\Rightarrow$ (ii). Suppose that (i) holds and

$$E(f - g_f) = \{t_1, \ldots, t_{n+1}\}. \tag{1.28}$$

Since G is weak Chebyshev, it follows from Theorem 1.8 that there exist points $a \le t_1 < \cdots < t_{n+1} \le b$ and a sign $\sigma \in \{-1, 1\}$ such that

$$\sigma \, (-1)^i \, (f(t_i) - g_f(t_i)) = \|f - g_f\|_\infty, \qquad i = 1, \ldots, n+1.$$

Suppose that there exists a set

$$\{t_1, \ldots, t_{i-1}, t_{i+1}, \ldots, t_{n+1}\},$$

which is not poised with respect to G. Then by Lemma 1.11 there exists a nontrivial function $g \in G$ such that

$$(f(t_i) - g_f(t_i)) \, g(t_i) \ge 0, \qquad i = 1, \ldots, n+1.$$

Then it follows from (1.28) and Theorem 3.17 in Chapter I that (i) fails, which is a contradiction. This proves Theorem 1.12.

Remark 1.13. The equivalence (i) $\Leftrightarrow$ (ii) in Theorem 1.12 is not true if the error $f - g_f$ has more than $n + 1$ extreme points. A complete characterization of property (i) in Theorem 1.12 was given by Nürnberger [1982a]. Further results on strong unicity in approximation and optimization can be found in Brosowski [1981], [1984] and Brosowski & Nürnberger [1988].

For a finite–dimensional subspace G of $C[a, b]$ we set

$$\mathrm{SU}\,(G) = \{f \in C[a, b] : \; f \text{ has a strongly unique best} \atop \text{uniform approximation from } G\}.$$

Since in practice functions are only known approximately, this leads to the following question of stability: which functions are from the interior of $\mathrm{SU}\,(G)$, denoted by $\mathrm{int}\,\mathrm{SU}\,(G)$?

The following characterization of those functions for weak Chebyshev spaces is a special case of more general results in Nürnberger [1983], [1985c].

Theorem 1.14. *Let G be an n–dimensional weak Chebyshev subspace of $C[a, b]$, $f \in C[a, b] \backslash G$, and $g_f \in G$ be a best uniform approximation of f. The following statements are equivalent:*

(i) $f \in \mathrm{int}\,SU\,(G)$

(ii) The error $f - g_f$ has at least $n + 1$ alternating extreme points $t_1 < \cdots < t_{n+1}$ in $[a, b]$ and for every such set $\{t_1, \ldots, t_{n+1}\}$, the set

$$\{t_1, \ldots, t_{i-1}, t_{i+1}, \ldots, t_{n+1}\}$$

is poised with respect to G, $i = 1, \ldots, n + 1$.

Proof. (i) $\Rightarrow$ (ii). Suppose that (i) holds. Then it follows from Theorem 1.8 that $f - g_f$ has at least $n + 1$ alternating extreme points. We now assume that (ii) fails, i.e. there exists a set $T = \{t_1, \ldots, t_{n+1}\}$ of alternating extreme points of $f - g_f$ such that the set

$$\{t_1, \ldots, t_{j-1}, t_{j+1}, \ldots, t_{n+1}\}$$

is not poised with respect to G for some $j \in \{1, \ldots, n+1\}$. By perturbing $f - g_f$ at the extreme points not contained in T, it is easy to see that there exists a sequence (h_m) in $C[a, b]$ such that $h_m \to f - g_f$, $h_m(t) = f(t) - g_f(t)$ for all $t \in T$ and $E(h_m) = T$ for all m. Then it follows from Theorem 1.9 and Theorem 1.12 that for all m, zero is a best uniform approximation of h_m from G but is not strongly unique. We set $f_m = h_m + g_f$ for all m. Then $f_m \to f$ and for all m, g_f is a best uniform approximation of f_m from G, but is not strongly unique. This shows that $f \notin \mathrm{int}\, \mathrm{SU}\,(G)$.

(ii) $\Rightarrow$ (i). Suppose that (ii) holds. Then it follows from Theorem 1.12 that $f \in \mathrm{SU}\,(G)$. We now assume that (i) fails, i.e. there exists a sequence (f_m) such that $f_m \to f$ and $f_m \notin \mathrm{SU}\,(G)$ for all m. By Theorem 1.8 for each m, there exists a best uniform approximation $g_m \in G$ of f_m and there exist points $t_{1,m} < \cdots < t_{n+1,m}$ in $[a, b]$ and a sign $\sigma_m \in \{-1, 1\}$ with

$$\sigma_m \, (-1)^i \, (f_m(t_{i,m}) - g_m(t_{i,m})) = \|f_m - g_m\|_\infty, \qquad i = 1, \ldots, n + 1.$$

Since $f_m \notin \mathrm{SU}\,(G)$ for all m, it follows from Theorem 1.12 that for each m, there exists an integer $j_m \in \{1, \ldots, n + 1\}$ such that the set

$$\{t_{1,m}, \ldots, t_{j_m-1,m}, t_{j_m+1,m}, \ldots, t_{n+1,m}\}$$

is not poised with respect to G. By going to a subsequence, we may assume that for all m, $\sigma_m = \sigma \in \{-1, 1\}$, $j_m = j \in \{1, \ldots, n + 1\}$, $g_m \to g_f \in G$ and $t_{i,m} \to t_i \in [a, b]$, $i = 1, \ldots, n + 1$. Taking limits, it follows that

$$\sigma \, (-1)^i \, (f(t_i) - g_f(t_i)) = \|f - g_f\|_\infty, \qquad i = 1, \ldots, n + 1,$$

and the set

$$\{t_1, \ldots, t_{j-1}, t_{j+1}, \ldots, t_{n+1}\}$$

is not poised with respect to G, with contradicts (ii). This proves Theorem 1.14.

1.3. Spline Spaces

We define spaces of polynomial splines and show that there exists a basis consisting of polynomials and truncated power functions. Spline spaces are prototypes of weak Chebyshev spaces.

We begin with the definition of polynomial splines.

Definition 1.15. Let points $a = x_0 < x_1 < \cdots < x_k < x_{k+1} = b$ and an integer $m \geq 1$ be given. We call

$$S_m(x_1, \ldots, x_k) = \{s \in C^{m-1}[a, b] : s|_{[x_i, x_{i+1}]} \in P_m, \ i = 0, \ldots, k\} \qquad (1.29)$$

the space of *polynomial splines of degree m with k fixed knots* $x_1, \ldots, x_k$. For a given spline space $S_m(x_1, \ldots, x_k)$, we always associate further points $x_{-m} < \cdots < x_{-1} < a$ and $b < x_{k+2} < \cdots < x_{k+m+1}$, where these points may be chosen arbitrarily.

We now show that there exists a basis of a given spline space consisting of polynomials and truncated power functions.

Definition 1.16. For a given point $x \in (a, b)$ the function

$$(t - x)_+^m = \begin{cases} 0, & \text{if } t \leq x \\ (t - x)^m, & \text{if } t > x \end{cases} \qquad (1.30)$$

is called the *truncated power function* of degree m with knot x.

Theorem 1.17. *The set of functions*

$$\{1, t, \ldots, t^m, (t - x_1)_+^m, \ldots, (t - x_k)_+^m\} \qquad (1.31)$$

forms a basis of $S_m(x_1, \ldots, x_k)$. *In particular, the dimension of* $S_m(x_1, \ldots, x_k)$ *is* $k + m + 1$.

Proof. It is easy to see that

$$\{1, t, \ldots, t^m, (t - x_1)_+^m, \ldots, (t - x_k)_+^m\}$$

is a subset of $S_m(x_1, \ldots, x_k)$. It remains to show that every $s \in S_m(x_1, \ldots, x_k)$ has a unique representation

$$s(t) = \sum_{i=0}^m a_i t^i + \sum_{i=1}^k b_i (t - x_i)_+^m, \qquad t \in [a, b]. \qquad (1.32)$$

Let a spline $s \in S_m(x_1, \ldots, x_k)$ be given. We set

$$p_i(t) = s(t), \qquad t \in [x_i, x_{i+1}], \quad i = 0, \ldots, k.$$

Then we have $p_0, \ldots, p_k \in P_m$. Therefore, $p_0 \in P_m$ has a unique representation

$$p_0(t) = \sum_{i=0}^m a_i t^i, \qquad t \in [x_0, x_1].$$

Moreover, since $s \in C^{(m-1)}[a, b]$, we have

$$p_1^{(i)}(x_1) = p_0^{(i)}(x_1), \qquad i = 0, \ldots, m - 1.$$

Since $p_1 - p_0 \in P_m$, this implies that $p_1 - p_0$ has a unique representation

$$p_1(t) - p_0(t) = b_1 (t - x_1)^m.$$

Therefore, we have

$$s(t) = \sum_{i=0}^{m} a_i t^i + b_1 (t - x_1)_+^m, \qquad t \in [x_0, x_2].$$

By continuing this method, we finally obtain (1.32). This proves Theorem 1.17.

The next theorem, due to Schoenberg & Whitney [1953], says that spline spaces are weak Chebyshev.

For proving this result, we need the following version of the well-known Rolle's Theorem which can be found in standard books on Analysis.

Theorem 1.18. *Let a function $f \in C^1[a, b]$ and points $a \leq t_1 < t_2 \leq b$ be given such that $f(t_1) = f(t_2) = 0$. Then the function f' has at least one zero in (t_1, t_2). If, in addition $f(t) \neq 0$ for some point $t \in (t_1, t_2)$, then f' has at least one sign change in (t_1, t_2).*

Theorem 1.19. *The space $S_m(x_1, \ldots, x_k)$ is a $(k + m + 1)$–dimensional weak Chebyshev subspace of $C[a, b]$.*

Proof. We will show that every $s \in S_m(x_1, \ldots, x_k)$ has at most $k + m$ sign changes. Then it follows from Theorem 1.6 that $S_m(x_1, \ldots, x_k)$ is weak Chebyshev. Suppose that a spline $s \in S_m(x_1, \ldots, x_k)$ has at least $k + m + 1$ sign changes. Then it follows from Theorem 1.18 that $s' \in S_{m-1}(x_1, \ldots, x_k)$ has at least $k + m$ sign changes. We consider further derivatives of s and finally get that $s^{(m-1)} \in S_1(x_1, \ldots, x_k)$ has at least $k + 2$ sign changes. This is a contradiction, since such a spline of degree one has at most $k + 1$ sign changes. This proves Theorem 1.19.

2. B–Splines

It is shown that the so–called B–splines form a basis of spline spaces. B–splines are splines which have smallest possible support, in other words, they are zero on a large set. For the evaluation of splines, it is desirable to have basis functions with this property. Moreover, a stable evaluation of B–splines with the aid of a recurrence relation is possible. It is shown that B–splines form a partition of unity and that the B–spline basis is variation diminishing. Finally, we give results on the differentiation and integration of splines.

2.1. Basic Properties

In this section we give the definition of B–splines and discuss their zero properties.

We need the following definition of splines on $(-\infty, \infty)$.

Definition 2.1. Let points $x_{-m} < \cdots < x_{-1} < a = x_0 < x_1 < \cdots < x_k < x_{k+1} = b < x_{k+2} < \cdots < x_{k+m+1}$ be given. A function $s : (-\infty, \infty) \to \mathbf{R}$ is called a *polynomial spline of degree m with knots* $x_{-m}, \ldots, x_{k+m+1}$ if s has $m - 1$ continuous derivatives at x_i, $i = -m, \ldots, k + m + 1$, and $s|_{(x_i, x_{i+1})} \in P_m$, $i = -m - 1, \ldots, k + m + 1$, where $x_{-m-1} = -\infty$ and $x_{k+m+2} = \infty$.

The first result on the existence and uniqueness of splines with certain zero properties is due to Curry & Schoenberg [1947], [1966].

Theorem 2.2. *For each $i \in \{-m, \ldots, k\}$, there exists a unique spline B_i^m of degree m with knots $x_{-m}, \ldots, x_{k+m+1}$ such that*

$$B_i^m(t) = 0, \qquad t \in (-\infty, x_i] \cup [x_{i+m+1}, \infty), \tag{2.1}$$

$$B_i^m(t) > 0, \qquad t \in (x_i, x_{i+m+1}), \tag{2.2}$$

and

$$\int_{x_i}^{x_{i+m+1}} B_i^m(t)\, dt = 1. \tag{2.3}$$

Proof. Every spline B_i^m of degree m satisfying (2.1) has the form

$$B_i^m(t) = \sum_{j=i}^{i+m+1} a_j\, (t - x_j)_+^m, \qquad t \in (-\infty, \infty).$$

It follows from (2.1) that

$$\sum_{j=i}^{i+m+1} a_j\, (t - x_j)^m = 0, \qquad t \in [x_{i+m+1}, \infty).$$

Then by the binomial theorem,

$$\sum_{j=i}^{i+m+1} \sum_{r=0}^{m+1} a_j\, (-1)^r \binom{m}{r}\, x_j^r\, t^{m-r} = 0, \qquad t \in [x_{i+m+1}, \infty).$$

Since the coefficients of the functions $1, t, \ldots, t^m$ must be zero, we get that

$$\sum_{j=i}^{i+m+1} a_j\, x_j^r = 0, \qquad r = 0, \ldots, m. \tag{2.4}$$

Moreover, it follows from (2.3) that

$$\sum_{j=i}^{i+m+1} a_j\, (x_{i+m+1} - x_j)^{m+1} = m + 1.$$

Then by the binomial theorem,

$$\sum_{j=i}^{i+m+1} \sum_{r=0}^{m+1} a_j\, (-1)^r \binom{m+1}{r}\, x_j^r\, x_{i+m+1}^{m+1-r} = m + 1. \tag{2.5}$$

It follows from (2.4) and (2.5) that

$$\sum_{j=i}^{i+m+1} a_j x_j^{m+1} = (-1)^{m+1} (m + 1). \tag{2.6}$$

The determinant corresponding to the linear system of equations (2.4) and (2.6) is the nonzero Vandermonde determinant. This shows that the unknowns $a_i, \ldots, a_{i+m+1}$ are uniquely determined and that there exists a unique spline B_i^m of degree m statisfying (2.1) and (2.3). Property (2.2) can be easily proved by induction on m with aid of the subsequent recurrence relation (2.17) which is independent of (2.2). This proves Theorem 2.2.

Definition 2.3. The spline B_i^m in Theorem 2.2 is called the *B–spline* of degree m with support $[x_i, x_{i+m+1}]$.

Remark 2.4. The proof of Theorem 2.2 shows that, if $i \in \{-m, \ldots, k\}$, $r \in \{1, \ldots, m\}$ and s is a spline of degree m with knots $x_{-m}, \ldots, x_{k+m+1}$ satisfying

$$s(t) = 0, \qquad t \in (-\infty, x_i] \cup [x_{i+r}, \infty), \tag{2.7}$$

then $s = 0$. Therefore, we may say that B–splines have "minimal" support.

Curry & Schoenberg [1966] proved the following result on the shape of B–splines.

Theorem 2.5. *Let an index $i \in \{-m, \ldots, k\}$ be given. Then for all $j \in \{1, \ldots, m - 1\}$, the spline $(B_i^m)^{(j)}$ has exactly j distinct zeros in (x_i, x_{i+m+1}) and changes sign at these zeros.*

Proof. Let an integer $i \in \{-m, \ldots, k\}$ be given. We first show that

$$(B_i^m)^{(j)} \text{ has at least } j \text{ sign changes in } (x_i, x_{i+1}), j = 1, \ldots, m - 1. \tag{2.8}$$

Since $B_i^m(x_i) = B_i^m(x_{i+m+1}) = 0$, it follows from (2.2) and Theorem 1.18 the spline $(B_i^m)'$ has at least one sign change. By applying Theorem 1.18 several times, we see that (2.8) holds. Next, we show that

$$(B_i^m)^{(j)} \text{ has only finitely many zeros in } (x_i, x_{i+m+1}), j = 1, \ldots, m - 1. \tag{2.9}$$

Indeed, if $(B_i^m)^{(j)}$ vanishes on a knot-interval in $[x_i, x_{i+m+1}]$, then $(B_i^m)^{(m-1)}$ vanishes on this interval. But then it is easy to see that the spline $(B_i^m)^{(m-1)}$ of degree one cannot have $m - 1$ sign changes, which contradicts (2.8). Finally, we show that

$$(B_i^m)^{(j)} \text{ has at most } j \text{ distinct zeros in } (x_i, x_{i+m+1}), j = 1, \ldots, m - 1. \tag{2.10}$$

Assume to the contrary that $(B_i^m)^{(j)}$ has at least $j+1$ distinct zeros in (x_i, x_{i+m+1}). Then, since in addition $(B_i^m)^{(j)}(x_i) = (B_i^m)^{(j)}(x_{i+m+1}) = 0$, it follows from Theorem 1.18 that $(B_i^m)^{(j)}$ has at least $j + 2$ sign changes. By applying Theorem 1.18 several times, we get that $(B_i^m)^{(m-1)}$ has at least m sign changes. This is a contradiction since $(B_i^m)^{(m-1)}$ is a spline of degree one with $(B_i^m)^{(m-1)}(x_i) = (B_i^m)^{(m-1)}(x_{i+m+1}) = 0$, and therefore has at most $m - 1$ sign changes. Now, the result follows from (2.8) and (2.10). This proves Theorem 2.5.

2.2. B–Spline Basis

It is shown that for a given spline space there exists a basis consisting of B–splines.

The result formulated in the next theorem is due to Curry & Schoenberg [1966].

Theorem 2.6. *The set of B–splines*

$$\{B_{-m}^m, \ldots, B_k^m\} \tag{2.11}$$

forms a basis of $S_m(x_1, \ldots, x_k)$ on $[a, b]$.

Proof. We will show that the B–splines $B_{-m}^m, \ldots, B_k^m$ are linearly independent on $[a, b]$. Suppose to the contrary that there exist real numbers $a_{-m}, \ldots, a_k$ such that

$$\sum_{i=-m}^k |a_i| \neq 0$$

and

$$\sum_{i=-m}^k a_i \, B_i^m(t) = 0, \qquad t \in [a, b].$$

We set $j = \min\{i \in \{-m, \ldots, k\} : a_i \neq 0\}$. Then by the properties of the B–splines

$$\sum_{i=-m}^k a_i \, B_i^m(t) = a_j \, B_j^m(t) \neq 0, \qquad t \in (x_j, x_{j+1}]. \tag{2.12}$$

This implies that $j < 0$, otherwise we get a contradiction to (2.12). We set

$$s(t) = \sum_{i=-m}^k a_i \, B_i^m(t), \qquad t \in (-\infty, x_{k+1}].$$

Then we have

$$s(t) = 0, \qquad t \in (-\infty, x_{-m}] \cup [x_0, x_{k+1}].$$

We can extend this spline s to $(-\infty, \infty)$ by setting

$$s(t) = 0, \qquad t \in (x_{k+1}, \infty).$$

Then it follows from Remark 2.4 that $s = 0$ which implies that

$$\sum_{i=-m}^k a_i \, B_i^m(t) = 0, \qquad t \in [x_{-m}, x_0].$$

Since $j \in \{-m, \ldots, -1\}$, this is a contradiction to (2.12). This proves Theorem 2.6.

Remark 2.7. It follows from Theorem 2.6 that every spline $s \in S_m(x_1, \ldots, x_k)$ has a unique representation

$$s(t) = \sum_{i=-m}^{k} a_i \, B_i^m(t), \qquad t \in [a, b]. \tag{2.13}$$

This representation has the desirable property that if we have to compute the value $s(t)$ for some $t \in [a, b]$, then only $m + 1$ values of the $k + m + 1$ values $B_{-m}^m(t), \ldots, B_k^m(t)$ are different from zero. (This follows from (2.1) in Theorem 2.2).

The next result shows that the linear hull of B–splines is a weak Chebyshev space.

Theorem 2.8. *For* $i \in \{-m, \ldots, k\}$ *and* $j \geq 1$, *the space*

$$\mathrm{span}\{B_i^m, \ldots, B_{i+j-1}^m\} \tag{2.14}$$

is a j*–dimensional weak Chebyshev subspace of* $C[x_i, x_{i+j+m}]$.

Proof. The claim can be proved by using arguments similar to those in the proof of Theorem 1.19 and Theorem 2.6.

2.3. Recurrence Relations

We show that B–splines can be represented as divided differences of truncated power functions and as complex contour integrals. Moreover, it is shown that a stable evaluation of B–splines is possible by using a recurrence relation. Finally, we prove that normalized B–splines form a partition of unity and give results on differentiation and integration of splines.

The first result shows that there is a fundamental relation between B–splines and divided differences.

Theorem 2.9. *For all* $t \in (-\infty, \infty)$,

$$B_i^m(t) = (-1)^{m+1} \, (m + 1) \, (t - x)_+^m \, [x_i, \ldots, x_{i+m+1}] \tag{2.15}$$

(i.e. $B_i^m(t)$ *is the divided difference of order* $m + 1$ *of the function* $x \rightarrow (-1)^{m+1}$ $(m + 1) \, (t - x)_+^m$, $x \in (-\infty, \infty)$, *with respect to the knots* $x_i, \ldots, x_{i+m+1}$).

Proof. We set

$$s(t) = (-1)^{m+1} \, (m + 1)(t - x)_+^m \, [x_i, \ldots, x_{i+m+1}], \qquad t \in (-\infty, \infty),$$

and show that s satisfies the B–spline properties (2.1) and (2.3). If $t \in (-\infty, x_i]$, then

$$(t - x_r)_+^m = 0, \qquad r = i, \ldots, i + m + 1,$$

and therefore $s(t) = 0$. If $t \in [x_{i+m+1}, \infty)$, then

$$(t - x)_+^m = (t - x)^m, \qquad x \in [x_i, x_{i+m+1}].$$

Since $x \to (t - x)^m$ is a polynomial of degree m, it follows from Remark 2.3 in Chapter I that $s(t) = 0$. This shows that s satisfies (2.1). Moreover, we have

$$
\begin{aligned}
\int_{x_i}^{x_{i+m+1}} s(t)\, dt &= \int_{x_i}^{x_{i+m+1}} (-1)^{m+1} (m + 1)(t - x)_+^m [x_i, \ldots, x_{i+m+1}]\, dt \\
&= (-1)^{m+1} (t - x)_+^{m+1} [x_i, \ldots, x_{i+m+1}] \Big|_{t=x_i}^{x_{i+m+1}} \\
&= (-1)^{m+1} (x_{i+m+1} - x)^{m+1} [x_i, \ldots, x_{i+m+1}] \\
&= x^{m+1} [x_i, \ldots, x_{i+m+1}] = 1.
\end{aligned}
$$

The last equality follows from Remark 2.3 in Chapter I. This shows that s satisfies (2.3). Then it follows from the proof of Theorem 2.2 that $s = B_i^m$. This proves Theorem 2.9.

The following complex integral representation of B–splines is due to Meinardus [1974]. (For details concerning integral representations of divided differences the reader is referred to Nörlund [1954].)

Theorem 2.10. *For all* $t \in (-\infty, \infty)$,

$$B_j^m(t) = \frac{1}{2\pi i} \int_{C_t} \frac{(m + 1)(z - t)^m}{(z - x_j) \cdots (z - x_{j+m+1})}\, dz, \qquad (2.16)$$

where C_t *is a simply closed rectifiable curve in the complex plane containing all knots* x_r *with* $t \le x_r \le x_{j+m+1}$ *and no others in its interior, and the integration is carried out in the positive direction.*

Proof. For all $t \in (-\infty, \infty)$, we denote the right hand side of (2.16) by $I_j^m(t)$. It follows form the residue theorem that the function $I_j^m : (-\infty, \infty) \to \mathbf{R}$ is a spline of degree m with knots $x_j, \ldots, x_{j+m+1}$. Since the numerator of the integrand in (2.16) is a polynomial of degree m and the denominator is a polynomial of degree $m + 2$, by the residue theorem we get that

$$I_j^m(t) = 0, \qquad t \in (-\infty, x_j].$$

Moreover, it follows from Cauchy's theorem that

$$I_j^m(t) = 0, \qquad t \in [x_{j+m+1}, \infty).$$

Furthermore, we have

$$
\begin{aligned}
\int_{x_j}^{x_{j+m+1}} I_j^m(t)\, dt &= -\frac{1}{2\pi i} \int_{C_{x_{j+m+1}}} \frac{(z - x_{j+m+1})^{m+1}}{(z - x_j) \cdots (z - x_{j+m+1})}\, dz \\
&\quad + \frac{1}{2\pi i} \int_{C_{x_j}} \frac{(z - x_j)^{m+1}}{(z - x_j) \cdots (z - x_{j+m+1})}\, dz.
\end{aligned}
$$

Again by Cauchy's theorem the first integral is zero. Moreover, it follows from the residue theorem that

$$\int_{C_{x_j}} \frac{(z - x_j) \cdots (z - x_{j+m}) - (z - x_j)^{m+1}}{(z - x_j) \cdots (z - x_{j+m+1})} dz = 0,$$

since the numerator of the integrand is a polynomial of degree m and the denominator is a polynomial of degree $m + 2$. This implies that

$$\int_{x_j}^{x_{j+m+1}} I_j^m(t) \, dt = \frac{1}{2\pi i} \int_{C_{x_j}} \frac{(z - x_j)^{m+1}}{(z - x_j) \cdots (z - x_{j+m+1})} dz$$

$$= \frac{1}{2\pi i} \int_{C_{x_j}} \frac{(z - x_j)^{m+1} + (z - x_j) \cdots (z - x_{j+m}) - (z - x_j)^{m+1}}{(z - x_j) \cdots (z - x_{j+m+1})} dz$$

$$= \frac{1}{2\pi i} \int_{C_{x_j}} \frac{1}{(z - x_{j+m+1})} dz = 1.$$

Therefore, it follows from Theorem 2.2 that $I_j^m = B_j^m$. This proves Theorem 2.10.

As has been shown by Meinardus [1974], the subsequent results on B–splines can also be derived from the representation (2.16) by using a simple decomposition technique for the rational integrand. Furthermore, by using (2.16) it is easy to compute derivatives of B–splines with respect to the knots. For example, if $m \geq 2$ and $r \in \{j, \ldots, j + m + 1\}$, then for all $t \in (-\infty, \infty)$,

$$\frac{\partial B_j^m(t)}{\partial x_r} = \frac{1}{2\pi i} \int_{C_t} \frac{(m + 1)(z - t)^m}{(z - x_j) \cdots (z - x_{r-1})(z - x_r)^2(z - x_{r+1}) \cdots (z - x_{j+m+1})} dz.$$

This expression can be reduced to B–splines of degree $m - 1$.

The next result, due to de Boor [1972] and Cox [1972], shows that B–splines can be evaluated with the aid of a recurrence relation.

Theorem 2.11. If $m \geq 2$, then for all $t \in (-\infty, \infty)$,

$$B_i^m(t) = \frac{m + 1}{m} \left(\frac{t - x_i}{x_{i+m+1} - x_i} B_i^{m-1}(t) + \frac{x_{i+m+1} - t}{x_{i+m+1} - x_i} B_{i+1}^{m-1}(t) \right). \qquad (2.17)$$

Proof. Let $m \geq 2$ and $t \in (-\infty, \infty)$ be given. We set $f_1(x) = (t - x)$ and $f_2(x) = (t - x)_+^{m-1}$ for all $x \in (-\infty, \infty)$. Then it follows from Theorem 2.9, Theorem 2.4 in Chapter I and Remark 2.3 in Chapter I, that

$$B_i^m(t) = (-1)^{m+1} (m + 1)(f_1 f_2) [x_i, \ldots, x_{i+m+1}]$$

$$= (-1)^{m+1} (m + 1) \sum_{r=i}^{i+m+1} f_1 [x_i, \ldots, x_r] f_2 [x_r, \ldots, x_{i+m+1}]$$

$$= (-1)^{m+1} (m + 1) \left(f_1 [x_i] f_2 [x_i, \ldots, x_{i+m+1}] + \right.$$

$$+ f_1 [x_i, x_{i+1}] f_2 [x_{i+1}, \ldots, x_{i+m+1}] \Big)$$

$$= (-1)^{m+1} \frac{m+1}{m} \Big(m f_2 [x_{i+1}, \ldots, x_{i+m+1}] +$$

$$+ (t - x_i) \frac{m f_2 [x_{i+1}, \ldots, x_{i+m+1}] - m f_2 [x_i, \ldots, x_{i+m}]}{x_{i+m+1} - x_i} \Big)$$

$$= \frac{m+1}{m} \Big(\frac{t - x_i}{x_{i+m+1} - x_i} (-1)^m m f_2 [x_i, \ldots, x_{i+m}] +$$

$$+ \frac{x_{i+m+1} - t}{x_{i+m+1} - x_i} (-1)^m m f_2 [x_{i+1}, \ldots, x_{i+m+1}] \Big)$$

$$= \frac{m+1}{m} \Big(\frac{t - x_i}{x_{i+m+1} - x_i} B_i^{m-1}(t) + \frac{x_{i+m+1} - t}{x_{i+m+1} - x_i} B_{i+1}^{m-1}(t) \Big).$$

This proves Theorem 2.11.

Remark 2.12. By using the recurrence relation (2.17) for B–splines, the value $B_i^m(t)$ can be easily computed according to the following scheme starting with the values of the linear B–splines in the first column.

$$
\begin{array}{ccccc}
B_i^1(t) & & & & \\
 & B_i^2(t) & & & \\
B_{i+1}^1(t) & & \ddots & & \\
 & & & B_i^{m-1}(t) & \\
 & & & & B_i^m(t) \\
\vdots & \vdots & & B_{i+1}^{m-1}(t) & \\
 & & & & \\
B_{i+m-2}^1(t) & & \mathinner{\mkern2mu\raise1pt\hbox{.}\mkern2mu\raise4pt\hbox{.}\mkern2mu\raise7pt\hbox{.}\mkern1mu} & & \\
 & B_{i+m-2}^2(t) & & & \\
B_{i+m-1}^1(t) & & & &
\end{array}
$$

Note that if $t \in [x_j, x_{j+1}]$, then all values in the first column are zero except $B_{j-1}^1(t)$ and $B_j^1(t)$.

We now introduce the so–called normalized B–splines which are used as a basis of spline spaces in practice.

Definition 2.13. The spline N_i^m, defined by

$$N_i^m(t) = \frac{1}{m+1} (x_{i+m+1} - x_i) B_i^m(t) \tag{2.18}$$

for all $t \in (-\infty, \infty)$, is called the *normalized B–spline* of degree m with support (x_i, x_{i+m+1}).

It follows immediately from (2.17) and (2.18) that a recurrence relation also holds for normalized B–splines.

Theorem 2.14. *If $m \geq 2$, then for all $t \in (-\infty, \infty)$,*

$$N_i^m(t) = \frac{t - x_i}{x_{i+m} - x_i} N_i^{m-1}(t) + \frac{x_{i+m+1} - t}{x_{i+m+1} - x_{i+1}} N_{i+1}^{m-1}(t). \qquad (2.19)$$

The next result due to Marsden & Schoenberg [1966] shows that normalized B–splines form a partition of unity.

Theorem 2.15. *The normalized B–splines $N_{-m}^m, \ldots, N_k^m$ form a partition of unity, i.e. for all $t \in [a, b]$,*

$$\sum_{i=-m}^{k} N_i^m(t) = 1. \qquad (2.20)$$

Proof. We first prove (2.20) for $m = 1$. Let a point $t \in [t_j, t_{j+1}]$ be given. Then it follows that for all $i \in \{-m, \ldots, k\} \setminus \{j-1, j\}$,

$$N_i^1(t) = 0.$$

Moreover, by (2.15) and (2.18) we have

$$N_{j-1}^1(t) = (x_{j+1} - x_{j-1})(t - x)_+ [x_{j-1}, x_j, x_{j+1}] = \frac{x_{j+1} - t}{x_{j+1} - x_j}$$

and

$$N_j^1(t) = (x_{j+2} - x_j)(t - x)_+ [x_j, x_{j+1}, x_{j+2}] = \frac{t - x_j}{x_{j+1} - x_j}.$$

This implies that

$$\sum_{i=-m}^{k} N_i^1(t) = 1.$$

We now proceed by induction as follows. Suppose that (2.20) is true for $m - 1$. Let a point $t \in [a, b]$ be given. Since $B_{-m}^{m-1}(t) = 0$ and $B_{k+1}^{m-1}(t) = 0$, it follows from (2.17) and (2.18) that

$$\sum_{i=-m}^{k} N_i^m(t) = \frac{1}{m} \sum_{i=-m}^{k} \left((t - x_i) B_i^{m-1}(t) + (x_{i+m+1} - t) B_{i+1}^{m-1}(t) \right)$$

$$= \frac{1}{m} \left(\sum_{i=-m+1}^{k} (t - x_i) B_i^{m-1}(t) + \sum_{i=-m+1}^{k} (x_{i+m} - t) B_i^{m-1}(t) \right)$$

$$= \sum_{i=-m+1}^{k} \frac{1}{m} (x_{i+m} - x_i) B_i^{m-1}(t) = \sum_{i=-m+1}^{k} N_i^{m-1}(t) = 1.$$

This proves Theorem 2.15.

In the following we give results on the differentiation and integration of splines. The first result due to de Boor [1972] shows that the derivative of a given B–spline can be computed by B–splines of lower degree.

Theorem 2.16. *If $m \geq 2$, then for all $t \in (-\infty, \infty)$,*

$$(N_i^m)'(t) = \frac{m}{x_{i+m} - x_i} N_i^{m-1}(t) - \frac{m}{x_{i+m+1} - x_{i+1}} N_{i+1}^{m-1}(t). \qquad (2.21)$$

Proof. It follows from (2.15) that for all $t \in (-\infty, \infty)$,

$$(N_i^m)'(t) = (-1)^{m+1} m \, (x_{i+m+1} - x_i) \, (t - x)_+^{m-1} \, [x_i, \ldots, x_{i+m+1}]$$

$$= (-1)^{m+1} m \, (x_{i+m+1} - x_i) \cdot$$

$$\cdot \frac{(t - x)_+^{m-1} \, [x_{i+1}, \ldots, x_{i+m+1}] - (t - x)_+^{m-1} \, [x_i, \ldots, x_{i+m}]}{x_{i+m+1} - x_i}$$

$$= \frac{m}{x_{i+m} - x_i} N_i^{m-1}(t) - \frac{m}{x_{i+m+1} - x_{i+1}} N_{i+1}^{m-1}(t).$$

This proves Theorem 2.16.

The next formula for computing the derivative of a given spline was proved by de Boor [1972].

Theorem 2.17. *Let a spline $s = \sum_{i=-m}^{k} a_i N_i^m$ be given. Then for all $j \in \{1, \ldots, m-1\}$ and all $t \in [a, b]$,*

$$s^{(j)}(t) = \sum_{i=-m+j}^{k} a_i^{(j)} N_i^{m-j}(t), \qquad (2.22)$$

where

$$a_i^{(j)} = \begin{cases} a_i, & \text{if } j = 0 \\ (m+1-j) \dfrac{a_i^{(j-1)} - a_{i-1}^{(j-1)}}{x_{i+m+1-j} - x_i}, & \text{if } j > 0 \end{cases} \qquad (2.23)$$

Proof. Let $j = 1$ and $t \in [a, b]$ be given. Since $N_{-m}^{m-1}(t) = 0$ and $N_{k+1}^{m-1}(t) = 0$, it follows from (2.21) that

$$s'(t) = \sum_{i=-m}^{k} a_i \, (N_i^m)'(t)$$

$$= \sum_{i=-m}^{k} a_i \left(\frac{m}{x_{i+m} - x_i} N_i^{m-1}(t) - \frac{m}{x_{i+m+1} - x_{i+1}} N_{i+1}^{m-1}(t) \right)$$

$$= \sum_{i=-m}^{k} a_i \frac{m}{x_{i+m} - x_i} N_i^{m-1}(t) - a_{i-1} \frac{m}{x_{i+m+1} - x_{i+1}} N_i^{m-1}(t)$$

$$= \sum_{i=-m+1}^{k} a_i^{(1)} N_i^{m-1}(t).$$

This proves (2.22) for $j = 1$. By repeating the same argument we obtain (2.22) for $j \in \{2, \ldots, m-1\}$. This proves Theorem 2.17.

We now derive a formula for computing the indefinite integral of a given spline due to de Boor, Lyche & Schumaker [1976].

Theorem 2.18. *Let a spline* $s = \sum_{i=-m}^{k} a_i N_i^m$ *be given. Then for all* $t \in [a, b]$,

$$\int_{x_{-m}}^{t} s(x)\,dx = \sum_{i=-m}^{k} \left(\sum_{j=-m}^{i} a_j \frac{x_{j+m+1} - x_j}{m+1} \right) N_i^{m+1}(t). \tag{2.24}$$

Proof. Since

$$s = \sum_{i=-m}^{k} a_i N_i^m,$$

the function $\tilde{s} : [a, b] \to \mathbf{R}$ defined by

$$\tilde{s}(t) = \int_{x_{-m}}^{t} s(x)\,dx, \qquad t \in [a, b],$$

is a spline of degree $m + 1$ and can be written as

$$\tilde{s} = \sum_{i=-m}^{k} \tilde{a}_i N_i^{m+1}.$$

It follows from (2.22) and (2.23) that for all $t \in [a, b]$,

$$\sum_{i=-m}^{k} a_i N_i^m(t) = s(t) = \tilde{s}'(t) = \sum_{i=-m}^{k} (m+1) \frac{\tilde{a}_i - \tilde{a}_{i-1}}{x_{i+m+1} - x_i} N_i^m(t),$$

where $\tilde{a}_{-m-1} = 0$. By comparing the basis coefficients, we get that

$$a_i = \frac{m+1}{x_{i+m+1} - x_i} \tilde{a}_i \qquad \text{for } i = -m$$

and

$$a_i = \frac{m+1}{x_{i+m+1} - x_i} \tilde{a}_i - \tilde{a}_{i-1}, \qquad i = -m+1, \ldots, k.$$

Then it follows that

$$\tilde{a}_i = \sum_{j=-m}^{i} a_j \frac{x_{j+m+1} - x_j}{m+1}, \qquad i = -m, \ldots, k.$$

This proves Theorem 2.18.

Finally, we discuss how to obtain the piecewise polynomial representation of a spline which is given as a linear combination of B–splines.

Let us assume that by applying an interpolation or best approximation method, we have computed a spline $s \in S_m(x_1, \ldots, x_k)$ in the following form

$$s(t) = \sum_{i=-m}^{k} a_i\, N_i^m(t), \qquad t \in [a, b]. \tag{2.25}$$

If we have to evaluate the spline s at a large number of points (e.g. if we want to draw the graph of s), then it is advantageous to determine the polynomials

$$p_j = s|_{[x_j, x_{j+1}]}, \qquad j = 0, \ldots, k, \tag{2.26}$$

and to evaluate these polynomials instead of the B–spline representation (2.25). Since the polynomials can be written as

$$p_j(t) = \sum_{r=0}^{m} \frac{1}{r!}\, s^{(r)}(x_j)\, (t - x_j)^r, \qquad t \in [x_j, x_{j+1}], \tag{2.27}$$

we have to compute the values $s^{(r)}(x_j)$, $r = 0, \ldots, m$, which can be done by using formula (2.22).

2.4. Variation Diminishing Property

It is shown that order complete weak Chebyshev spaces have a basis which is variation diminishing. In particular, spline spaces are order complete weak Chebyshev spaces and the B–spline basis has the variation diminishing property.

Definition 2.19. An n–dimensional subspace G of $C[a, b]$ is called an *order complete weak Chebyshev subspace* if there exists a basis $\{g_1, \ldots, g_n\}$ of G such that for all $p \in \{1, \ldots, n\}$ and all $i_1 < \cdots < i_p$ in $\{1, \ldots, n\}$, the subspace $\mathrm{span}\{g_{i_1}, \ldots, g_{i_p}\}$ is an p–dimensional weak Chebyshev subspace.

Example 2.20. The space $S_m(x_1, \ldots, x_k)$ is an order complete weak Chebyshev subspace of $C[a, b]$. This can be seen as follows. Since by Remark 3.10 below all minors of the B–spline determinant

$$D\left(\begin{matrix} B_{-m}^m & , \ldots, & B_k^m \\ t_1 & , \ldots, & t_{k+m+1} \end{matrix} \right)$$

are nonnegative for all points $t_1 < \cdots < t_{k+m+1}$ in $[a, b]$, it follows that

$$\mathrm{span}\{B_{i_1}^m, \ldots, B_{i_p}^m\}$$

is a p–dimensional weak Chebyshev subspace of $C[a, b]$ for all $p \in \{1, \ldots, k + m + 1\}$ and all $i_1 < \cdots < i_p$ in $\{1, \ldots, k + m + 1\}$. Therefore, $\{B_{-m}^m, \ldots, B_k^m\}$ is the desired basis of $S_m(x_1, \ldots, x_k)$.

Order complete weak Chebyshev spaces G have the important property that there exists a basis of G which is *variation diminishing* in the following sense (compare Theorem 1.21 in Chapter I).

Theorem 2.21. *Let G be an n–dimensional order complete weak Chebyshev subspace of $C[a,b]$. Then there exists a basis $\{g_1,\ldots,g_n\}$ of G with the following property: For every function $g = \sum_{i=1}^n a_i\, g_i$ in G, the number of sign changes of g is less than or equal to the number of sign changes in the sequence $a_1,\ldots,a_n$.*

Proof. The proof of Theorem 2.21 is completely analogous to the proof of Theorem 1.21 in Chapter I.

Corollary 2.22. *For every spline $s = \sum_{i=-m}^k a_i\, B_i^m$ in $S_m(x_1,\ldots,x_k)$, the number of sign changes of s is less than or equal to the number of sign changes in the sequence $a_{-m},\ldots,a_k$.*

Proof. Since by Example 2.20 the desired basis for the order complete weak Chebyshev space $S_m(x_1,\ldots,x_k)$ is the B–spline basis $\{B_{-m}^m,\ldots,B_k^m\}$ the claim follows from Theorem 2.21. This proves Corollary 2.22.

3. Interpolation by Splines

In Section 1 of Chapter I we have seen that Lagrange interpolation by Chebyshev spaces and Hermite interpolation by extended Chebyshev spaces is always possible. This is not the case for spline spaces. We give a complete characterization of those sets of points for which the Hermite interpolation problem for splines has a unique solution. Moreover, we investigate interpolation at the knots, where the interpolating splines satisfy certain boundary conditions. In particular, it is shown that the interpolation problem for complete splines, periodic splines and natural splines has a unique solution and that the interpolating splines satisfy certain optimality properties. Moreover, we give some results on the norm of spline operators resulting from Lagrange interpolation and interpolation by periodic splines. Finally, we discuss the method of quasi–interpolation.

3.1. Lagrange and Hermite Interpolation by Splines

At the beginning of this section we give a result on the number of zeros of splines. This theorem will be used to prove a complete characterization of those sets of points for which the Hermite interpolation problem for splines is uniquely solvable. Finally, we investigate the norm of Lagrange interpolation operators in the case of linear, quadratic and cubic splines.

To prove our first result on the number of zeros of splines we need the following definitions.

Definition 3.1. A spline $s \in S_m(x_1,\ldots,x_k)$ is said to have a *zero of multiplicity* (at least) r at a point $t \in [a,b]$, where $r \le m$, if

$$s^{(i)}(t) = 0, \quad i = 0,\ldots,r-1.$$

Definition 3.2. Let a function $f \in C[a,b]$ and points $a \leq t_1 < \cdots < t_p \leq b$ with $f(t_i) = 0$, $i = 1, \ldots, p$, be given, where $p \geq 2$. The zeros $t_1, \ldots, t_p$ of f are called *isolated* if for each $i \in \{1, \ldots, p-1\}$, there exists a point $\tilde{t}_i \in (t_i, t_{i+1})$ with $f(\tilde{t}_i) \neq 0$.

We now give a result on the number of zeros of splines. A general result on zeros of splines can be found in Schumaker [1976a].

Theorem 3.3. *If a spline s from the j-dimensional space $\mathrm{span}\{B_i^m, \ldots, B_{i+j-1}^m\}$ has only finitely many zeros in (x_i, x_{i+m+j}), $j \geq 1$, then s has at most $j-1$ zeros (counting multiplicities) in (x_i, x_{i+m+j}).*

Proof. Let a spline $s \in \mathrm{span}\{B_i^m, \ldots, B_{i+j-1}^m\}$ be given such that s has only finitely many zeros. If $m = 1$, than the claim is obvious. Therefore, we consider the case that $m \geq 2$. Let $x_i = z_0 < z_1 < \cdots < z_r < z_{r+1} = x_{i+m+j}$ be the zeros of s with corresponding multiplicities $m_1, \ldots, m_r$. We set $p = \sum_{i=1}^{r} m_i$. Then it follows from Theorem 1.18 that the spline s' of degree $m-1$ has at least one sign change in every interval (z_i, z_{i+1}), $i = 0, \ldots, r$, and a zero of multiplicity $m_i - 1$ in z_i, if $m_i \geq 2$, $i = 0, \ldots, r+1$. This implies that s' has at least $p+1$ zeros (counting multiplicities) in (x_i, x_{i+m+j}) and these zeros are isolated. We continue this method by considering $s'', \ldots, s^{(m-1)}$. By the same arguments as above, we finally obtain a spline $s^{(m-1)}$ of degree one which has at least $p + (m-1)$ isolated zeros. It is obvious that such a spline has at most $m + j - 2$ isolated zeros which implies that $p + (m-1) \leq m + j - 2$, i.e. $p \leq j - 1$. This proves Theorem 3.3.

The next result is an immediate consequence of Theorem 3.3 and Theorem 2.6.

Corollary 3.4. *If a spline $s \in S_m(x_1, \ldots, x_k)$ has only finitely many zeros in $[a,b]$, then s has at most $k+m$ zeros (counting multiplicities) in $[a,b]$.*

Remark 3.5. It follows from Theorem 1.17 that every nontrivial function from an n–dimensional extended Chebyshev space has at most $n-1$ zeros (counting multiplicities up to $n-1$). By comparing this result with Corollary 3.4, we see that every spline in $S_m(x_1, \ldots, x_k)$ with only finitely many zeros has at most $n-1$ zeros (counting multiplicities), where n is the dimension of $S_m(x_1, \ldots, x_k)$.

In the following we investigate Hermite interpolation by splines.

Definition 3.6. Let a function $f \in C[a,b]$ which is sufficiently differentiable and points $t_1 \leq \cdots \leq t_{k+m+1}$ in $[a,b]$ be given. We set for all $j \in \{1, \ldots, k+m+1\}$,

$$d_j = \max\{i : t_j = \cdots = t_{j-i}\}, \tag{3.1}$$

and assume that

$$\begin{aligned} d_j &\leq m, && \text{if } t_j \notin \{x_1, \ldots, x_k\} \\ d_j &\leq m-1, && \text{if } t_j \in \{x_1, \ldots, x_k\}. \end{aligned} \tag{3.2}$$

The *Hermite interpolation problem* is to determine a function $s \in S_m(x_1, \ldots, x_k)$ such that

$$s_f^{(d_j)}(t_j) = f^{(d_j)}(t_j), \qquad j = 1, \ldots, k+m+1. \tag{3.3}$$

If $t_1 < \cdots < t_{k+m+1}$, then we obtain the *Lagrange interpolation problem* to determine a function $s \in S_m(x_1, \ldots, x_k)$ such that

$$s_f(t_j) = f(t_j), \qquad j = 1, \ldots, k+m+1. \tag{3.4}$$

If we use the B–spline basis $\{B_{-m}^m, \ldots, B_k^m\}$ of $S_m(x_1, \ldots, x_k)$, then the determinant which corresponds to the Hermite interpolation problem (3.3) is

$$D \left(\begin{array}{ccc} B_{-m}^m & , \cdots , & B_k^m \\ t_1 & , \cdots , & t_{k+m+1} \end{array} \right)$$

$$= \left| \begin{array}{ccc} (B_{-m}^m)^{(d_1)}(t_1) & \cdots & (B_k^m)^{(d_1)}(t_1) \\ \vdots & & \vdots \\ (B_{-m}^m)^{(d_{k+m+1})}(t_{k+m+1}) & \cdots & (B_k^m)^{(d_{k+m+1})}(t_{k+m+1}) \end{array} \right|.$$

The next theorem is a characterization of those sets of points for which the Hermite interpolation problem (3.3) has a unique solution. In the case of Lagrange interpolation, this result was proved by Schoenberg & Whitney [1953]. The generalization to Hermite interpolation was obtained by Karlin & Ziegler [1966]. Extensions of these results were given by Nürnberger, Schumaker, Sommer & Strauß [1983], [1984].

Theorem 3.7. *Let points $t_1 \leq \cdots \leq t_{k+m+1}$ in $[a, b]$ be given. The following statements are equivalent:*
(i) *For all sufficiently differentiable functions $f \in C[a, b]$ the Hermite interpolation problem (3.3) has a unique solution from $S_m(x_1, \ldots, x_k)$.*
(ii)

$$D \left(\begin{array}{ccc} B_{-m}^m & , \cdots , & B_k^m \\ t_1 & , \cdots , & t_{k+m+1} \end{array} \right) \neq 0. \tag{3.5}$$

(iii)

$$t_j < x_j < t_{j+m+1}, \qquad j = 1, \ldots, k. \tag{3.6}$$

(iv) *Every interval $(x_i, x_{i+m+j}) \subset (x_{-m}, x_{k+m+1})$, $j \geq 1$, contains at least j (not necessarily distinct) points from $\{t_1, \ldots, t_{k+m+1}\}$.*

Proof. (i) $\Longleftrightarrow$ (ii). This is a standard result in Linear Algebra.
   (i) $\Rightarrow$ (iii). Suppose that (iii) fails.
Case 1. $t_{r+m+1} \leq x_r$ for some $r \in \{1, \ldots, k\}$.
The Hermite interpolation problem is to determine real numbers $a_{-m}, \ldots, a_k$ such that

$$\sum_{i=-m}^{k} a_i \, (B_i^m)^{(d_j)}(t_j) = f^{(d_j)}(t_j), \qquad j = 1, \ldots, k+m+1. \tag{3.7}$$

Since $(B_i^m)^{(d_j)}(t_j) = 0$, $j = 1,\ldots,r+m+1$, $i = r,\ldots,k$, it follows from (3.7) that

$$\sum_{i=-m}^{r-1} a_i \,(B_i^m)^{(d_j)}(t_j) = f^{(d_j)}(t_j), \qquad j = 1,\ldots,r+m+1. \qquad (3.8)$$

Since (3.8) is a linear system of equations with $r+m+1$ equations, but only $r+m$ unknowns $a_{-m},\ldots,a_{r-1}$, this system does not have a solution for all values $f^{(d_j)}(t_j)$, $j = 1,\ldots,r+m+1$, i.e. (i) fails.

Case 2. $x_r \le t_r$ for some $r \in \{1,\ldots,k\}$.

Analogously as in Case 1 we obtain a linear system of equations with more equations than unknowns which again shows that (i) fails.

(iii) $\Rightarrow$ (i). We assume that (iii) holds, but (i) fails. Since (i) fails, there exist real numbers $a_{-m},\ldots,a_k$ such that

$$\sum_{i=-m}^{k} |a_i| \ne 0$$

and

$$\sum_{i=-m}^{k} a_i \,(B_i^m)^{(d_j)}(t_j) = 0, \qquad j = 1,\ldots,k+m+1. \qquad (3.9)$$

We define the spline s by

$$s(t) = \sum_{i=-m}^{k} a_i \, B_i^m(t), \qquad t \in (-\infty,\infty).$$

Moreover, we choose additional knots x_{-m-1} and x_{k+m+2} such that $x_{-m-1} < x_{-m}$ and $x_{k+m+1} < x_{k+m+2}$. Since s is a nontrivial spline, by Remark 2.4 there exists an interval $(x_i, x_{i+m+j}) \subset (x_{-m}, x_{k+m+1})$, $j \ge 1$, such that

$$s(t) = 0, \qquad t \in [x_{i-1}, x_i] \cup [x_{i+m+j}, x_{i+m+j+1}]$$

and s has only finitely many zeros in (x_i, x_{i+m+j}). It follows from (iii) that

$$x_i < t_{i+m+1} \le \cdots \le t_{i+m+j} < x_{i+m+j}.$$

By (3.9) the spline s has at least j zeros (counting multiplicities) in (x_i, x_{i+m+j}). On the other hand, since $s \in \text{span}\{B_i^m,\ldots,B_{i+j-1}^m\}$ on $[x_i, x_{i+m+j}]$ and s has only finitely many zeros, it follows from Theorem 3.3 that s has at most $j-1$ zeros (counting multiplicities) in (x_i, x_{i+m+j}) which is a contradiction.

(iii) $\Rightarrow$ (iv). Let an interval (x_i, x_{i+m+j}) as in (iv) be given. It follows from (iii) that

$$x_i < t_{i+m+1} \le \cdots \le t_{i+m+j} < x_{i+m+j}.$$

This shows that (x_i, x_{i+m+j}) contains at least j points from $\{t_1,\ldots,t_{k+m+1}\}$.

(iv) $\Rightarrow$ (iii). Let $j \in \{1,\ldots,k\}$ be given. It follows from (iv) that (x_{-m}, x_j) contains at least j points from $\{t_1,\ldots,t_{k+m+1}\}$ which implies that $t_j < x_j$. Moreover, it follows from (iv) that (x_j, x_{k+m+1}) contains at least $k-j+1$ points from $\{t_1,\ldots,t_{k+m+1}\}$ which implies that $x_j < t_{j+m+1}$. This proves Theorem 3.7.

Remark 3.8. In order to solve the interpolation problem (3.3) in the case that $t_j < x_j < t_{j+m+1}$, $j = 1, \ldots, k$, (see Theorem 3.7) it is convenient to use the B-spline basis of $S_m(x_1, \ldots, x_k)$ (in practice the normalized B–spline basis), since the matrix

$$
\begin{pmatrix}
(B_{-m}^m)^{(d_1)}(t_1) & \cdots & (B_k^m)^{(d_1)}(t_1) \\
\vdots & & \vdots \\
(B_{-m}^m)^{(d_{k+m+1})}(t_{k+m+1}) & \cdots & (B_k^m)^{(d_{k+m+1})}(t_{k+m+1})
\end{pmatrix}
$$

which corresponds to problem (3.3) has the property that all elements are zero except at the diagonal, m superdiagonals and m subdiagonals. (This follows from Theorem 2.2). Such a matrix is called *banded*.

The following result shows that all determinants corresponding to the Hermite interpolation problem (3.3) have the same sign.

Theorem 3.9. *For all points $t_1 \leq \cdots \leq t_{k+m+1}$ in $[a, b]$ with $t_i < t_{i+m}$ for all i,*

$$
D\begin{pmatrix} B_{-m}^m & , & \cdots & , & B_k^m \\ t_1 & , & \cdots & , & t_{k+m+1} \end{pmatrix} \geq 0. \tag{3.10}
$$

Proof. By Theorem 2.8 the space $\operatorname{span}\{B_{-m}^m, \ldots, B_k^m\}$ is a weak Chebyshev subspace of $C[x_{-m}, x_{k+m+1}]$, i.e. there exists a sign $\sigma \in \{-1, 1\}$ such that for all $t_1 < \cdots < t_{k+m+1}$ in $[x_{-m}, x_{k+m+1}]$,

$$
\sigma D\begin{pmatrix} B_{-m}^m & , & \cdots & , & B_k^m \\ t_1 & , & \cdots & , & t_{k+m+1} \end{pmatrix} \geq 0. \tag{3.11}
$$

To determine the sign σ, we choose points $t_1 < \cdots < t_{k+m+1}$ in $[x_{-m}, x_{k+m+1}]$ such that
$$
t_{m+1-i} \in (x_i, x_{i+1}), \qquad i = -m, \ldots, k.
$$
Then it follows from Theorem 2.2 that the diagonal elements of the above determinant are positive and all elements above the diagonal are zero. This shows that the determinant is positive and therefore $\sigma = 1$. Then it follows from (3.11) and the proof of Theorem 1.8 in Chapter I, (i) $\Rightarrow$ (ii) that (3.10) holds for all $t_1 \leq \cdots \leq t_{k+m+1}$ in $[x_{-m}, x_{k+m+1}]$. This proves Theorem 3.9.

Remark 3.10. The following stronger version of Theorem 3.7 and Theorem 3.9 was proved by de Boor [1976a]:
For all $i_1 < \cdots < i_p$ in $\{-m, \ldots, k\}$ and all $t_1 \leq \cdots \leq t_p$ with $t_i < t_{i+m}$ for all i,

$$
D\begin{pmatrix} B_{i_1}^m & , & \cdots & , & B_{i_p}^m \\ t_1 & , & \cdots & , & t_p \end{pmatrix} \geq 0. \tag{3.12}
$$

and the determinant is strictly positive if and only if

$$
t_j \in (x_{i_j}, \ldots, x_{i_j+m+1}), \qquad j = 1, \ldots, p. \tag{3.13}
$$

This implies that for given points $t_1 < \cdots < t_{k+m+1}$ in $[a, b]$ all minors of the determinant

$$D \begin{pmatrix} B_{-m}^m & , & \cdots & , & B_k^m \\ t_1 & , & \cdots & , & t_{k+m+1} \end{pmatrix}$$

are nonnegative. Matrices with this property are called *totally positive*. This property was proved by Karlin [1968]. It was shown by de Boor & Pinkus [1977] that a linear system with an invertible totally positive matrix can be solved in a stable way by Gauss elimination without pivoting.

Up to now we have investigated the solvability of spline interpolation problems. We now give results on the norm of spline interpolation operators.

Interpolation is a relatively simple approximation method. Let $L : C[a, b] \to S_m(x_1, \ldots, x_k)$ be a projection which results from interpolation by splines in $S_m(x_1, \ldots, x_k)$. Then it follows from Theorem 2.17 in Chapter I that for all $f \in C[a, b]$,

$$\|f - L(f)\|_\infty \leq (1 + \|L\|_\infty) \inf_{s \in S_m(x_1, \ldots, x_k)} \|f - s\|_\infty.$$

This shows that if the norm of L is small, then this operator yields nearly best approximations.

In the following we will show that in the case of arbitrary knot location Lagrange interpolation with linear and quadratic splines has this property.

We consider the following interpolation problem.

Definition 3.11. Let arbitrary knots $a = x_0 < x_1 < \cdots < x_k < x_{k+1} = b$ be given. Moreover, we set

$$x_{-m+1} = \cdots = x_{-1} = a$$

and

$$x_{k+2} = \cdots = x_{k+m} = b.$$

We consider Lagrange interpolation by splines from $S_m(x_1, \ldots, x_k)$ at the following points, defined as averages of m consecutive knots:

$$t_i = \frac{x_{i-m} + \cdots + x_{i-1}}{m}, \qquad i = 1, \ldots, k + m + 1. \tag{3.14}$$

Let a function $f \in C[a, b]$ be given. Then the Lagrange interpolation problem is to determine a spline $s_f \in S_m(x_1, \ldots, x_k)$ such that

$$s_f(t_i) = f(t_i), \qquad i = 1, \ldots, k + m + 1. \tag{3.15}$$

The next result on the unique solvability of the interpolation problem (3.15) is a simple observation.

Lemma 3.12. *The Lagrange interpolation problem (3.15) has a unique solution from $S_m(x_1, \ldots, x_k)$.*

Proof. It follows from (3.14) that

$$t_j \leq x_{j-1} \quad \text{and} \quad x_{j+1} \leq t_{j+m+1}, \qquad j = 1, \ldots, k,$$

which implies

$$t_j < x_j < t_{j+m+1}, \qquad j = 1, \ldots, k.$$

Therefore, the claim follows from Theorem 3.7. This proves Lemma 3.12.

We now define the operator which results from the interpolation problem (3.15).

Definition 3.13. Let $L_{S_m(x_1,\ldots,x_k)} : C[a,b] \to S_m(x_1,\ldots,x_k)$ be the mapping which associates to each function $f \in C[a,b]$ the unique solution $L_{S_m(x_1,\ldots,x_k)}(f) = s_f$ from $S_m(x_1,\ldots,x_k)$ of the Lagrange interpolation problem (3.15).

In the following we give results on the norm of $L_{S_m(x_1,\ldots,x_k)}$ for $m \in \{1,2,3\}$. The first result on linear splines is a simple observation.

Theorem 3.14. *We have*

$$\|L_{S_1(x_1,\ldots,x_k)}\|_\infty = 1. \tag{3.16}$$

In particular, for every $f \in C[a,b]$,

$$\|f - L_{S_1(x_1,\ldots,x_k)}(f)\|_\infty \leq 2 \inf_{s \in S_1(x_1,\ldots,x_k)} \|f - s\|_\infty. \tag{3.17}$$

Proof. Let $f \in C[a,b]$ be given. Then it is easy to see that

$$\|L_{S_1(x_1,\ldots,x_k)}(f)\|_\infty = \max_{i=0,\ldots,k+1} |L_{S_1(x_1,\ldots,x_k)}(f)(t_i)|$$

$$= \max_{i=0,\ldots,k+1} |f(t_i)| \leq \|f\|_\infty.$$

This implies that

$$\|L_{S_1(x_1,\ldots,x_k)}(f)\|_\infty = 1.$$

Statement (3.17) follows directly from Theorem 2.17 in Chapter I. This proves Theorem 3.14.

The next result, due to Marsden [1974a], shows that for quadratic splines the Lagrange interpolation method provides a nearly optimal error.

Theorem 3.15. *We have*

$$\|L_{S_2(x_1,\ldots,x_k)}\|_\infty \leq 2. \tag{3.18}$$

In particular, for every $f \in C[a,b]$,

$$\|f - L_{S_2(x_1,\ldots,x_k)}(f)\|_\infty \leq 3 \inf_{s \in S_2(x_1,\ldots,x_k)} \|f - s\|_\infty. \tag{3.19}$$

Proof. Let $f \in C[a, b]$ and $s_f = L_{S_2(x_1, \ldots, x_k)}(f)$ be given. For each $i \in \{0, \ldots, k\}$ we set $s_i = s_f|_{[x_i, x_{i+1}]}$ and $h_i = x_{i+1} - x_i$. Now, let $i \in \{0, \ldots, k-1\}$ be given. Then $s_i \in P_2$ (respectively $s_{i+1} \in P_2$) is uniquely determined by the function values $s_f(x_i)$, $f(t_{i+2})$ and $s_f(x_{i+1})$ (respectively $s_f(x_{i+1})$, $f(t_{i+3})$ and $s_f(x_{i+2})$). A simple computation shows that

$$s_i'(x_{i+1}) = \frac{1}{h_i} \left(s_f(x_i) - 4 f(t_{i+2}) + 3 s_f(x_{i+1}) \right)$$

and

$$s_{i+1}'(x_{i+1}) = \frac{1}{h_{i+1}} \left(-3 s_f(x_{i+1}) + 4 f(t_{i+3}) - s_f(x_{i+2}) \right).$$

Since $s_f \in C^1[a, b]$, we have

$$s_i'(x_{i+1}) = s_{i+1}'(x_{i+1}),$$

which implies that

$$h_{i+1} s_f(x_i) + 3(h_i + h_{i+1}) s_f(x_{i+1}) + h_i s_f(x_{i+2})$$
$$= 4 h_{i+1} f(t_{i+2}) + 4 h_i f(t_{i+3}). \tag{3.20}$$

We set
$$K = \max\{ |s_f(x_i)| : i = 0, \ldots, k+1 \}.$$

Then it follows from (3.20) that

$$3(h_i + h_{i+1}) |s_f(x_i)|$$
$$\leq 4 h_{i+1} |f(t_{i+2})| + 4 h_i |f(t_{i+3})| + h_{i+1} |s_f(x_i)| + h_i |s_f(x_{i+2})|$$
$$\leq (4 \|f\|_\infty + K)(h_i + h_{i+1}).$$

Since $s_f(x_0) = f(x_0)$ and $s_f(x_{k+1}) = f(x_{k+1})$, it follows that

$$K \leq 2 \|f\|_\infty. \tag{3.21}$$

By using (3.21) we will show that

$$\|s_f\|_\infty \leq 2 \|f\|_\infty. \tag{3.22}$$

We consider $[x_i, x_{i+1}]$ for some $i \in \{0, \ldots, k\}$. For simplicity of notation we assume that $[x_i, x_{i+1}] = [0, 1]$. First, let $t \in [0, \frac{1}{2}]$ be given. By the Lagrange interpolation formula (2.17) in Chapter I and (3.21) we have

$$\begin{aligned} |s_f(t)| &= |2(t - \tfrac{1}{2})(t - 1) s_f(0) - 4t(t - 1) f(\tfrac{1}{2}) + 2t(t - \tfrac{1}{2}) s_f(1)| \\ &\leq 4\|f\|_\infty \left[(\tfrac{1}{2} - t)(1 - t) + t(1 - t) + t(t - \tfrac{1}{2}) \right] \\ &= 4\|f\|_\infty (\tfrac{1}{2} - t^2) \leq 2\|f\|_\infty. \end{aligned}$$

It follows analogously that

$$|s_f(t)| \leq 2\|f\|_\infty \qquad \text{for all } t \in [\tfrac{1}{2}, 1].$$

This shows (3.22) and thus we obtain

$$\|L_{S_2(x_1,\dots,x_k)}\| \leq 2.$$

Then statement (3.19) follows from Theorem 2.17 in Chapter I. This proves Theorem 3.15.

We now state without proof a result on the norm of the Lagrange interpolation operator for cubic splines which was proved by de Boor [1975].

Theorem 3.16. *We have*

$$\|L_{S_3(x_1,\dots,x_k)}\|_\infty \leq 27. \tag{3.23}$$

The above results show that for $m \leq 3$ the norm of $L_{S_m(x_1,\dots,x_k)}$ is bounded by a constant which does not depend on the knot set $\{x_1,\dots,x_k\}$. On the other hand, it was proved by Jia [1988] that this is not true for $m \geq 19$, i.e.

$$\sup_{x_1,\dots,x_k} \|L_{S_m(x_1,\dots,x_k)}\|_\infty = \infty.$$

Demko [1985] established the existence of interpolation operators for which the norms are bounded independent of the knot set $\{x_1,\dots,x_k\}$ for all m. However, the corresponding interpolation points are not known explicitly but have to be computed by best approximation methods.

3.2. Interpolation by Complete Splines, Periodic Splines and Natural Splines

In this section we study interpolation at the knots, where the interpolating splines satisfy certain boundary conditions. This leads to the so–called complete splines, periodic splines and natural splines. It is shown that the interpolation problems for these types of splines have a unique solution and that the interpolating splines satisfy certain optimality properties. Moreover, we give results on the norm of the interpolation operator for periodic cubic splines.

We first consider the following minimization problem. Let a function $f \in C[a, b]$ and points $a = x_0 < x_1 < \cdots < x_k < x_{k+1} = b$ be given. Determine a function $g \in C^2[a, b]$ with

$$g(x_i) = f(x_i), \qquad i = 0, \dots, k+1,$$

which minimizes the integral

$$\int_a^b (g''(t))^2 \, dt.$$

Holladay [1957] showed that there exists a unique so–called natural spline $s \in S_3(x_1,\ldots,x_k)$ with $s''(a) = s''(b) = 0$ which solves this problem.

There is a connection between this result and the following mechanical problem. Architects use a thin rod of some elastic material, called a "spline", to draw a smooth curve through given points, say $(x_i, f(x_i))$, $i = 0,\ldots,k+1$. It is known from mechanics that the resulting curve $g : [a,b] \to \mathbf{R}$ of the rod minimizes the strain energy

$$\int_a^b \frac{(g''(t))^2}{(1 + (g'(t))^2)^3}\, dt.$$

(Note, that the above integral is equal to $\int_a^b (c(t))^2\, dt$, where

$$c(t) = \frac{g''(t)}{(1 + (g'(t))^2)^{\frac{3}{2}}}$$

is the curvature of g.) It follows from Holladay's result that g is approximately a natural spline, however only for the special case when g' is nearly constant on $[a,b]$.

Holladay's result was generalized by Schoenberg [1964c] to natural splines of arbitrary degree.

In the following we give results of this type also for complete splines and periodic splines.

There is a vast literature on this topic including generalizations to minimization problems for linear operators. The interested reader is referred e.g. to the books of Ahlberg, Nilson & Walsh [1967], Böhmer [1974] and Schumaker [1981].

We begin with interpolation by splines, where the knots are chosen as interpolation points. Let a function $f \in C[a,b]$ and points $a = x_0 < x_1 < \cdots < x_k < x_{k+1} = b$ be given. The problem is to determine a spline $s \in S_m(x_1,\ldots,x_k)$ such that

$$s(x_i) = f(x_i), \qquad i = 0,\ldots,k+1.$$

Since the dimension of $S_m(x_1,\ldots,x_k)$ is $k + m + 1$ and there are only $k + 2$ interpolation conditions, we require that the spline s satisfies additional boundary conditions. This leads to various types of splines.

We first consider interpolation by complete splines.

Definition 3.17. Let a function $f \in C^r[a,b]$ be given, where $r \geq 1$. The *complete spline interpolation problem* is to determine a spline $s_f \in S_{2r+1}(x_1,\ldots,x_k)$ such that

$$s_f(x_i) = f(x_i), \qquad i = 0,\ldots,k+1, \tag{3.24}$$

$$s_f^{(j)}(a) = f^{(j)}(a), \qquad j = 1,\ldots,r, \tag{3.25}$$

$$s_f^{(j)}(b) = f^{(j)}(b), \qquad j = 1,\ldots,r, \tag{3.26}$$

(It is easy to see that the number of conditions is equal to the dimension of $S_{2r+1}(x_1,\ldots,x_k)$).

We first show that the complete spline interpolation problem has a unique solution.

Theorem 3.18. *The complete spline interpolation problem (3.24)–(3.26) has a unique solution from $S_{2r+1}(x_1, \ldots, x_k)$.*

Proof. It suffices to show that the homogeneous problem

$$s(x_i) = 0, \qquad i = 0, \ldots, k+1, \tag{3.27}$$

$$s^{(j)}(a) = 0, \qquad j = 1, \ldots, r, \tag{3.28}$$

$$s^{(j)}(b) = 0, \qquad j = 1, \ldots, r, \tag{3.29}$$

has only the trivial solution $s = 0$. Therefore, let $s \in S_{2r+1}(x_1, \ldots, x_k)$ satisfy (3.27)–(3.29). Applying integration by parts several times, we obtain

$$\int_a^b (s^{(r+1)}(t))^2 \, dt = \sum_{j=0}^{r-1} (-1)^j s^{(r+1+j)}(t) s^{(r-j)}(t) \Big|_{t=a}^b + (-1)^r \int_a^b s^{(2r+1)}(t) s'(t) \, dt.$$

It follows from (3.28) and (3.29) that

$$\sum_{j=0}^{r-1} (-1)^j s^{(r+1+j)}(t) s^{(r-j)}(t) \Big|_{t=a}^b = 0.$$

Therefore, we get by (3.27) that

$$\begin{aligned}
\int_a^b (s^{(r+1)}(t))^2 \, dt &= (-1)^r \int_a^b s^{(2r+1)}(t) s'(t) \, dt \\
&= (-1)^r \sum_{i=0}^k s^{(2r+1)}(t_i) \int_{x_i}^{x_{i+1}} s'(t) \, dt \\
&= (-1)^r \sum_{i=0}^k \{ s^{(2r+1)}(t_i) (s(x_{i+1}) - s(x_i)) \} = 0,
\end{aligned}$$

where t_i is an arbitrary point in (x_i, x_{i+1}), $i = 0, \ldots, k$. Note, that $s^{(2r+1)}$ is constant on each interval (x_i, x_{i+1}), since $s|_{(x_i, x_{i+1})} \in P_{2r+1}$, $i = 0, \ldots, k$. Since $\int_a^b (s^{(r+1)}(t))^2 \, dt = 0$, it follows that $s^{(r+1)} = 0$ and $s \in P_r$. Moreover, since by Example 1.13 in Chapter I the space P_r is an extended Chebyshev space and

$$s^{(j)}(a) = 0, \qquad j = 0, \ldots, r,$$

it follows from Theorem 1.8 in Chapter I that $s = 0$. This proves Theorem 3.18.

We now show that the complete spline which solves the interpolation problem (3.24)–(3.26) has the following optimality property.

Theorem 3.19. *Let $s_f \in S_{2r+1}(x_1, \ldots, x_k)$ be the unique solution of the complete spline interpolation problem (3.24)–(3.26). Then for every function $g \in C^{r+1}[a,b]$, $g \neq s_f$, with*

$$g(x_i) = f(x_i), \qquad i = 0, \ldots, k+1, \tag{3.30}$$

$$g^{(j)}(a) = f^{(j)}(a), \qquad j = 1, \ldots, r, \tag{3.31}$$

$$g^{(j)}(b) = f^{(j)}(b), \qquad j = 1, \ldots, r, \tag{3.32}$$

we have

$$\|s_f^{(r+1)}\|_2 < \|g^{(r+1)}\|_2. \tag{3.33}$$

Proof. Let a function $g \in C^{r+1}[a,b]$ satisfy (3.24)–(3.26). We have

$$\int_a^b (s_f^{(r+1)}(t))^2 \, dt \;=\; \int_a^b (g^{(r+1)}(t))^2 \, dt$$

$$- \int_a^b (g^{(r+1)}(t) - s_f^{(r+1)}(t))^2 \, dt$$

$$- 2 \int_a^b s_f^{(r+1)}(t) \, (g^{(r+1)}(t) - s_f^{(r+1)}(t)) \, dt.$$

It suffices to show that

$$\int_a^b s_f^{(r+1)}(t) \, (g^{(r+1)}(t) - s_f^{(r+1)}(t)) \, dt = 0. \tag{3.34}$$

We argue analogously as in the proof of Theorem 3.18. Applying integration by parts, we obtain

$$\int_a^b s_f^{(r+1)}(t) \, (g^{(r+1)}(t) - s_f^{(r+1)}(t)) \, dt$$

$$= \sum_{j=0}^{r-1} (-1)^j \, s_f^{(r+1+j)}(t) \, (g^{(r-j)}(t) - s_f^{(r-j)}(t)) \, \Big|_{t=a}^{b}$$

$$+ (-1)^r \int_a^b s_f^{(2r+1)}(t) \, (g'(t) - s_f'(t)) \, dt.$$

Since by (3.25), (3.26), (3.31) and (3.32) the above sum is zero and by (3.24) and (3.30) the last integral is zero, we get (3.34). This proves Theorem 3.19.

The next result shows that complete splines possess a further optimality property.

Theorem 3.20. *Let a function $f \in C^{r+1}[a,b]$ be given and $s_f \in S_{2r+1}(x_1, \ldots, x_k)$ be the unique solution of the corresponding complete spline interpolation problem (3.24)–(3.26). Then for every spline $s \in S_{2r+1}(x_1, \ldots, x_k)$, we have*

$$\|f^{(r+1)} - s_f^{(r+1)}\|_2 \leq \|f^{(r+1)} - s^{(r+1)}\|_2. \tag{3.35}$$

The equality sign in (3.35) holds if and only if $s - s_f \in P_r$.

Proof. Let a function $f \in C^{r+1}[a, b]$ be given. We have

$$\int_a^b (f^{(r+1)}(t) - s_f^{(r+1)}(t))^2 \, dt$$

$$= \int_a^b (f^{(r+1)}(t) - s^{(r+1)}(t))^2 \, dt - \int_a^b (s^{(r+1)}(t) - s_f^{(r+1)}(t))^2 \, dt$$

$$- 2 \int_a^b (s^{(r+1)}(t) - s_f^{(r+1)}(t)) \, (f^{(r+1)}(t) - s_f^{(r+1)}(t)) \, dt.$$

We will show that

$$\int_a^b (s^{(r+1)}(t) - s_f^{(r+1)}(t)) \, (f^{(r+1)}(t) - s_f^{(r+1)}(t)) \, dt = 0. \tag{3.36}$$

Then statement (3.35) follows. Moreover, the equality sign in (3.25) holds if and only if

$$\int_a^b (s^{(r+1)}(t) - s_f^{(r+1)}(t))^2 \, dt = 0,$$

i.e. $s - s_f \in P_r$.

In order to prove (3.36), we argue analogously as in the proof of Theorem 3.18. Applying integration by parts, we obtain

$$\int_a^b (s^{(r+1)}(t) - s_f^{(r+1)}(t)) \, (f^{(r+1)}(t) - s_f^{(r+1)}(t)) \, dt$$

$$= \sum_{j=0}^{r-1} (-1)^j \, (s^{(r+1+j)}(t) - s_f^{(r+1+j)}(t))(f^{(r-j)}(t) - s_f^{(r-j)}(t)) \, \Big|_{t=a}^b$$

$$+ (-1)^r \int_a^b (s^{(2r+1)}(t) - s_f^{(2r+1)}(t)) \, (f'(t) - s_f'(t)) \, dt.$$

Since by (3.25) and (3.26) the above sum is zero and by (3.24) the last integral is zero, we get (3.36). This proves Theorem 3.20.

In the following we investigate interpolation by periodic splines. This class of splines is the natural class to approximate periodic functions.

Definition 3.21. A function $s \in S_{2r+1}(x_1, \ldots, x_k)$ is called a *periodic spline* of degree $2r + 1$ with knots $x_1, \ldots, x_k$, if

$$s^{(j)}(a) = s^{(j)}(b), \qquad j = 0, \ldots, 2r. \tag{3.37}$$

Remark 3.22. A function $s \in S_{2r+1}(x_1, \ldots, x_k)$ is a periodic spline if and only if there exists an extension $\tilde{s} : (-\infty, \infty) \to \mathbf{R}$ of s such that $\tilde{s}$ has $2r$ continuous derivatives and $\tilde{s}$ is a periodic function with period $p = b - a$ (i.e. $\tilde{s}(t + p) = \tilde{s}(t)$ for all $t \in (-\infty, \infty)$).

We now formulate the interpolation problem for perodic splines.

Definition 3.23. Let a periodic function $f \in C[a, b]$ (i.e. $f(a) = f(b)$) be given. The *periodic spline interpolation problem* is to determine a periodic spline $s_f \in S_{2r+1}(x_1, \ldots, x_k)$, where $r \geq 1$, such that

$$s_f(x_i) = f(x_i), \qquad i = 0, \ldots, k+1, \tag{3.38}$$

(It is easy to see that the number of conditions is equal to the dimension of $S_{2r+1}(x_1, \ldots, x_k)$).

The next result shows that the periodic spline interpolation problem has a unique solution.

Theorem 3.24. *The periodic spline interpolation problem (3.38) has a unique solution from $S_{2r+1}(x_1, \ldots, x_k)$.*

Proof. It suffices to show that the homogeneous problem

$$s(x_i) = 0, \qquad i = 0, \ldots, k+1, \tag{3.39}$$

has only the trivial solution $s = 0$ in the space of periodic splines. The proof is completely analogous to the proof of Theorem 3.18. We note that by (3.37)

$$\sum_{j=0}^{r-1}(-1)^j \, s^{(r+1+j)}(t) \, s^{(r-j)}(t) \, \Big|_{t=a}^{b} = 0.$$

Using this fact, it follows that

$$\int_a^b (s^{(r+1)}(t))^2 \, dt = 0$$

which implies that $s \in P_r$. Then it follows from (3.37) that s is a constant function. Finally, by (3.39) we get $s = 0$. This proves Theorem 3.24.

The periodic spline which solves the interpolation problem (3.38) has the following optimality property.

Theorem 3.25. *Let $s_f \in S_{2r+1}(x_1, \ldots, x_k)$ be the unique solution of the periodic spline interpolation problem (3.38). Then for every function $g \in C^{r+1}[a, b]$, $g \neq s_f$, with*

$$g(x_i) = f(x_i), \qquad i = 0, \ldots, k+1, \tag{3.40}$$
$$g^{(j)}(a) = g^{(j)}(b), \qquad j = 0, \ldots, r, \tag{3.41}$$

we have

$$\|s_f^{(r+1)}\|_2 < \|g^{(r+1)}\|_2. \tag{3.42}$$

Proof. The proof is completely analogous to the proof of Theorem 3.19. We only note that by (3.37) and (3.41)

$$\sum_{j=0}^{r-1}(-1)^j\, s_f^{(r+1+j)}(t)\,(g^{(r-j)}(t) - s_f^{(r-j)}(t))\,\Big|_{t=a}^b = 0$$

and by (3.38) and (3.40)

$$\int_a^b s_f^{(2r+1)}(t)\,(g'(t) - s_f'(t))\,dt = 0.$$

This proves Theorem 3.25.

Interpolating periodic splines possess a further optimality property.

Theorem 3.26. *Let a function* $f \in C^{r+1}[a, b]$ *with*

$$f^{(j)}(a) = f^{(j)}(b), \qquad j = 0, \ldots, r, \tag{3.43}$$

be given and $s_f \in S_{2r+1}(x_1, \ldots, x_k)$ *be the unique solution of the corresponding periodic spline interpolation problem (3.38). Then for every periodic spline* $s \in S_{2r+1}(x_1, \ldots, x_k)$, *we have*

$$\|f^{(r+1)} - s_f^{(r+1)}\|_2 \le \|f^{(r+1)} - s^{(r+1)}\|_2. \tag{3.44}$$

The equality sign in (3.44) holds if and only if $s - s_f \in P_r$.

Proof. The proof is completely analogous to the proof of Theorem 3.20. We only note that by (3.43) and the fact that s and s_f are periodic splines,

$$\sum_{j=0}^{r-1}(-1)^j\,(s^{(r+1+j)}(t) - s_f^{(r+1+j)}(t))\,(f^{(r-j)}(t) - s_f^{(r-j)}(t))\,\Big|_{t=a}^b = 0$$

and by (3.38)

$$\int_a^b (s^{(2r+1)}(t) - s_f^{(2r+1)}(t))\,(f'(t) - s_f'(t))\,dt = 0.$$

This proves Theorem 3.26.

In the following we investigate the optimality property of interpolating natural splines which we discussed at the beginning of this section.

However, we mention that the approximation power of natural splines near the endpoints of the considered interval is bad compared with splines not satisfying the boundary conditions (3.45). For details see e.g. de Boor [1978].

Definition 3.27. A function $s \in S_{2r+1}(x_1, \ldots, x_k)$, where $r \ge 1$, is called a *natural spline* of degree $2r + 1$ with knots $x_1, \ldots, x_k$, if

$$s^{(j)}(a) = s^{(j)}(b) = 0, \qquad j = r + 1, \ldots, 2r. \tag{3.45}$$

Remark 3.28. A function $s \in S_{2r+1}(x_1, \ldots, x_k)$ is a natural spline if and only if there exists an extension $\tilde{s} : (-\infty, \infty) \to \mathbf{R}$ of s such that $\tilde{s}$ has $2r$ continuous derivatives, $\tilde{s}|_{(-\infty,a)} \in P_r$ and $\tilde{s}|_{(b,\infty)} \in P_r$.

We now formulate the interpolation problem for natural splines.

Definition 3.29. Let a function $f \in C[a, b]$ be given. The *natural spline interpolation problem* is to determine a natural spline $s_f \in S_{2r+1}(x_1, \ldots, x_k)$, where $r \geq 1$, such that

$$s_f(x_i) = f(x_i), \qquad i = 0, \ldots, k+1, \tag{3.46}$$

(It is easy to see that the number of conditions is equal to the dimension of $S_{2r+1}(x_1, \ldots, x_k)$).

The next result shows that the natural spline interpolation problem has a unique solution.

Theorem 3.30. *If $k \geq r-1$, then the natural spline interpolation problem (3.46) has a unique solution from $S_{2r+1}(x_1, \ldots, x_k)$.*

Proof. Let $k \geq r - 1$ be given. It suffices to show that the homogeneous problem

$$s(x_i) = 0, \qquad i = 0, \ldots, k+1, \tag{3.47}$$

has only the trivial solution $s = 0$ in the space of natural splines. The proof is completely analogously to the proof of Theorem 3.18. We note that by (3.45)

$$\sum_{j=0}^{r-1}(-1)^j s^{(r+1+j)}(t) s^{(r-j)}(t) \Big|_{t=a}^{b} = 0.$$

Using this fact, it follows that

$$\int_a^b (s^{(r+1)}(t))^2 \, dt = 0$$

which implies that $s \in P_r$. Since $k \geq r - 1$, it follows from (3.47) that $s = 0$. This proves Theorem 3.30.

The natural spline which solves the interpolation problem (3.46) has the following optimality property compared with arbitrary functions from $C^{r+1}[a, b]$ which interpolate at the knots.

Theorem 3.31. *Let $k \geq r - 1$ and $s_f \in S_{2r+1}(x_1, \ldots, x_k)$ be the unique solution of the natural spline interpolation problem (3.46). Then for every function $g \in C^{r+1}[a, b]$, $g \neq s_f$, with*

$$g(x_i) = f(x_i), \quad i = 0, \ldots, k+1 \tag{3.48}$$

we have

$$\|s_f^{(r+1)}\|_2 < \|g^{(r+1)}\|_2. \tag{3.49}$$

Proof. The proof is completely analogous to the proof of Theorem 3.19. We only note that by (3.45)

$$\sum_{j=0}^{r-1}(-1)^j\, s_f^{(r+1+j)}(t)\,(g^{(r-j)}(t) - s_f^{(r-j)}(t))\,\Big|_{t=a}^{b} = 0$$

and by (3.46) and (3.48)

$$\int_a^b s_f^{(2r+1)}(t)\,(g'(t) - s_f'(t))\,dt = 0.$$

This proves Theorem 3.31.

Interpolating natural splines possess a further optimality property.

Theorem 3.32. *Let* $k \geq r - 1$, $f \in C^{r+1}[a, b]$ *and* $s_f \in S_{2r+1}(x_1, \ldots, x_k)$ *be the unique solution of the corresponding natural spline interpolation problem (3.46). Then for every natural spline* $s \in S_{2r+1}(x_1, \ldots, x_k)$, *we have*

$$\|f^{(r+1)} - s_f^{(r+1)}\|_2 \leq \|f^{(r+1)} - s^{(r+1)}\|_2. \tag{3.50}$$

The equality sign in (3.50) holds if and only if $s - s_f \in P_r$.

Proof. The proof is completely analogous to the proof of Theorem 3.20. We only note that since s and s_f are natural splines,

$$\sum_{j=0}^{r-1}(-1)^j\,(s^{(r+1+j)}(t) - s_f^{(r+1+j)}(t))\,(f^{(r-j)}(t) - s_f^{(r-j)}(t))\,\Big|_{t=a}^{b} = 0$$

and by (3.46)

$$\int_a^b (s^{(2r+1)}(t) - s_f^{(2r+1)}(t))\,(f'(t) - s_f'(t))\,dt = 0.$$

This proves Theorem 3.32.

The above results show that interpolation at the knots is always possible for splines which satisfy certain boundary conditions. In the following we describe the computation of such splines in the cubic case.

Let a function $f \in C[a, b]$ be given and let $s_f \in S_3(x_1, \ldots, x_k)$ be a spline which satisfies

$$s_f(x_i) = f(x_i), \qquad i = 0, \ldots, k + 1, \tag{3.51}$$

and additional boundary conditions as in Definition 3.17 (complete splines), Definition 3.23 (periodic splines) or Definition 3.29 (natural splines). It will be shown that the spline s_f can be computed in two steps. We first compute the derivatives $s_f'(x_i)$, $i = 0, \ldots, k + 1$, by solving a system of linear equations. Then the spline s_f is uniquely determined on each knot–interval $[x_i, x_{i+1}]$ by the values $f(x_i), f(x_{i+1})$, $s_f'(x_i)$ and $s_f'(x_{i+1})$ $(i = 0, \ldots, k + 1)$ and the polynomial pieces can be computed in the Newton–form (see Theorem 2.6 in Chapter I.)

Therefore, it remains to show how to compute the derivatives $s_f'(x_i)$, $i = 0, \ldots, k + 1$. To do this we first give a result on these derivatives for arbitrary cubic splines.

Lemma 3.33. *Let a spline* $s \in S_3(x_1, \ldots, x_k)$ *be given. Then for all indices* $i \in \{1, \ldots, k\}$,

$$(x_{i+1} - x_i)\, s'(x_{i-1}) + 2\,(x_{i+1} - x_{i-1})\, s'(x_i) + (x_i - x_{i-1})\, s'(x_{i+1})$$
$$= 3\,(x_{i+1} - x_i)\, s[x_{i-1}, x_i] + 3\,(x_i - x_{i-1})\, s[x_i, x_{i+1}]. \tag{3.52}$$

Proof. Let an arbitrary polynomial $p \in P_3$ be given. We will show that for all $u, v \in \mathbf{R}$ with $u \neq v$,

$$p''(u) = \frac{2}{u - v}\,(2p'(u) - 3p[u, v] + p'(v)). \tag{3.53}$$

By setting
$$q(t) = p((u - v)\, t + v)$$

for all $t \in \mathbf{R}$, we see that equation (3.53) is equivalent to

$$q''(1) = 4q'(1) - 6q(1) + 6q(0) + 2q'(0).$$

It is easy to verify that this relation holds for the basis functions 1, t, t^2, t^3 of P_3 and therefore, it holds for all $q \in P_3$. This shows (3.53).
Now, let an index $i \in \{1, \ldots, k\}$ be given. By setting $u = x_i$ and $v = x_{i-1}$, it follows from (3.53) that

$$s''(x_i) = \frac{2}{x_i - x_{i-1}}\,(2s'(x_i) - 3s[x_{i-1}, x_i] + s'(x_{i-1})). \tag{3.54}$$

Moreover, by setting $u = x_i$ and $v = x_{i+1}$, we get

$$s''(x_i) = \frac{2}{x_i - x_{i+1}}\,(2s'(x_i) - 3s[x_i, x_{i+1}] + s'(x_{i+1})). \tag{3.55}$$

By comparing (3.54) and (3.55), we obtain (3.52). This proves Lemma 3.33.

By using Lemma 3.33, we will show how to compute the derivatives $s'_f(x_i)$, $i = 0, \ldots, k + 1$, for splines $s \in S_3(x_1, \ldots, x_k)$ satisfying (3.51). We distinguish three cases of boundary conditions.
If s_f is a complete spline, then the values $s'_f(x_0) = f'(x_0)$ and $s'_f(x_{k+1}) = f'(x_{k+1})$ are known. Therefore, by inserting these values in (3.52) for $s = s_f$, we obtain a system of k linear equations for the k unknowns $s'_f(x_i)$, $i = 1, \ldots, k$.
If s_f is a periodic spline, then s_f can be extended to $(-\infty, \infty)$ with period $b - a$ (see Remark 3.22). Therefore, by setting $x_{k+2} = x_1 + (b - a)$, equation (3.52) also holds for $i = k + 1$. Since $s'_f(x_{k+1}) = s'_f(x_0)$ and $s'_f(x_{k+2}) = s'_f(x_1)$, equation (3.52) considered for $i = 1, \ldots, k + 1$ yields a system of $k + 1$ linear equations for the $k + 1$ unknowns $s'_f(x_i)$, $i = 0, \ldots, k$.
If s_f is a natural spline, then we have $s''(x_0) = 0$ and $s''(x_{k+1}) = 0$. Therefore, it follows from (3.55) that

$$2s'_f(x_0) + s'_f(x_1) = 3s_f[x_0, x_1] \tag{3.56}$$

and from (3.54) that

$$s'_f(x_k) + 2s'_f(x_{k+1}) = 3s_f[x_k, x_{k+1}]. \tag{3.57}$$

Finally, the equations (3.52) for $s = s_f$, (3.56) and (3.57) yield a system of $k + 2$ linear equations for the $k + 2$ unknowns $s'_f(x_i)$, $i = 0, \ldots, k + 1$.

It is easy to verify that in all three cases the systems of linear equations have a unique solution.

At the end of this section we give results on the norm of interpolation operators for cubic periodic splines.

Definition 3.34. The space of periodic continuous functions on $[a, b]$ is denoted by

$$P[a, b] = \{f \in C[a, b] : f(a) = f(b)\} \tag{3.58}$$

and the space of periodic splines of degree $2r + 1$ with knots $x_1, \ldots, x_k$ is denoted by

$$P_{2r+1}(x_1, \ldots, x_k) =$$
$$\{s \in S_{2r+1}(x_1, \ldots, x_k) : s^{(j)}(a) = s^{(j)}(b), \quad j = 0, \ldots, 2r\}. \tag{3.59}$$

Let $L_{P_{2r+1}(x_1, \ldots, x_k)} : P[a, b] \to P_{2r+1}(x_1, \ldots, x_k)$ be the mapping which associates to each function $f \in P[a, b]$ the unique solution

$$L_{P_{2r+1}(x_1, \ldots, x_k)}(f) = s_f$$

from $P_{2r+1}(x_1, \ldots, x_k)$ of the interpolation problem (3.38). The norm of the operator is defined by

$$\|L_{P_{2r+1}(x_1, \ldots, x_k)}\|_\infty =$$
$$\sup\{\|L_{P_{2r+1}(x_1, \ldots, x_k)}(f)\|_\infty : f \in P[a, b], \|f\|_\infty \le 1\}. \tag{3.60}$$

The following result, due to Schurer & Cheney [1968], which we state without proof shows that interpolation by cubic periodic splines at equidistant knots provides a nearly optimal error.

Theorem 3.35. *Let equidistant knots* $a = x_0 < x_1 < \cdots < x_k < x_{k+1} = b$ *be given, i.e.*

$$x_i = x_0 + i \frac{b - a}{k + 1}, \quad i = 0, \ldots, k + 1.$$

Then we have

$$\|L_{P_3(x_1, \ldots, x_k)}\|_\infty < \frac{1}{4}(1 + 3\sqrt{3}). \tag{3.61}$$

In particular, for every $f \in P[a, b]$,

$$\|f - L_{P_3(x_1, \ldots, x_k)}(f)\|_\infty < \frac{1}{4}(5 + 3\sqrt{3}) \inf_{s \in P_3(x_1, \ldots, x_k)} \|f - s\|_\infty. \tag{3.62}$$

Schurer & Cheney [1968] proved that the number $\frac{1}{4}(1 + 3\sqrt{3})$ in Theorem 3.35 is the smallest possible bound for $\|L_{P_3(x_1,\ldots,x_k)}\|_\infty$.

It is natural to ask for which distributions of the knots $x_1, \ldots, x_k$ the norm $\|L_{P_3(x_1,\ldots,x_k)}\|_\infty$ is minimal when k is fixed. It is conjectured that this is the case if and only if the knots are equidistant. But this problem is unsolved at present. Meinardus & Taylor [1978] proved the analogous conjecture for quadratic periodic splines.

Many authors investigated the question for which distributions of knots the norm $\|L_{P_3(x_1,\ldots,x_k)}\|_\infty$ is bounded. (A survey of norms of spline interpolation operators can be found in Merz [1979]). The boundedness of the norm depends on bounds of the *local mesh ratio*, defined by

$$\alpha(x_1, \ldots, x_k) = \max\left\{\frac{x_i - x_{i-1}}{x_j - x_{j-1}} : |i - j| = 1, \, i, j = 1, \ldots, k+1\right\}.$$

It was proved by de Boor [1976b] and Meinardus [1976] that there exists a constant $K > 0$ such that for every positive integer k and every set of knots $\{x_1, \ldots, x_k\}$ with $\alpha(x_1, \ldots, x_k) < \frac{1}{2}(3 + \sqrt{5})$,

$$\|L_{P_3(x_1,\ldots,x_k)}\|_\infty \leq K.$$

This theorem together with a result of Marsden [1974b] shows that the number $\frac{1}{2}(3 + \sqrt{5})$ is the largest bound of the local mesh ratio such that the above statement holds.

Finally, we give a brief description of a further minimization problem (compare Theorem 3.31).

Let a function $f \in C[a, b]$, points $a = x_0 < x_1 < \cdots < x_k < x_{k+1} = b$ and an integer r with $1 \leq r \leq k$ be given. Determine a function $g : [a, b] \to \mathbf{R}$ with

$$g(x_i) = f(x_i), \qquad i = 0, \ldots, k+1, \tag{3.63}$$

which minimizes the norm $\|g^{(r+1)}\|_\infty$.

This minimization problem leads to the following type of splines.

A spline $s \in S_m(x_1, \ldots, x_k)$ is called a *perfect spline* if

$$|s^{(m)}(t)| = \|s^{(m)}\|_\infty, \qquad t \in [a, b].$$

We say that x_i is an *active knot* if $s^{(m)}$ changes sign at x_i, $i = 1, \ldots, k$. (Here we consider $s^{(m)}$ only on $[a, b] \setminus \{x_1, \ldots, x_k\}$.)

It was independently proved by de Boor [1974a], [1976c] and Karlin [1975] that a function $g : [a, b] \to \mathbf{R}$ solves the minimization problem (3.63) if and only if there exists an interval $[x_p, x_{p+r+j}]$, $j \geq 1$, such that g is a perfect spline of degree $r + 1$ on $[x_p, x_{p+r+j}]$ which satisfies the following conditions:

(i) $|g^{(r+1)}(t)| = \|g^{(r+1)}\|_\infty$ for all $t \in (x_p, x_{p+r+j})$, where $\|g^{(r+1)}\|_\infty = \sup_{t \in [a,b]} |g^{(r+1)}(t)|$.

(ii) g has at most $j - 1$ active knots in (x_p, x_{p+r+j}).

3.3. Quasi–Interpolation

In this section we describe methods for constructing spline approximations which — in contrast to interpolation — do not require the numerical solution of a system of linear equations. In particular, we discuss the method of quasi–interpolation for approximating functions which are m-times continuously differentiable. The corresponding linear operator is local in a certain sense, reproduces splines of degree m and the resulting error is of order $m + 1$.

Let points $a = x_0 < x_1 < \cdots < x_k < x_{k+1} = b$, $x_{-m} < \cdots < x_{-1} < a$ and $b < x_{k+2} < \cdots < x_{k+m+1}$ be given. For a given function f we will consider spline approximations of the form

$$A(f) = \sum_{i=-m}^{k} \mu_i(f)\, N_i^m, \tag{3.64}$$

where $\mu_{-m}, \ldots, \mu_k$ are certain linear functionals and $N_{-m}^m, \ldots, N_k^m$ are normalized B–splines of degree m (see Definition 2.13). Obviously, the approximation $A(f)$ is a spline from $S_m(x_1, \ldots, x_k)$.

The simplest type of approximations (3.64) is given if $\mu_{-m}, \ldots, \mu_k$ are point functionals, i.e. for all $f \in C[a, b]$,

$$A(f) = \sum_{i=-m}^{k} f(t_i)\, N_i^m, \tag{3.65}$$

where

$$t_i \in (x_i, x_{i+m+1}) \cap [a, b], \qquad i = -m, \ldots, k.$$

A typical choice of the points $t_{-m}, \ldots, t_k$ would be

$$t_i = \begin{cases} \frac{1}{2}(x_i + x_{i+m+1}), & \text{if } \frac{1}{2}(x_i + x_{i+m+1}) \in [a, b] \\ a, & \text{if } \frac{1}{2}(x_i + x_{i+m+1}) < a \\ b, & \text{if } \frac{1}{2}(x_i + x_{i+m+1}) > b \end{cases} \tag{3.66}$$

for all $i \in \{-m, \ldots, k\}$. The proof of Theorem 4.27 below shows that in this case the following error estimate holds: for all $f \in C[a, b]$,

$$\|f - A(f)\|_\infty \le \frac{1}{2}(m + 1)\omega(f; \delta), \tag{3.67}$$

where $\delta = \max\{|x_{i+1} - x_i| : i = 0, \ldots, k\}$. It follows from (3.67) that for all $f \in C^1[a, b]$,

$$\|f - A(f)\|_\infty \le \frac{1}{2}(m + 1)\delta \|f'\|_\infty. \tag{3.68}$$

This shows that the error behaves up to some constant like the maximal length of the knot–intervals. If we choose in (3.65) for all $i \in \{-m, \ldots, k\}$,

$$t_i = \frac{1}{m}(y_{i+1} + \cdots + y_{i+m}), \tag{3.69}$$

where $y_i = x_i$, $i = 0, \ldots, k+1$, $y_{-m+1} = \cdots = y_{-1} = a$ and $y_{k+2} = \cdots y_{k+m} = b$, then we obtain an operator introduced by Schoenberg [1967]. In this case the error can be estimated as follows: there exists a constant $K > 0$ such that for all $f \in C^2[a, b]$,

$$\|f - A(f)\|_\infty \leq K \, \delta^2 \, \|f''\|_\infty. \tag{3.70}$$

However, even for functions $f \in C^r[a, b]$ with $r > 2$, the exponent 2 in (3.70) cannot be increased. Results of this type can be found in de Boor [1978].

On the other hand, the subsequent Theorem 4.27 shows that the minimal deviation satisfies the following estimation: there exists a constant $K > 0$ such that for all $f \in C^{m+1}[a, b]$,

$$d_\infty(f, S_m(x_1, \ldots, x_k)) \leq K \, \delta^{m+1} \, \|f^{(m+1)}\|_\infty. \tag{3.71}$$

In view of this result, the problem arises to construct spline approximations which yield an error as in (3.71). Obviously, best spline approximations have this property. Moreover, it follows from Theorem 2.17 in Chapter I that interpolation operators (compare Section 3.2) also have this property, since these operators have bounded norm.

In order to obtain an error as in (3.71) for approximations of the form (3.64), the corresponding operator has to be more complicated than (3.65). We will describe the quasi–interpolant of de Boor & Fix [1973] which has this property. Similar approximation methods were investigated by Lyche & Schumaker [1975].

Definition 3.36. Let points $t_{-m} \leq \cdots \leq t_k$ be given such that

$$t_i \in (x_i, x_{i+m+1}) \cap [a, b], \qquad i = -m, \ldots, k. \tag{3.72}$$

We define for all $i \in \{-m, \ldots, k\}$ and all $f \in C^m[a, b]$,

$$\mu_i(f) = \sum_{j=0}^{m} \frac{1}{m!} (-1)^j \, w_i^{(m-j)}(t_i) \, f^{(j)}(t_i) \tag{3.73}$$

and

$$Q(f) = \sum_{i=-m}^{k} \mu_i(f) \, N_i^m, \tag{3.74}$$

where

$$w_i(t) = \frac{1}{m!} (t - x_{i+1}) \cdots (t - x_{i+m}).$$

The linear operator $Q : C^m[a, b] \to S_m(x_1, \ldots, x_k)$ is called a *quasi–interpolant*.

A typical choice of the points $t_{-m}, \ldots, t_k$ in (3.72) would be

$$t_i = \begin{cases} x_{i + \frac{m+1}{2}}, & \text{if } x_{i + \frac{m+1}{2}} \in [a, b] \\ a, & \text{if } x_{i + \frac{m+1}{2}} < a \\ b, & \text{if } x_{i + \frac{m+1}{2}} > b \end{cases} \tag{3.75}$$

for all $i \in \{-m, \ldots, k\}$, where $x_{j+\frac{1}{2}} = \frac{1}{2}(x_j + x_{j+1})$.

We first note that the operator Q is *local* in the following sense. Let a function $f \in C^m[a, b]$ be given. Since for all $i \in \{-m, \ldots, k\}$,

$$t_i \in (x_i, x_{i+m+1}) \cap [a, b]$$

and

$$N_i^m(t) = 0, \qquad t \in (-\infty, x_i] \cup [x_{i+m+1}, \infty),$$

it follows that for $j \in \{-m, \ldots, k\}$ the values of $Q(f)$ in (x_j, x_{j+1}) depend only on the values of f in $(x_{j-m}, x_{j+m+1}) \cap [a, b]$.

The following result, due to de Boor & Fix [1973], shows that the quasi-interpolant Q reproduces the functions from $S_m(x_1, \ldots, x_k)$. Here we use the convention that in the definition of $Q(s)$ for $s \in S_m(x_1, \ldots, x_k)$ we denote by $s^{(m)}(x_i)$, $i = 1, \ldots, k$, the left derivative of order m.

Theorem 3.37. *For every spline $s \in S_m(x_1, \ldots, x_k)$, we have*

$$Q(s) = s.$$

Proof. We will show that for all $i, j \in \{-m, \ldots, k\}$,

$$\mu_i(N_j^m) = \begin{cases} 1, & \text{if } i = j \\ 0, & \text{if } i \neq j \end{cases}. \tag{3.76}$$

Then it follows that for all

$$s = \sum_{j=-m}^{k} \alpha_j N_j^m \in S_m(x_1, \ldots, x_k),$$

$$
\begin{aligned}
Q(s) &= \sum_{i=-m}^{k} \mu_i \left(\sum_{j=-m}^{k} \alpha_j N_j^m \right) N_i^m \\
&= \sum_{i=-m}^{k} \left(\sum_{j=-m}^{k} \alpha_j \mu_i(N_j^m) \right) N_i^m = \sum_{i=-m}^{k} \alpha_i N_i^m = s.
\end{aligned}
$$

Therefore, it suffices to prove (3.76). It follows from Theorem 2.9 that for all $t \in (-\infty, \infty)$,

$$N_j^m(t) = (-1)^{m+1}(x_{j+m+1} - x_j)(t - x)_+^m[x_j, \ldots, x_{j+m+1}]. \tag{3.77}$$

In order to compute $\mu_i(N_j^m)$, we first have to compute

$$\mu_i \left((\cdot - x)_+^m \right), \qquad i = -m, \ldots, k.$$

Let an index $i \in \{-m, \ldots, k\}$ be given. If $t_i \leq x$, then there exists a neighborhood U of t_i such that for all $t \in U \cap (-\infty, x]$,

$$(t - x)_+^m = 0.$$

This implies that

$$\mu_i\left((\cdot - x)_+^m\right) = 0.$$

If $t_i > x$, then there exists a neighborhood U of t_i such that for all $t \in U$,

$$(t - x)_+^m = (t - x)^m.$$

Therefore, it follows that

$$\mu_i\left((\cdot - x)_+^m\right) = \sum_{j=0}^{m} \frac{1}{m!} (-1)^j w_i^{(m-j)}(t_i) \left((\cdot - x)^m\right)^{(j)}(t_i)$$

$$= \sum_{j=0}^{m} \frac{1}{m!} (-1)^j w_i^{(m-j)}(t_i)\, m \cdots (m - j + 1)(t_i - x)^{m-j}$$

$$= (-1)^m \sum_{j=0}^{m} \frac{1}{(m - j)!} (x - t_i)^{m-j} w_i^{(m-j)}(t_i) = (-1)^m w_i(x).$$

The last equation is a consequence of Taylor's theorem. This shows that

$$\mu_i\left((\cdot - x)_+^m\right) = (-1)^m w_i(x)(t_i - x)_+^0. \tag{3.78}$$

It follows from (3.77) and (3.78) that

$$\mu_i(N_j^m) = (x_j - x_{j+m+1})\, z_i\,[x_j, \ldots, x_{j+m+1}],$$

where

$$z_i(x) = w_i(x)(t_i - x)_+^0, \qquad x \in (-\infty, \infty).$$

In the subsequent arguments we use the fact that

$$t_r \in (x_r, x_{r+m+1}), \qquad r = -m, \ldots, k.$$

If $j > i$, then

$$z_i(x_r) = 0, \qquad r = j, \ldots, j + m + 1,$$

which implies that

$$\mu_i(N_j^m) = 0.$$

If $j < i$, then

$$w_i(x_r) = z_i(x_r), \qquad r = j, \ldots, j + m + 1.$$

Since $w_i \in P_m$, it follows from the definition of the divided differences (Definition 2.1 in Chapter I) that

$$z_i[x_j, \ldots, x_{j+m+1}] = 0,$$

which implies that

$$\mu_i(N_j^m) = 0.$$

Finally, we consider the case when $j = i$. It is easy to see that the polynomial

$$p(x) = \frac{x - x_{j+m+1}}{x_j - x_{j+m+1}} w_i(x)$$

satisfies

$$p(x_r) = z_i(x_r), \qquad r = j, \ldots, j + m + 1.$$

Since $p \in P_{m+1}$ can be written as

$$p(x) = \sum_{r=0}^{m+1} a_r x^r,$$

where $a_{m+1} = 1/(x_j - x_{j+m+1})$, it follows from the definition of the divided differences that

$$z_i[x_j, \ldots, x_{j+m+1}] = 1/(x_j - x_{j+m+1}),$$

which implies that

$$\mu_i(N_j^m) = 1.$$

This proves Theorem 3.37.

Finally, we state without proof a result of de Boor & Fix [1973] which shows that for the quasi-interpolant Q an error estimate as in (3.71) holds.

Theorem 3.38. *There exists a constant $K > 0$ (depending only on m) such that for every function $f \in C^m[a, b]$ and for every set of knots $\{x_1, \ldots, x_k\}$, we have*

$$\|f - Q(f)\|_\infty \leq K \, \delta^m \, \omega(f^{(m)}; \delta), \tag{3.79}$$

where $\delta = \max\{|x_{i+1} - x_i| : i = 0, \ldots, k\}$. In particular, if $f \in C^{m+1}[a, b]$, then

$$\omega(f^{(m)}; \delta) \leq \|f^{(m+1)}\|_\infty \, \delta. \tag{3.80}$$

4. Best Uniform Approximation by Splines

Spline spaces are prototypes of weak Chebyshev spaces, but they do not belong to the class of Chebyshev spaces. Therefore, the theory of best approximation by Chebyshev spaces is not applicable.

We give results on characterization, unicity and strong unicity of best uniform approximations from spline spaces and discuss their approximation power. Moreover, we describe a method for computing best spline approximations with fixed knots. We also give an algorithm for best piecewise polynomial approximations with free knots. A combination of these methods yields good spline approximations with free knots.

4.1. Characterization, Unicity and Strong Unicity of Best Uniform Approximations

It will be shown that best uniform approximations and strongly unique best uniform approximations from spline spaces can be characterized by alternation properties of the error, however this is not the case for unique best uniform approximations. Moreover, we give stability results for unique best approximations and a formula for computing strong unicity constants.

We first recall the definition of alternating extreme points which plays a fundamental role in the following investigations.

Definition 4.1. Let a function $h \in [a, b]$ and a subset T of $[a, b]$ be given. Points $t_1 < \cdots < t_p$ in T are called *alternating extreme points*, briefly denoted by *A–points*, of h in T if there exists a sign $\sigma \in \{-1, 1\}$ such that

$$\sigma(-1)^i h(t_i) = \|h\|_\infty, \qquad i = 1, \ldots, p. \tag{4.1}$$

The maximal number of A–points of h in T is denoted by $A(h)|_T$.

We begin with a characterization of best spline approximations which was independently proved by Rice [1967] and Schumaker [1968a]. Extensions of this result were given by Sommer [1980a] and Nürnberger, Schumaker, Sommer & Strauß [1985].

Theorem 4.2. *Let a function $f \in C[a, b]$ and a spline $s_f \in S_m(x_1, \ldots, x_k)$ be given. The following statements are equivalent:*
(i)   The spline s_f is a best uniform approximation of f from $S_m(x_1, \ldots, x_k)$.
(ii) There exists an interval $[x_p, x_{p+q}] \subset [a, b]$, $q \geq 1$, such that

$$A(f - s_f)\big|_{[x_p, x_{p+q}]} \geq q + m + 1.$$

Proof. (ii) $\Rightarrow$ (i). Suppose that (ii) holds, i.e. there exist points $t_1 < \cdots < t_{q+m+1}$ in $[x_p, x_{p+q}]$ and a sign $\sigma \in \{-1, 1\}$ such that

$$\sigma(-1)^i \left(f(t_i) - s_f(t_i) \right) = \|f - s_f\|_\infty, \qquad i = 1, \ldots, q + m + 1. \tag{4.2}$$

If (i) fails, then there exists a spline $s \in S_m(x_1, \ldots, x_k)$ such that

$$\|f - s\|_\infty < \|f - s_f\|_\infty. \tag{4.3}$$

If follows from (4.2) and (4.3) that

$$\begin{aligned}
\sigma(-1)^i \left(f(t_i) - s(t_i) \right) &\leq \|f - s\|_\infty < \|f - s_f\|_\infty \\
&= \sigma(-1)^i \left(f(t_i) - s_f(t_i) \right), \qquad i = 1, \ldots, q + m + 1.
\end{aligned}$$

This implies that

$$\sigma(-1)^i \left(s_f(t_i) - s(t_i) \right) < 0, \qquad i = 1, \ldots, q + m + 1.$$

Therefore, the spline $s - s_f$ has at least $q + m$ sign changes on $[x_p, x_{p+q}]$. This is a contradiction, since by Theorem 1.19 the space $S_m(x_1, \ldots, x_k)\big|_{[x_p, x_{p+q}]}$ is a $(q + m)$–dimensional weak Chebyshev space.

(i) $\Rightarrow$ (ii). Let $f \in C[a, b]$ be given. Since by Theorem 1.19 the space $S_m(x_1, \ldots, x_k)$ is weak Chebyshev, it follows from Theorem 1.8 that there exists a best uniform approximation $s_0 \in S_m(x_1, \ldots, x_k)$ of f such that $f - s_0$ has at least $k + m + 2$ A–points in $[x_0, x_{k+1}]$. Going to subintervals, if necessary, it follows that there exists an interval $[x_p, x_{p+q}]$ such that $f - s_0$ has $q + m + 1$ A–points $t_1 < \cdots < t_{q+m+1}$ in $[x_p, x_{p+q}]$, but

$$\begin{aligned} &\text{every proper subinterval } [x_i, x_{i+j}] \text{ of } [x_p, x_{p+q}] \\ &\text{contains at most } j + m \text{ points from } T = \{t_1, \ldots, t_{q+m+1}\}. \end{aligned} \tag{4.4}$$

We will show that

$$\begin{aligned} &\text{for all } i \in \{1, \ldots, q + m + 1\}, \text{ the set } \{t_1, \ldots, t_{i-1}, t_{i+1}, \ldots, t_{q+m+1}\} \\ &\text{is poised with respect to } S_m(x_1, \ldots, x_k)\big|_{[x_p, x_{p+q}]}. \end{aligned} \tag{4.5}$$

Suppose that (4.5) fails, i.e. there exist points $u_1 < \cdots < u_{q+m}$ in T such that $\{u_1, \ldots, u_{q+m}\}$ is not poised with respect to $S_m(x_1, \ldots, x_k)\big|_{[x_p, x_{p+q}]}$. Then it follows from Theorem 3.7 that the condition

$$u_j < x_{p+j} < u_{j+m+1}, \qquad i = 1, \ldots, q - 1,$$

is not satisfied. If $u_{j+m+1} \leq x_{p+j}$, then $[x_p, x_{p+j}]$ contains at least $j + m + 1$ points from T which contradicts (4.4). If $x_{p+j} \leq u_j$, then $[x_{p+j}, x_{p+q}]$ contains at least $q - j + m + 1$ points from T which again contradicts (4.4). This proves (4.5).

Since $S_m(x_1, \ldots, x_k)\big|_{[x_p, x_{p+q}]}$ is a $(m + q)$–dimensional weak Chebyshev space, it follows from (4.5) and Theorem 1.12 that s_0 is a strongly unique best uniform approximation of f on $[x_p, x_{p+q}]$. Now, let s_f be any best approximation of f from $S_m(x_1, \ldots, x_k)$. Then we have

$$\big\|(f - s_0)\big|_{[x_p, x_{p+q}]}\big\|_\infty = \|f - s_0\|_\infty = \|f - s_f\|_\infty \geq \big\|(f - s_f)\big|_{[x_p, x_{p+q}]}\big\|_\infty .$$

This shows that s_f is also a best uniform approximation of f from $S_m(x_1, \ldots, x_k)$ on $[x_p, x_{p+q}]$. This implies that $s_f = s_0$ on $[x_p, x_{p+q}]$. Since $A(f - s_f)\big|_{[x_p, x_{p+q}]} \geq q + m + 1$, condition (ii) follows. This proves Theorem 4.2.

We briefly discuss the alternation condition in Theorem 4.2. It follows from Theorem 1.8 that for every function $f \in C[a, b]$, there exists a best uniform approximation $s_f \in S_m(x_1, \ldots, x_k)$ such that the error $f - s_f$ has $n+1$ alternating extreme points, where n is the dimension of $S_m(x_1, \ldots, x_k)$. Theorem 4.2 shows that not all best spline approximations satisfy this classical alternation property (compare Theorem 3.12 in Chapter I on Chebyshev spaces). The number $q+m+1$ in condition (ii) of Theorem 4.2 is only equal to $\tilde{n} + 1$, where $\tilde{n}$ is the dimension of $S_m(x_1, \ldots, x_k)$ restricted to $[x_p, x_{p+q}]$.

A close inspection of the proof of Theorem 4.2, (i) $\Rightarrow$ (ii) shows that the following result holds.

Theorem 4.3. *Let a function $f \in C[a,b]$ be given and $s_f \in S_m(x_1, \ldots, x_k)$ be a best uniform approximation of f. Then there exists an interval $[x_p, x_{p+q}] \subset [a,b]$, $q \geq 1$, such that s_f is a strongly unique best uniform approximation of f from $S_m(x_1, \ldots, x_k)$ on $[x_p, x_{p+q}]$. In particular, all best uniform approximations of f from $S_m(x_1, \ldots, x_k)$ coincide on $[x_p, x_{p+q}]$.*

In addition to Theorem 4.3, it was proved by Berens & Nürnberger [1987] that for every function $f \in C[a,b]$, the set of points on which all best uniform approximations of f from $S_m(x_1, \ldots, x_k)$ coincide is the union of knot–intervals and that this set is a single knot–interval, if $k \leq m+1$.

Since spline spaces do not belong to the class of Chebyshev spaces, best uniform approximations from spline spaces are not always (strongly) unique (compare Theorem 3.18 in Chapter I). Therefore, in the following we investigate the unicity and strong unicity of best spline approximations. Schaback [1978] gave a sufficient condition for strongly unique best spline approximations. The following characterization, proved by Nürnberger [1982a], was deduced from a general theorem for weak Chebyshev spaces. Extensions of this result were given by Nürnberger, Schumaker, Sommer & Strauß [1985].

Theorem 4.4. *Let a function $f \in C[a,b]$ and a spline $s_f \in S_m(x_1, \ldots, x_k)$ be given. The following statements are equivalent:*
(i)   *The spline s_f is a strongly unique best uniform approximation of f from $S_m(x_1, \ldots, x_k)$.*
(ii) *For every interval $(x_i, x_{i+m+j}) \subset (x_{-m}, x_{k+m+1})$, $j \geq 1$, we have*
$$A(f - s_f)\big|_{(x_i, x_{i+m+j})} \geq j + 1.$$

Proof. (i) $\Rightarrow$ (ii). Suppose that (ii) fails, i.e. there exists an interval (x_i, x_{i+m+j}) with $A(f - s_f)\big|_{(x_i, x_{i+m+j})} \leq j$. We choose a maximal number of A–points $t_1 < \ldots < t_p$ of $f - s_f$ in (x_i, x_{i+m+j}). Then we have $p \leq j$. Moreover, we set for all $i \in \{2, \ldots, p\}$,

$$z_{i-1} = \min\{t \in (t_{i-1}, t_i] : f(t) - s_f(t) = f(t_i) - s_f(t_i)\}$$

and $z_0 = x_i$, $z_p = x_{i+m+j}$. Since by Theorem 2.8

$$S = \text{span}\{B_i^m, \ldots, B_{i+j-1}^m\}$$

is a j–dimensional weak Chebyshev space and $p \leq j$, by Corollary 1.7 there exists a nontrivial $s \in S$ such that

$$(-1)^i s(t) \geq 0, \qquad t \in [z_{i-1}, z_i], \quad i = 1, \ldots, p.$$

By the choice of the points $z_0, \ldots, z_p$ and by replacing s by $-s$, if necessary, we have

$$(f(t) - s_f(t)) s(t) \geq 0, \qquad t \in E(f - s_f). \tag{4.6}$$

Note, that by Theorem 2.2 the B–splines $B_i^m, \ldots, B_{i+j-1}^m$ vanish on $\mathbf{R}\backslash(x_i, x_{i+m+j})$. Now, it follows from (4.6) and Theorem 3.17 in Chapter I that s_f is not a strongly unique best uniform approximation of f.

(ii) $\Rightarrow$ (i). Suppose that (ii) holds, but (i) fails. Since (i) fails, it follows from Theorem 3.17 in Chapter I that there exists a nontrivial $s \in S_m(x_1, \ldots, x_k)$ such that

$$(f(t) - s_f(t))\, s(t) \geq 0, \qquad t \in E(f - s_f). \tag{4.7}$$

We distinguish the following cases.

Case 1. s has only finitely many zeros in $[a, b]$.

It follows from (ii) that $A(f - s_f)\big|_{[a,b]} \geq k + m + 2$. By the proof of Theorem 4.2 there exists an interval $[x_p, x_{p+q}]$ containing $m+q+1$ A-points $t_1 < \ldots < t_{m+q+1}$ of $f - s_f$ in $[x_p, x_{p+q}]$ such that for all $i \in \{1, \ldots, q+m+1\}$, the set

$$\{t_1, \ldots, t_{i-1}, t_{i+1}, \ldots, t_{q+m+1}\}$$

is poised with respect to $S_m(x_1, \ldots, x_k)\big|_{[x_p, x_{p+q}]}$. \tag{4.8}

Then there exists a sign $\sigma \in \{-1, 1\}$ such that

$$\sigma(-1)^i\left(f(t_i) - s_f(t_i)\right) = \|f - s_f\|_\infty, \quad i = 1, \ldots, m+q+1.$$

It follows from (4.7) that

$$\sigma(-1)^i \|f - s_f\|_\infty s(t_i) = (f(t_i) - s_f(t_i))s(t_i) \geq 0,$$

i.e.

$$\sigma(-1)^i s(t_i) \geq 0, \quad i = 1, \ldots, m+q+1. \tag{4.9}$$

Finally, by (4.8),(4.9) and Lemma 1.11 we have s=0 on $[x_p, x_{p+q}]$ which is a contradiction.

Case 2. There exists an interval $[x_v, x_w]$ such that $s = 0$ on $[x_v, x_w]$ and s has only finitely many zeros on $[a, b]\backslash[x_v, x_w]$.

We may assume that $v > 0$. (The other case follows analogously.) It follows form (ii) that $[x_0, x_v)$ contains $v + 1$ A-points $t_1 < \ldots < t_{v+1}$ of $f - s_f$.

We choose $m + 1$ additional points $t_{v+2} < \ldots < t_{v+m+2}$ in $[x_v, x_{v+1}]$. As in Case 1 there exists a sign $\sigma \in \{-1, 1\}$ such that

$$\sigma(-1)^i s(t_i) \geq 0, \quad i = 1, \ldots, v+m+2.$$

Analogously to the proof of Theorem 4.2, there exists an interval $[x_p, x_{p+q}] \subset [x_0, x_{v+1}]$ and points $u_1 < \ldots < u_{q+m+1}$ in $\{t_1, \ldots, t_{v+m+2}\}$ such that for all $i \in \{1, \ldots, q + m + 1\}$, the set $\{u_1, \ldots, u_{i-1}, u_{i+1}, \ldots, u_{q+m+1}\}$ is poised with respect to $S_m(x_1, \ldots, x_k)\big|_{[x_p, x_{p+q}]}$ and

$$\sigma(-1)^i s(u_i) \geq 0, \quad i = 1, \ldots, q+m+1.$$

Then it follows from Lemma 1.11 that $s = 0$ on $[x_p, x_{p+q}]$. Since $[x_p, x_{p+q}] \neq [x_v, x_{v+1}]$ this is a contradiction.

Case 3. There exist intervals $[x_u, x_v]$ and $[x_w, x_z]$, $v < w$, such that $s = 0$ on $[x_u, x_v] \cup [x_w, x_z]$ and s has only finitely many zeros in (x_v, x_w).

By assumption we have $w - v \geq m + 1$, otherwise by Remark 2.4, $s(t) = 0$ for all $t \in (x_v, x_w)$. It follows from (ii) that (x_v, x_w) contains $w - v - m + 1$ A–points of $f - s_f$ denoted by $t_{m+2} < \cdots < t_{w-v+2}$. We choose $2m + 2$ additional points $t_1 < \cdots < t_{m+1}$ in $[x_{v-1}, x_v]$ and $t_{w-v+3} < \cdots < t_{w-v+m+3}$ in $[x_w, x_{w+1}]$.

As in Case 2, it follows that there exists an interval $[x_p, x_{p+q}] \subset [x_{v-1}, x_{w+1}]$ with $[x_p, x_{p+q}] \neq [x_{v-1}, x_v]$ and $[x_p, x_{p+q}] \neq [x_w, x_{w+1}]$ such that $s = 0$ on $[x_p, x_{p+q}]$, which is a contradiction. This proves Theorem 4.4.

By comparing Theorem 4.2 and Theorem 4.4 we see that best spline approximations are characterized by the existence of sufficiently many alternating extreme points in *one* knot–interval, while strongly unique best spline approximations are characterized by the existence of sufficiently many alternating extreme points in *every* knot–interval.

Unique and strongly unique best uniform spline approximations are not the same in general. For the next characterization of unique spline approximations we need the following notation introduced by Nürnberger & Singer [1982].

Definition 4.5. A function $h \in C[a, b]$ is called *flat of order m from the left* (respectively *from the right*) at a given point $t_0 \in (a, b)$, if

$$\liminf_{\substack{t \to t_0 \\ t < t_0}} \frac{|h(t_0) - h(t)|}{|t_0 - t|^m} = 0$$

(respectively

$$\liminf_{\substack{t \to t_0 \\ t > t_0}} \frac{|h(t_0) - h(t)|}{|t_0 - t|^m} = 0).$$

Building on earlier work of Rice [1967], Schumaker [1968a] and Strauß [1975b], the following characterization of unique best spline approximations was proved by Nürnberger & Singer [1982]. For extensions of this result see Nürnberger, Schumaker, Sommer & Strauß [1985]. The following result shows that in contrast to best approximations and strongly unique best approximations from spline spaces, unique best spline approximation depend not only on alternation properties of the error, but also on the flatness of the error near the knots.

Theorem 4.6. *Let a function* $f \in C[a, b]$ *and a spline* $s_f \in S_m(x_1, \ldots, x_k)$ *be given. The following statements are equivalent:*

(i) *The spline* s_f *is a unique best uniform approximation of the function f from* $S_m(x_1, \ldots, x_k)$.

(ii) *The following three conditions hold simultaneously:*

(a) *For every interval* $[x_i, x_{i+m+j}] \subset [x_{-m}, x_{k+m+1}]$, $j \geq 1$, *we have*

$$A(f - s_f)\big|_{[x_i, x_{i+m+j}]} \geq j + 1.$$

(b) *If $A(f - s_f)\big|_{(x_i, x_{i+m+j}]} = j$ (respectively $A(f - s_f)\big|_{[x_i, x_{i+m+j})} = j$), then $f - s_f$ is flat of order m from the right at x_i (respectively from the left at x_{i+m+j}).*

(c) *If $A(f - s_f)\big|_{(x_i, x_{i+m+j})} = j$, then $f - s_f$ is flat of order m from the right at x_i or from the left at x_{i+m+j}.*

Proof. (i) $\Rightarrow$ (ii). We first show that (i) $\Rightarrow$ (a). Suppose that (a) fails, i.e. there exists an interval $[x_i, x_{i+m+j}]$ with $A(f - s_f)\big|_{[x_i, x_{i+m+j}]} \leq j$. We choose a maximal number of A–points $t_1 < \cdots < t_p$ of $f - s_f$ in $[x_i, x_{i+m+j}]$. Then we have $p \leq j$. We may assume that $f(t_1) - s_f(t_1) = \|f - s_f\|_\infty$. Since p is the maximal number of A–points of $f - s_f$ in $[x_i, x_{i+m+j}]$, there exist points $x_i = z_0 < z_1 < \cdots < z_{p-1} < z_p = x_{i+m+j}$ satisfying $t_i < z_i < t_{i+1}$, $i = 1, \ldots, p-1$, and a constant $K > 0$ such that

$$(-1)^i (f(t) - s_f(t)) \leq \|f - s_f\| - K, \qquad t \in [z_{i-1}, z_i], \quad i = 1, \ldots, p. \qquad (4.10)$$

Since by Theorem 2.8

$$S = \text{span}\{B_i^m, \ldots, B_{i+j-1}^m\}$$

is a j–dimensional weak Chebyshev space and $p \leq j$, by Corollary 1.7 there exists a nontrivial spline $s \in S$ such that

$$(-1)^{i+1} s(t) \geq 0, \qquad t \in [z_{i-1}, z_i], \quad i = 1, \ldots, p. \qquad (4.11)$$

By multiplying s with an appropiate factor, we may assume that $\|s\|_\infty \leq K$. Then by (4.10) and (4.11) we have for all $i \in \{1, \ldots, p\}$ and all $t \in [z_{i-1}, z_i]$,

$$\begin{aligned} -\|f - s_f\|_\infty &\leq (-1)^i (f(t) - s_f(t)) \\ &\leq (-1)^i (f(t) - s_f(t)) - (-1)^i s(t) \\ &\leq \|f - s_f\| - K + \|s\|_\infty \leq \|f - s_f\|_\infty. \end{aligned}$$

Moreover, by Theorem 2.2 we have for all $t \in \mathbf{R} \setminus [x_i, x_{i+m+j}]$, $s(t) = 0$. This implies that

$$\|f - (s_f + s)\|_\infty \leq \|f - s_f\|_\infty,$$

i.e. (i) fails.

We now show that (i) $\Rightarrow$ (b) and (i) $\Rightarrow$ (c). Suppose that (i) holds but (b) or (c) fails. We only consider the following case. (The other case can be proved analogously.) Suppose that there exists an interval $[x_i, x_{i+m+j}]$ such that $A(f - s_f)\big|_{(x_i, x_{i+m+j}]} = j$ and $f - s_f$ is not flat of order m from the right at x_i, i.e. there exists a real number $\varepsilon > 0$ and a neighborhood U of x_i such that for all $t \in [x_i, x_{i+m+j}] \cap U$,

$$|(f(x_i) - s_f(x_i)) - (f(t) - s_f(t))| \geq \varepsilon |x_i - t|^m. \qquad (4.12)$$

It follows from the proof of (i) $\Rightarrow$ (a) that there exists a nontrivial spline $s \in S = \text{span}\{B_i^m, \ldots, B_{i+j-1}^m\}$ such that for all $t \in [x_i, x_{i+m+j}] \setminus U$,

$$|f(t) - (s_f(t) + s(t))| \leq \|f - s_f\|_\infty.$$

Since $s \in S$, there exists a real number α such that for all $t \in [x_i, x_{i+m+j}] \cap U$, $s(t) = \alpha(t - x_i)_+^m$. By multiplying s with an appropriate factor, we may assume that $|\alpha| \leq \varepsilon$. Then by (4.12) we have for all $t \in [x_i, x_{i+m+j}] \cap U$,

$$
\begin{aligned}
&|f(t) - (s_f(t) + s(t))| \\
&= |(f(x_i) - s_f(x_i)) + s(t) - [(f(x_i) - s_f(x_i)) - (f(t) - s_f(t))]| \\
&= \|f - s_f\|_\infty + \alpha|t - x_i|_+^m - [(f(x_i) - s_f(x_i)) - (f(t) - s_f(t))] \\
&\leq \|f - s_f\|_\infty.
\end{aligned}
$$

Moreover, by Theorem 2.2 we have for all $t \notin [x_i, x_{i+m+j}]$, $s(t) = 0$. This implies that

$$\|f - (s_f + s)\|_\infty \leq \|f - s_f\|_\infty,$$

which contradicts (i).

(ii) $\Rightarrow$ (i). Suppose that (ii) holds, but (i) fails, i.e. there exists a spline $s \in S_m(x_1, \ldots, x_k)$ such that $s \neq s_f$ and $\|f - s\|_\infty \leq \|f - s_f\|_\infty$. Then we have

$$(f(t) - s_f(t))(s(t) - s_f(t)) \geq 0, \qquad t \in E(f - s_f), \tag{4.13}$$

otherwise $\|f - s\|_\infty > \|f - s_f\|_\infty$, a contradiction. Since (i) fails, it follows from Theorem 4.4 that condition (ii) in Theorem 4.4 cannot be satisfied. Then it follows from (a) and (b) that there exist knots where $f - s_f$ is flat of order m from the left respectively from the right. We will show that there exists a knot x_i and a neighborhood U of x_i such that $f - s_f$ is flat of order m from the right (respectively from the left) and for all $t \in (x_i, b] \cap U$ (respectively $t \in [a, x_i) \cap U$),

$$(f(t) - s_f(t))(s(t) - s_f(t)) < 0. \tag{4.14}$$

If this is not true, then it follows from (4.13) and (ii) that there exists a subset T of $[a, b]$ such that every interval $(x_i, x_{i+m+j}) \subset (x_{-m}, x_{k+m+1})$, $j \geq 1$, contains at least $j + 1$ points $t_1 < \cdots < t_{j+1}$ from T with

$$\sigma(-1)^i \left(s(t_i) - s_f(t_i) \right) \geq 0, \quad i = 1, \ldots, j + 1,$$

for some sign $\sigma \in \{-1, 1\}$. The proof of (ii) $\Rightarrow$ (i) in Theorem 4.4 shows that this is not possible. This shows that (4.14) holds.

We consider only the case when $f - s_f$ is flat of order m from the right of x_i. The other case follows analogously. It follows from (4.14) that

$$(f(x_i) - s_f(x_i))(s(x_i) - s_f(x_i)) \leq 0.$$

Then $(f(x_i) - s_f(x_i))(s(x_i) - s_f(x_i)) = 0$; otherwise $\|f - s\|_\infty > \|f - s_f\|_\infty$, since $|(f(x_i) - s_f(x_i))| = \|f - s_f\|_\infty$. This implies that $s(x_i) - s_f(x_i) = 0$. Therefore, we have for all $t \in [x_i, x_{i+m+j}] \cap U$,

$$s(t) - s_f(t) = \alpha(t - x_i)^q \, p(t),$$

where $\alpha \neq 0$ is a real number and p is a polynomial with $p(x_i) \neq 0$. Moreover, there exists a real number $\varepsilon > 0$ such that for all $t \in [x_i, x_{i+m+j}] \cap U$,

$$|s(t) - s_f(t)| = |\alpha(t - x_i)^q \, p(t)| \geq |\varepsilon(t - x_i)^q|. \qquad (4.15)$$

Since $f - s_f$ is flat of order m from the right at x_i, there exists a sequence (t_n) such that $t_n \to x_i$, $t_n > x_i$ for all n, and

$$|(f(x_i) - s_f(x_i)) - (f(t_n) - s_f(t_n))| < \varepsilon |t_n - x_i|^m. \qquad (4.16)$$

Therefore, it follows from (4.14)–(4.16) that for sufficiently large n,

$$
\begin{aligned}
\|f - s\|_\infty &\geq |f(t_n) - s(t_n)| \\
&= |(f(t_n) - s_f(t_n)) - (s(t_n) - s_f(t_n))| \\
&= |f(t_n) - s_f(t_n)| + |s(t_n) - s_f(t_n)| \\
&\geq |f(t_n) - s_f(t_n)| + |\varepsilon(t_n - x_i)^q| \\
&\geq |f(t_n) - s_f(t_n)| + |\varepsilon(t_n - x_i)^m| \\
&> |f(x_i) - s_f(x_i)| \\
&= \|f - s_f\|_\infty,
\end{aligned}
$$

which is a contradiction. This proves Theorem 4.6.

We now consider stability questions for unique best spline approximations (compare Theorem 1.14). The following sets are defined:

$$U(S_m(x_1, \ldots, x_k)) = \{\, f \in C[a, b] : f \text{ has a unique best} \\ \text{uniform approximation from } S_m(x_1, \ldots, x_k)\}$$

and

$$SU(S_m(x_1, \ldots, x_k)) = \{\, f \in C[a, b] : f \text{ has a strongly unique best} \\ \text{uniform approximation from } S_m(x_1, \ldots, x_k)\}.$$

We consider the following stability problem: which functions belong to the interior (briefly denoted by int) of these sets ?

The next result, due to Nürnberger [1983], characterizes these functions.

Theorem 4.7. *Let a function $f \in C[a, b]$ be given and s_f be a best uniform approximation of f from $S_m(x_1, \ldots, x_k)$. The following statements are equivalent:*

(i) $f \in \operatorname{int} U(S_m(x_1, \ldots, x_k))$.

(ii) $f \in \operatorname{int} SU(S_m(x_1, \ldots, x_k))$.

(iii) $A(f - s_f)\big|_{[a,b]} \geq k + m + 2$ *and for every interval* $[x_p, x_{p+q}] \subsetneq [a, b]$, $q \geq 1$, *we have* $A(f - s_f)\big|_{[x_p, x_{p+q}]} < q + m + 1$.

Proof. (i) $\Rightarrow$ (iii). Suppose that (i) holds. Then it follows from Theorem 4.6 that $A(f - s_f)\big|_{[a,b]} \geq k + m + 2$. Suppose that (iii) fails, i.e. there exists an interval $[x_p, x_{p+q}] \underset{\neq}{\subsetneq} [a, b]$ which contains a set $T = \{t_1, \ldots, t_{q+m+1}\}$ of $q + m + 1$ A–points of $f - s_f$. As in the proof of (i) $\Rightarrow$ (ii) in Theorem 1.14, there exists a sequence (h_n) in $C[a, b]$ such that $h_n \rightarrow f - s_f$, $h_n(t) = f(t) - s_f(t)$ for all $t \in T$ and $E(h_n) = T$ for all n. Then it follows from Theorem 4.2 and Theorem 4.6 that for all n, zero is a best uniform approximation of h_n from $S_m(x_1, \ldots, x_k)$, but is not unique. We set $f_n = h_n + s_f$ for all n. Then $f_n \rightarrow f$ and for all n, s_f is best uniform approximation of f_n but is not unique. This shows that $f \notin \mathrm{int}\, U(S_m(x_1, \ldots, x_k))$, which contradicts (i).

(iii) $\Rightarrow$ (ii). Suppose that (iii) holds. We will show that condition (ii) in Theorem 1.14 is satisfied. If this is not true, then there exists a set $\{t_1, \ldots, t_{k+m+2}\}$ of A–points of $f - s_f$ such that the set

$$\tilde{T} = \{t_1, \ldots, t_{r-1}, t_{r+1}, \ldots, t_{k+m+2}\}$$

is not poised with respect to $S_m(x_1, \ldots, x_k)$ for some $r \in \{1, \ldots, k + m + 2\}$. Then by Theorem 3.7 there exists an interval $(x_i, x_{i+m+j}) \subset (x_{-m}, x_{k+m+1})$, $j \geq 1$, which contains at most $j - 1$ points from $\tilde{T}$. But then $[x_0, x_i]$ contains at least $i + m + 1$ points from $\tilde{T}$ or $[x_{i+m+j}, x_{k+1}]$ contains at least $k - i + j + 2$ points from $\tilde{T}$, which contradicts (iii).

Since the implication (ii) $\Rightarrow$ (i) is obvious, this proves Theorem 4.7. $\blacksquare$

In the following we derive from Theorem 4.7 a uniqueness result for special functions.

The set

$$K(S_m(x_1, \ldots, x_k))$$
$$= \{ f \in C[a, b] : \quad \mathrm{span}(S_m(x_1, \ldots, x_k) \cup \{f\}) \text{ is weak Chebyshev}\}$$

is called the *convexity cone* of $S_m(x_1, \ldots, x_k)$.

As an application of Theorem 4.7 we obtain the following unicity result for functions from the convexity cone due to Zwick [1987].

Theorem 4.8. *If $m \geq 2$, then the set $K(S_m(x_1, \ldots, x_k)) \cap C^1[a, b]$ is a subset of $\mathrm{int}\, SU(S_m(x_1, \ldots, x_k))$.*

Proof. Let $m \geq 2$ and a function $f \in K(S_m(x_1, \ldots, x_k)) \cap C^1[a, b]$ be given. We will show that condition (iii) in Theorem 4.7 holds which proves the claim. Since by Theorem 1.19 the space $S_m(x_1, \ldots, x_k)$ is weak Chebyshev, it follows from Theorem 1.8 that there exists a best uniform approximation $s_f \in S_m(x_1, \ldots, x_k)$ of f such that $A(f - s_f)\big|_{[a,b]} \geq m + k + 2$. Now, suppose that there exists an interval $[x_p, x_{p+q}] \underset{\neq}{\subsetneq} [a, b]$ such that

$$A(f - s_f)\big|_{[x_p, x_{p+q}]} \geq q + m + 1.$$

Since $f \in K(S_m(x_1, \ldots, x_k))$, it follows from Zwick [1987] that $\text{span}(S_m(x_1, \ldots, x_k) \cup \{f\})$ is also a weak Chebyshev subspace on $[x_p, x_{p+q}]$ and that therefore $f - s_f$ has exactly $q + m + 1$ extreme points

$$x_p = t_1 < t_2 < \cdots < t_{q+m} < t_{q+m+1} = x_{p+q}$$

such that

$$f'(t_1) - s'_f(t_1) \neq 0$$

and

$$f'(t_{q+m+1}) - s'_f(t_{q+m+1}) \neq 0.$$

But this is a contradiction, since t_1 and t_{q+m+1} are extreme points of $f - s_f$ and one of them is contained in (a, b). This proves Theorem 4.8.

In view of Theorem 4.8, we are interested in a description of functions belonging to the convexity cone of spline spaces. The following result on those functions is due to Micchelli [1977]. A more general characterization was given by Zwick [1986].

Theorem 4.9. *For a function $f \in C^{m+1}[a, b]$ the following statements are equivalent:*

(i) $f \in K(S_m(x_1, \ldots, x_k))$

(ii) *There exists a sign $\sigma \in \{-1, 1\}$ such that*

$$\sigma(-1)^i f^{(m+1)}(t) \geq 0, \qquad t \in [x_i, x_{i+1}], \quad i = 0, \ldots, k.$$

Proof. Let a function $f \in C^{m+1}[a, b]$ be given. By the Taylor expansion we have for all $t \in [a, b]$,

$$
\begin{aligned}
f(t) &= \sum_{i=0}^{m} \frac{f^{(i)}(a)}{i!}(t - a)^i + \frac{1}{m!} \int_a^b (t - u)_+^m f^{(m+1)}(u)\, du \\
&= \sum_{i=0}^{m} \frac{f^{(i)}(a)}{i!}(t - a)^i + \frac{1}{m!} \sum_{j=0}^{k} \int_{x_j}^{x_{j+1}} (t - u)_+^m f^{(m+1)}(u)\, du.
\end{aligned}
$$

We set for all $j \in \{0, \ldots, k\}$ and all $t \in [a, b]$,

$$f_j(t) = \int_{x_j}^{x_{j+1}} (t - u)_+^m f^{(m+1)}(u)\, du.$$

We define the following basis functions of $S_m(x_1, \ldots, x_k)$

$$s_j(t) = t^{j-1}, \qquad j = 1, \ldots, m + 1,$$

and

$$s_{m+1+j}(t) = (t - x_j)_+^m, \qquad j = 1, \ldots, k.$$

It is easy to see that for all points $t_1 < \cdots < t_{k+m+2}$ in $[a, b]$,

$$D\left(\begin{matrix} s_1, \ldots, s_{k+m+1}, f \\ t_1, \quad \ldots\ldots \quad , t_{k+m+2} \end{matrix}\right)$$

$$= \frac{1}{m!} \sum_{j=0}^{k} D\left(\begin{matrix} s_1, \ldots, s_{k+m+1}, f_j \\ t_1, \quad \ldots\ldots \quad , t_{k+m+2} \end{matrix}\right)$$

$$= \frac{1}{m!} \sum_{j=0}^{k} \int_{x_j}^{x_{j+1}} D\left(\begin{matrix} s_1, \ldots, s_{j+m+1}, (t-u)_+^m, s_{j+m+2}, \ldots, s_{k+m+1} \\ t_1, \quad\quad\quad \ldots\ldots \quad\quad\quad\quad , t_{k+m+2} \end{matrix}\right)$$

$$\cdot (-1)^{k+j} f^{(m+1)}(u)\, du.$$

Since by Theorem 1.19 spline spaces are weak Chebyshev, the determinant below the integral has the same sign for all $t_1 < \cdots < t_{k+m+2}$ in $[a, b]$. Therefore,

$$D\left(\begin{matrix} s_1, \ldots, s_{k+m+1}, f \\ t_1, \quad \ldots\ldots \quad , t_{k+m+2} \end{matrix}\right)$$

has the same sign for all $t_1 < \cdots < t_{k+m+2}$ in $[a, b]$ if and only if (ii) holds. This proves Theorem 4.9.

Having proved several results on strongly unique best spline approximations, we now turn to the question of how to determine the corresponding strong unicity constant.

A general formula for computing strong unicity constants for splines was proved by Nürnberger [1982b]. Extensions of this result were obtained by Nürnberger [1987a]. We state a special case of these results.

Theorem 4.10. *Let a function $f \in C[a, b]$ be given and s_f be a strongly unique best uniform approximation of f from $S_m(x_1, \ldots, x_k)$ such that $E(f - s_f) = \{t_1, \ldots, t_{k+m+2}\}$. For each $j \in \{1, \ldots, k+m+2\}$, let $s_j \in S_m(x_1, \ldots, x_k)$ be the uniquely determined spline with*

$$s_j(t_i) = \operatorname{sgn}(f(t_i) - s_f(t_i)), \qquad i = 1, \ldots, k+m+2, \quad i \neq j.$$

Then we have

$$K(f) = \min\{1/\|s_j\|_\infty : j = 1, \ldots, k+m+2\}.$$

Proof. Since $s_f \in S_m(x_1, \ldots, x_k)$ is a strongly unique best uniform approximation of f and $E(f - s_f) = \{t_1, \ldots, t_{k+m+2}\}$, it follows from Theorem 1.12 that for all $i \in \{1, \ldots, k+m+2\}$, the set

$$\{t_1, \ldots, t_{i-1}, t_{i+1}, \ldots, t_{k+m+2}\}$$

is poised with respect to $S_m(x_1, \ldots, x_k)$. Therefore, the splines $s_1, \ldots, s_{k+m+2} \in S_m(x_1, \ldots, x_k)$ as defined in the theorem are uniquely determined. Since the above sets are poised, the proof of Theorem 3.20 in Chapter I for Chebyshev spaces also applies to the case of $S_m(x_1, \ldots, x_k)$. This proves Theorem 4.10.

Theorem 4.6 shows that functions from $U(S_m(x_1, \ldots, x_k))$ cannot be characterized by alternation properties alone. On the other hand, it was proved by Berens & Nürnberger [1987] that such a characterization holds for functions from the closure of $U(S_m(x_1, \ldots, x_k))$.

Theorem 4.11. *For a function $f \in C[a, b]$, the following statements are equivalent:*

(i) $f \in \overline{U(S_m(x_1, \ldots, x_k))}$.
(ii) There exists a best uniform approximation $s_f \in S_m(x_1, \ldots, x_k)$ of f such that for every interval $[x_i, x_{i+m+j}] \subset [x_{-m}, x_{k+m+1}]$, $j \geq 1$, we have

$$A(f - s_f)\big|_{[x_i, x_{i+m+j}]} \geq j + 1.$$

Proof. Theorem 4.11 can be proved analogously as Theorem 4.4 by using Theorem 3.24 in Chapter I instead of Theorem 3.17 in Chapter I.

4.2. Algorithm (Fixed Knots)

In this section we describe an iterative method for computing best uniform approximations from spline spaces which was developed by Nürnberger & Sommer [1983b].

The method is similar to that of Remez [1934a] for Chebyshev spaces (see Section 3.4 in Chapter I). However, there are two fundamental differences. In applying the exchange rule of Remez [1934a] to splines, it may happen that the systems of linear equations which appear cannot be solved. Therefore, a modified rule has to be used. Moreover, since best uniform approximations from spline spaces are not unique in general, the convergence proof is more involved than that for Chebyshev spaces.

In the following we describe the algorithm. Let a spline space $S_m(x_1, \ldots, x_k)$ and a function $f \in C[a, b] \backslash S_m(x_1, \ldots, x_k)$ be given. Moreover, let $\{h_1, \ldots, h_n\}$ be a basis of $S_m(x_1, \ldots, x_k)$, where $n = k + m + 1$ is the dimension of $S_m(x_1, \ldots, x_k)$. In practice we use the basis of normalized B–splines $N_{-m}^m, \ldots, N_k^m$ (see Theorem 2.6 and Definition 2.13).

As in the algorithm of Remez [1934a], we iteratively compute a sequence of functions (s_p) in $S_m(x_1, \ldots, x_k)$ converging to a best uniform approximation $s_f \in S_m(x_1, \ldots, x_k)$ of f on I_e (see the definition of I_e below) with the property that there exist points $a \leq t_1 < \cdots < t_{n+1} \leq b$ and a sign $\sigma \in \{-1, 1\}$ such that

$$\sigma (-1)^i \left(f(t_i) - s_f(t_i) \right) = \| (f - s_f) \big|_{I_e} \|_\infty, \qquad i = 1, \ldots, n + 1.$$

Before describing the method, we need some results on nonzero determinants. Let a subset

$$M = \{t_1, \ldots, t_{n+1}\}$$

of $[a, b]$ be given. We set

$$D(M) = \begin{vmatrix} h_1(t_1) & \cdots & h_n(t_1) & (-1)^1 \\ \vdots & & \vdots & \vdots \\ h_1(t_{n+1}) & \cdots & h_n(t_{n+1}) & (-1)^{n+1} \end{vmatrix}.$$

Such determinants correspond to systems of linear equations which have to be solved during the algorithm.

Lemma 4.12. *The following statements are equivalent:*
(i) $D(M) \neq 0$.
(ii) *Every interval* $(x_i, x_{i+m+j}) \subset (x_{-m}, x_{k+m+1})$, $j \geq 1$, *contains at least* j *points from* M.
(iii) *There exists a unique maximal interval* $I_M = [x_u, x_{u+v}] \subset [a, b]$, $v \geq 1$, *with the following property: If a point from* $M \cap I_M$ *is replaced by an arbitrary point in* $[a, b]$, *then* $D(\tilde{M}) \neq 0$ *for the resulting set* $\tilde{M}$.

Proof. We first note that, since $S_m(x_1, \ldots, x_k)$ is a weak Chebyshev space,

$$|D(M)| = \sum_{i=1}^{n+1} |D_i|, \tag{4.17}$$

where

$$D_i = D\begin{pmatrix} h_1, & \cdots\cdots & , h_n \\ t_1, \ldots, t_{i-1}, t_{i+1}, \ldots, t_{n+1} \end{pmatrix}.$$

(i) $\Rightarrow$ (ii). Suppose that (i) holds. Then by (4.17) there exists an index $j \in \{1, \ldots, n+1\}$ such that $D_j \neq 0$. It follows from Theorem 3.7 that (ii) holds.

(ii) $\Rightarrow$ (iii). Suppose that (ii) holds. Then every interval $[a, x_i)$, $i \geq 1$ (respectively $(x_{k+1-j}, b]$, $j \geq 1$) contains at least i (respectively j) points from M. We set $x_u = x_i$, where i is the largest index such that $[a, x_i)$, $i \geq 1$, contains exactly i points from M, and $x_u = a$, if no such interval exists. Moreover, we set $x_{u+v} = x_{k+1-j}$, where j is the largest index such that $(x_{k+1-j}, b]$, $j \geq 1$ contains exactly j points from M, and $x_{u+v} = b$, if no such interval exists. Since $[a, x_{k+1})$ contains at least $k+2$ points from M, we have $x_u < x_{k+1}$. Moreover, since M consists of $k+m+2$ points, the interval $(x_u, b]$ contains at least $k+2-u$ points from M. This implies that $x_u < x_{u+v}$. Now, let a point $t_r \in M \cap I_M$ be given.

Claim. Every interval $(x_i, x_{i+m+j}) \subset (x_{-m}, x_{k+m+1})$, $j \geq 1$, contains at least j points from $M_r = \{t_1, \ldots, t_{r-1}, t_{r+1}, \ldots, t_{n+1}\}$.

If the claim is true, then it follows from Theorem 3.7 that $D_r \neq 0$. Therefore, by (4.17) we have $D(\tilde{M}) \neq 0$, where $\tilde{M} = M_r \cup \{t\}$ and t is an arbitrary point in $[a, b]$. This shows that (iii) holds. It remains to prove the claim.

Case 1. $(x_i, x_{i+m+j}) \not\subset (x_1, x_k)$.

Suppose that $x_i < x_1$. Since $t_r \in I_M$, the interval $[a, x_{i+m+j})$ contains at least $i+m+j$ points from M_r. If $x_i = x_0$, then (x_i, x_{i+m+j}) contains at least $m+$

$j - 1 \geq j$ points from M_r. If $x_{-m} \leq x_i < x_0$, then (x_i, x_{i+m+j}) contains at least $i + m + j \geq j$ points from M_r. Now, suppose that $x_k < x_{i+m+j}$. Since $t_r \in M_r$, the interval $(x_i, b]$ contains at least $k + 1 - i$ points from M_r. If $x_{k+1} = x_{i+m+j}$ then (x_i, x_{i+m+j}) contains at least $m + j - 1 \geq j$ points from M_r. If $x_{k+1} < x_{i+m+j} \leq x_{k+m+1}$, then (x_i, x_{i+m+j}) contains at least $k + 1 - i \geq j$ points from M_r.

Case 2. $(x_i, x_{i+m+j}) \subset (x_1, x_k)$.

Suppose that (x_i, x_{i+m+j}) contains at most $j - 1$ points from M_r. Since the interval $[a, x_{i+m+j})$ contains at least $i + m + j$ points from M_r, the interval $[a, x_i)$ contains at least $i + m + 1$ points from M_r. This implies that $(x_i, b]$ contains at most $(k + m + 1) - (i + m + 1) = k - i$ points from M_r, which is a contradiction. This shows that (ii) $\Rightarrow$ (iii).

(iii) $\Rightarrow$ (i). The implication is obvious. This proves Lemma 4.12.

The interval I_M in Lemma 4.12 plays an important role in the subsequent algorithm. The next result which follows from the proof of Lemma 4.12 shows that the interval I_M can be easily determined.

Lemma 4.13. *If $D(M) \neq 0$, then the interval $I_M = [x_u, x_{u+v}]$ in Lemma 4.12 can be determined as follows:*

(i)   *$x_u = x_i$, where i is the largest index such that $[a, x_i)$, $i \geq 1$, contains exactly i points from M, and $x_u = a$, if no such interval exists.*

(ii)   *$x_{u+v} = x_{k+1-j}$, where j is the largest index such that $(x_{k+1-j}, b]$, $j \geq 1$, contains exactly j points from M, and $x_{u+v} = b$, if no such interval exists.*

We are now in position to describe the algorithm. Subsequently, we will discuss some details concerning the structure of the method.

Description of the Algorithm

We first choose sufficiently small real numbers $\varepsilon_1, \ldots, \varepsilon_k > 0$ and set $\varepsilon = (\varepsilon_1, \ldots, \varepsilon_k)$ and

$$I_\varepsilon = [a, b] \setminus \bigcup_{i=1}^{k} [(x_i - \varepsilon_i, x_i) \cup (x_i, x_i + \varepsilon_i)]$$

(compare Remark 4.14). In the following we only work on I_ε in order to get a stable algorithm.

In the first step by using Lemma 4.12, we choose a subset

$$M_1 = \{t_{1,1}, \ldots, t_{n+1,1}\}$$

of I_ε such that

$$a \leq t_{1,1} < \cdots < t_{n+1,1} \leq b$$

and $D(M_1) \neq 0$. Then we determine the unique spline $s_1 \in S_m(x_1, \ldots, x_k)$ and the unique real number $\lambda_1 \in R$ satisfying

$$(-1)^i \left(f(t_{i,1}) - s_1(t_{i,1}) \right) = \lambda_1, \qquad i = 1, \ldots, n+1. \qquad (4.18)$$

If we set $s_1 = \sum_{j=1}^{n} a_{j,1} h_j$, then (4.18) is equivalent to

$$\sum_{j=1}^{n} a_{j,1}\, g_j(t_{i,1}) + (-1)^i \lambda_1 = f(t_{i,1}), \qquad i = 1, \ldots, n+1. \qquad (4.19)$$

This is a system of $n+1$ linear equations with $n+1$ unknowns $a_{1,1}, \ldots, a_{n,1}$ and λ_1. Since the corresponding determinant $D(M_1) \neq 0$, the system has a unique solution.

For $p \geq 1$, we proceed by induction as follows. We determine a point $t_p \in I_\varepsilon$ such that

$$|\lambda_p| < |f(t_p) - s_p(t_p)| \approx \|(f - s_p)\big|_{I_\varepsilon}\|_\infty \qquad (4.20)$$

and replace a point from

$$M_p = \{t_{1,p}, \ldots, t_{n+1,p}\},$$

where

$$a \leq t_{1,p} < \cdots < t_{n+1,p} \leq b,$$

such that a new set

$$M_{p+1} = \{t_{1,p+1}, \ldots, t_{n+1,p+1}\},$$

where

$$a \leq t_{1,p+1} < \cdots < t_{n+1,p+1} \leq b$$

is obtained.

In the following we describe which point from M_p will be replaced by t_p.

Rule of Exchange

By using Lemma 4.13, we determine the unique interval I_{M_p} corresponding to the set M_p. Let $t_{\alpha,p}$ (respectively $t_{\beta,p}$) be the first (respectively last) point from $M_p \cap I_{M_p}$. Moreover, we set $t_{0,p} = -\infty$ and $t_{n+2,p} = \infty$. Then there exists an integer $j \in \{0, \ldots, n+1\}$ such that $t_{j,p} < t_p < t_{j+1,p}$.

Case 1. $j \in \{1, \ldots, n\}$
If $\operatorname{sgn}(f(t_{j,p}) - s_p(t_{j,p})) = \operatorname{sgn}(f(t_p) - s_p(t_p))$ (respectively $\operatorname{sgn}(f(t_{j+1,p}) - s_p(t_{j+1,p}))$ $= \operatorname{sgn}(f(t_p) - s_p(t_p)))$, then we replace $t_{j,p}$ (respectively $t_{j+1,p}$) by t_p, if $D(M_{p+1}) \neq 0$ for the resulting set M_{p+1}. Otherwise, we replace $t_{\alpha,p}$ (respectively $t_{\beta,p}$) by t_p, if $t_p > t_{\beta,p}$ (respectively $t_p < t_{\alpha,p}$). It follows from Lemma 4.12 that $D(M_{p+1}) \neq 0$.

Case 2. $j = 0$
If $\operatorname{sgn}(f(t_{1,p}) - s_p(t_{1,p})) = \operatorname{sgn}(f(t_p) - s_p(t_p))$, then we replace $t_{1,p}$ by t_p. Otherwise, we replace $t_{\beta,p}$ by t_p. It follows from Lemma 4.12 that $D(M_{p+1}) \neq 0$.

Case 3. $j = n+1$
If $\operatorname{sgn}(f(t_{n+1,p}) - s_p(t_{n+1,p})) = \operatorname{sgn}(f(t_p) - s_p(t_p))$, then we replace $t_{n+1,p}$ by t_p. Otherwise, we replace $t_{\alpha,p}$ by t_p. It follows from Lemma 4.12 that $D(M_{p+1}) \neq 0$.

Since in all cases $D(M_{p+1}) \neq 0$, we determine the unique spline $s_{p+1} \in S_m(x_1, \ldots, x_k)$ and the unique real number $\lambda_{p+1} \in \mathbf{R}$ satisfying

$$(-1)^i \left(f(t_{i,p+1}) - s_{p+1}(t_{i,p+1}) \right) = \lambda_{p+1}, \qquad i = 1, \ldots, n+1. \qquad (4.21)$$

Moreover, since there is some freedom in choosing the point $t_p \in I_e$ with property (4.20), the following *additional rule* should be applied.

In the p-th step of the algorithm we determine all points $t_p \in I_e$ with property (4.20). Then for each $j \in \{1, \ldots, n+1\}$ with $t_{j,p} \notin \{x_1, \ldots, x_k\}$ we choose a small neighborhood $U_j \subset I_e$ of $t_{j,p}$ such that

$$U_j \subset (x_r, x_{r+1}), \quad \text{if } t_{j,p} \in (x_r, x_{r+1}), \tag{4.22}$$

$$U_j \subset [a, x_1), \quad \text{if } t_{j,p} = a, \tag{4.23}$$

$$U_j \subset (x_k, b], \quad \text{if } t_{j,p} = b. \tag{4.24}$$

Now, if there exists a point $t_p \in I_e$ satisfying (4.20) which is contained in some U_j, we replace $t_{j,p}$ by t_p. In this case we replace all such points $t_{j,p}$, but no other point.

We now discuss some details of the above algorithm.

Remark 4.14. (i) We first note that, since $S_m(x_1, \ldots, x_k)$ is a weak Chebyshev space,

$$|D(M_p)| = \sum_{i=1}^{n+1} |D_{i,p}|,$$

where

$$D_{i,p} = D \begin{pmatrix} h_1, & \cdots\cdots & , h_n \\ t_{1,p}, \ldots, & t_{i-1,p}, t_{i+1,p}, & \ldots, t_{n+1,p} \end{pmatrix}.$$

(ii) Suppose that there exists an index $j \in \{1, \ldots, n+1\}$ such that for all p, $t_{j,p} \notin I_{M_p}$ and $t_{j,p} \in (x_r, x_{r+1})$. Then we have $\lim_{p \to \infty} D(M_p) = 0$, if $\lim_{p \to \infty} t_{j,p} = x_r$. In order to avoid such a situation, we work on I_e which ensures that

$$\inf_p |D(M_p)| > 0.$$

Therefore, the real numbers $\varepsilon_1, \ldots, \varepsilon_k > 0$ should be choosen small, but big enough so that the linear systems (4.21) can be solved numerically.

(iii) The exchange rule of the algorithm of Remez (see Section 3.4 in Chapter I) applied to splines does not guarantee that $D(M_p) \neq 0$ for all p. We give a simple example. Let $m = 1$, $k = 1$,

$$a \leq t_{1,p} < t_{2,p} < t_{3,p} < t_p < x_1 < t_{4,p} \leq b$$

and

$$\text{sgn}(f(t_{4,p}) - s_p(t_{4,p})) = \text{sgn}(f(t_p) - s_p(t_p)).$$

Then according to the exchange rule of Remez, the point $t_{4,p}$ has to be replaced by t_p which implies that $D(M_{p+1}) = 0$ for the resulting set M_{p+1}. Therefore, the algorithm cannot be continued.

(iv) In contrast to the algorithm of Remez, the property

$$\text{sgn}(f(t_{i,p+1}) - s_p(t_{i,p+1}))$$
$$= \sigma_p \, \text{sgn}(f(t_{i,p}) - s_p(t_{i,p})), i = 1, \ldots, n+1, \tag{4.25}$$

where $\sigma_p \in \{-1, 1\}$, does not hold in general. However, property (4.25) is satisfied for all points $t_{i,p+1} \in I_{M_p} \cup I_{M_{p+1}}$. Moreover, property (4.25) holds for all points $t_{i,p+1}$, $i = 1, \ldots, n+1$, after finitely many steps (see (v)).

(v) The rule that we replace a point $t_{j,p} \in M_p$ by a point $t_p \in I_\varepsilon$ with property (4.20) in a small neighborhood of $t_{j,p}$, whenever this is possible, guarantees the convergence of the algorithm. It can be shown that such an exchange is always possible after finitely many steps (see Nürnberger & Sommer [1983b]). Then the location of the points from M_p with respect to the knot intervals and also the interval I_{M_p} remains unchanged for all p.

In the following we give results on the convergence of the algorithm which were proved by Nürnberger & Sommer [1983b]. The results are given without proofs, since a complete convergence proof would be beyond the scope of this book.

The first result shows that the sequence $(|\lambda_p|)$ is monotonely increasing.

Lemma 4.15. *For all $p = 1, 2, \ldots$ the following properties hold:*
(i) $|\lambda_p| \leq |\lambda_{p+1}|$.
(ii) $|\lambda_p| < |\lambda_{p+1}|$ *if and only if $D_{j,p+1} \neq 0$, where $t_{j,p+1} = t_p$.*
(iii) There exists a subsequence $(|\lambda_{p_r}|)$ of $(|\lambda_p|)$ such that for all r,

$$|\lambda_{p_r}| < |\lambda_{p_r+1}|.$$

Since the sequence constructed in the algorithm depends on I_ε, to be more precise in the following we set $(s_p^{(\varepsilon)}) = (s_p)$ and $(\lambda_p^{(\varepsilon)}) = (\lambda_p)$.

The next result shows that the sequence $(s_p^{(\varepsilon)})$ converges to a best uniform approximation $s_f^{(\varepsilon)} \in S_m(x_1, \ldots, x_k)$ of f on I_ε. In particular, property (iii) in Theorem 4.16 shows that in practice the spline $s_f^{(\varepsilon)}$ can be considered as a best approximation on the whole interval $[a, b]$. Moreover, the sequence $(|\lambda_p^{(\varepsilon)}|)$ converges to $d(f, S_m(x_1, \ldots, x_k))$.

Theorem 4.16. *The following statements hold:*
(i) $\lim_{p \to \infty} |\lambda_p^{(\varepsilon)}| = d_\varepsilon(f, S_m(x_1, \ldots, x_k)) = d(f, S_m(x_1, \ldots, x_k))$, *where*

$$d_\varepsilon(f, S_m(x_1, \ldots, x_k)) = \inf_{s \in S_m(x_1, \ldots, x_k)} \| (f - s) |_{I_\varepsilon} \|_\infty.$$

(ii) $\lim_{p \to \infty} \| s_p^{(\varepsilon)} - s_f^{(\varepsilon)} \|_\infty = 0$, *where $s_f^{(\varepsilon)} \in S_m(x_1, \ldots, x_k)$ is a best uniform approximation of f from $S_m(x_1, \ldots, x_k)$ on I_ε.*
(iii) $\lim_{\varepsilon \to 0} \| f - s^{(\varepsilon)} \|_\infty = d(f, S_m(x_1, \ldots, x_k))$.
(iv) If f has a strongly unique best uniform approximation $s_f \in S_m(x_1, \ldots, x_k)$, then $\lim_{p \to \infty} \| s_p^{(\varepsilon)} - s_f \|_\infty = 0$.

Remark 4.17. The spline $s_f^{(\varepsilon)}$ in Theorem 4.16 has the property that there exist points $t_1 < \cdots < t_{n+1}$ in I_ε and a sign $\sigma \in \{-1, 1\}$ such that

$$\sigma(-1)^i\left(f(t_i) - s_f^{(\varepsilon)}(t_i)\right) = \left\|(f - s_f^{(\varepsilon)})\big|_{I_\varepsilon}\right\|_\infty, \quad i = 1, \ldots, n+1,$$

and $D(M) \neq 0$, where $M = \{t_1, \ldots, t_{n+1}\}$. If the corresponding interval I_M and the sign σ are fixed, then such a spline is uniquely determined (see Nürnberger & Sommer [1983a], [1983b]).

Remark 4.18. (Simultaneous exchange method)

(i) In order to obtain faster convergence of the algorithm, one should exchange not only one point $t_p \in I_\varepsilon$ satisfying (4.20) in the p–th step but several such points according to the above exchange rule (if possible). In particular, the resulting set M_{p+1} must satisfy $D(M_{p+1}) \neq 0$ and

$$\operatorname{sgn}(f(t_{i,p+1}) - s_p(t_{i,p+1})) = \sigma_p \operatorname{sgn}(f(t_{i,p}) - s_p(t_{i,p}))$$

for all $t_{i,p+1} \in I_{M_p} \cup I_{M_{p+1}}$, where $\sigma_p \in \{-1, 1\}$.

(ii) Under analogous assumptions to those in Theorem 3.28 of Chapter I, and the additional assumption that f has a strongly unique best uniform approximation from $S_m(x_1, \ldots, x_k)$, the simultaneous exchange method converges quadratically (Nürnberger [1983]).

Remark 4.19. (Error estimation). It is easy to verify that for all $p = 1, 2, \ldots$ the following properties hold:

(i) $\|f - s_p^{(\varepsilon)}\|_\infty - \|f - s_f^{(\varepsilon)}\|_\infty \leq \|f - s_p^{(\varepsilon)}\|_\infty - |\lambda_p^{(\varepsilon)}|.$

(ii) $\|s_p^{(\varepsilon)} - s_f\|_\infty \leq \frac{1}{K(f)}\left(\|f - s_p^{(\varepsilon)}\|_\infty - |\lambda_p^{(\varepsilon)}|\right),$

if f has a strongly unique best uniform approximation $s_f \in S_m(x_1, \ldots, x_k)$ and $K(f)$ is the strong unicity constant of f.

Since the values $\|f - s_p^{(\varepsilon)}\|_\infty$ and $|\lambda_p^{(\varepsilon)}|$ are computed during the algorithm (and also the strong unicity constant $K(f)$ can be computed (compare Theorem 4.10)), we stop the algorithm if $\|f - s_p^{(\varepsilon)}\|_\infty - |\lambda_p^{(\varepsilon)}| \leq \delta$, where δ is the desired accuracy.

We can compare the effort to compute a best spline approximation with the construction of an interpolating spline. In order to obtain a spline approximation by interpolation, we have to solve a system of linear equations (see Definition 3.6 and Theorem 3.7) and, of course, we want to know the norm of the resulting error function. Therefore, the computations in spline interpolation are approximately the same as the computations in one step of the above iterative method.

It is interesting that the approximations computed by this method are also interpolating splines, as follows from (4.21). However, in contrast to usual spline interpolation, with the aid of the above inequality (i), we have a control on how efficient these approximations are compared with the best approximation. The algorithm can be continued until the desired accuracy is obtained, and according to our numerical experience, this is the case after very few steps, in general.

When we tested our algorithm, we saw that most of the standard functions are from $\mathrm{int\,SU}\,(S_m(x_1,\ldots,x_k))$ (compare Theorem 4.7) and that the corresponding error has exactly $n + 1 = k + m + 2$ alternating extreme points. Numerical examples are given at the end of Section 4.3.

Finally, we note that Nürnberger & Sommer [1983b] actually gave a more detailed description of the above algorithm. Our numerical tests show that the somewhat simpler variant in this section suffices for practical purposes.

Schumaker [1969a] already observed that the idea of the Remez algorithm for Chebysev spaces can also be used for spline spaces. The application of the Remez type method was also studied by Blatter [1986]. Esch & Eastman [1969] used a different optimization approach for computing best uniform spline approximations. An algorithm for strict spline approximations was developed by Strauß [1984b], [1984c]. Optimization methods for computing best uniform approximations from arbitrary finite–dimensional spaces can be found in Carasso & Laurent [1978], Hettich & Zencke [1982] and Blatt [1984].

4.3. Algorithm (Free Knots)

In this section we describe an algorithm for computing best piecewise polynomial approximations with free knots in the uniform norm which was developed by Nürnberger, Sommer & Strauß [1986]. The piecewise polynomial constructed in the algorithm is discontinuous, in general. However, if we take the corresponding optimal knots as fixed knots and apply the algorithm of Section 4.2, then we obtain a best uniform spline approximation which is differentiable. Moreover, our numerical results show that this function is a good spline approximation for free knots.

We begin with the definition of piecewise polynomials with free knots.

Definition 4.20. Let integers $m \geq 1$ and $k \geq 1$ be given. We call

$$
\begin{aligned}
PP_{m,k} \;=\; \{ s : [a,b] \to \mathbf{R} : \ & \text{there exist knots} \\
& a = x_0 < x_1 < \cdots < x_k < x_{k+1} = b \\
& \text{such that } s|_{[x_i,x_{i+1})} \in P_m, \ i = 0,\ldots,k-1, \\
& \text{and } s|_{[x_k,x_{k+1}]} \in P_m \}
\end{aligned}
\tag{4.26}
$$

the set of *piecewise polynomials of degree m with k free knots*. (Note, that for functions from $PP_{m,k}$ no continuity is required.)

The problem of best approximation is defined as follows.

Definition 4.21. Let a function $f \in C[a,b]$ be given. A piecewise polynomials $s_f \in PP_{m,k}$ is called a *best uniform approximation* of f from $PP_{m,k}$, if

$$
\|f - s_f\|_\infty = \inf_{s \in PP_{m,k}} \|f - s\|_\infty.
\tag{4.27}
$$

The set of knots of the piecewise polynomial s_f is called an *optimal set of knots* for f. The *minimal deviation* of f from $PP_{m,k}$ is defined by

$$d(f, PP_{m,k}) = \inf_{s \in PP_{m,k}} \|f - s\|_\infty. \tag{4.28}$$

For a subinterval I of $[a, b]$, we set

$$d(f, P_m, I) = \inf_{p \in P_m} \|(f - p)|_I\|_\infty. \tag{4.29}$$

A set of knots $\{x_1, \ldots, x_k\}$ with $a = x_0 \leq x_1 \leq \cdots \leq x_k \leq x_{k+1} = b$ is called a *leveled set of knots* for f if

$$d(f, P_m, [x_{i-1}, x_i]) = d(f, P_m, [x_i, x_{i+1}]), \qquad i = 1, \ldots, k. \tag{4.30}$$

We first give a result, due to Lawson [1964], which will be needed in the algorithm.

Theorem 4.22. *For a function $f \in C[a, b]$, the following statements hold:*
(i)   For every set of knots $\{x_1, \ldots, x_k\}$ with

$$a = x_0 \leq x_1 \leq \cdots \leq x_k \leq x_{k+1} = b,$$

we have

$$\min_{i=0,\ldots,k} d(f, P_m, [x_i, x_{i+1}]) \leq d(f, PP_{m,k}) \leq \max_{i=0,\ldots,k} d(f, P_m, [x_i, x_{i+1}]).$$

(ii) There exists an optimal set of knots for f which is leveled.
(iii) Every leveled set of knots for f is optimal.

Proof. Let a function $f \in C[a, b]$ be given. We first show that there exists an optimal set of knots for f. To do this, we set

$$M = \{(x_1, \ldots, x_k) : a = x_0 \leq x_1 \leq \cdots \leq x_k \leq x_{k+1} = b\}$$

and define the mapping $m : M \to \mathbf{R}$ by

$$m(x_1, \ldots, x_k) = \max_{i=0,\ldots,k} d(f, P_m, [x_i, x_{i+1}])$$

for all $(x_1, \ldots, x_k) \in M$. Since the set M is compact and the mapping m is continuous, it follows that m attains its minimum. Therefore, there exists an optimal set of knots for f.
   (i). It is obvious that

$$d(f, PP_{m,k}) \leq \max_{i=0,\ldots,k} d(f, P_m, [x_i, x_{i+1}]).$$

We now assume that

$$\min_{i=0,\ldots,k} d(f, P_m, [x_i, x_{i+1}]) > d(f, PP_{m,k}).$$

Let $\{y_1, \ldots, y_k\}$ be a set of optimal knots for f such that

$$a = y_0 \leq y_1 \leq \cdots \leq y_k \leq y_{k+1} = b.$$

Then there exists an integer $j \in \{0, \ldots, k\}$ such that $[x_j, x_{j+1}] \subset [y_j, y_{j+1}]$. This implies that

$$d(f, PP_{m,k}) \geq d(f, P_m, [y_j, y_{j+1}]) \geq d(f, P_m, [x_j, x_{j+1}]) > d(f, PP_{m,k}),$$

which is a contradiction. This proves (i).

(iii). This statement follows directly from (i).

(ii). We prove this statement by induction on k. Let $k = 1$ and set $x_0 = a$ and $x_2 = b$. We define the function $c_1 : [a, b] \to \mathbf{R}$ by

$$c_1(x) = d(f, P_m, [x_0, x]) - d(f, P_m, [x, x_2]), \qquad x \in [a, b].$$

It follows from the intermediate value theorem that there exists a point $x_1 \in [a, b]$ such that $c_1(x_1) = 0$, i.e. $d(f, P_m, [x_0, x_1]) = d(f, P_m, [x_1, x_2])$. We now assume that the claim is true for $k - 1$. We define functions $c_2 : [a, b] \to \mathbf{R}$ and $c_3 : [a, b] \to \mathbf{R}$ by

$$c_2(x) = \min\Big\{ \max_{i=0,\ldots,k-1} d(f, P_m, [x_i, x_{i+1}]) :$$
$$a = x_0 \leq x_1 \leq \cdots \leq x_{k-1} \leq x_k = x, \ x \in [a, b]\Big\}$$

and

$$c_3(x) = c_2(x) - d(f, P_m, [x, x_{k+1}]), \qquad x \in [a, b].$$

Analogously to above, there exists a point $x_k \in [a, b]$ such that

$$c_2(x_k) = d(f, P_m, [x_k, x_{k+1}]).$$

Moreover, by the induction hypothesis there exist points

$$a = x_0 \leq x_1 \leq \cdots \leq x_{k+1} \leq x_k$$

such that

$$d(f, P_m, [x_i, x_{i+1}]) = c_2(x_k), \qquad i = 0, \ldots, k - 1.$$

Therefore, it follows that

$$d(f, P_m, [x_{i-1}, x_i]) = d(f, P_m, [x_i, x_{i+1}]), \qquad i = 1, \ldots, k.$$

This proves Theorem 4.22.

In the following we describe the algorithm. Let a function $f \in C[a, b] \setminus PP_{m,k}$ be given. The method is to compute a sequence of real numbers which converges to the minimal deviation of f from $PP_{m,k}$. Simultaneously, we obtain a leveled set of knots which by Theorem 4.22 is optimal for f.

Description of the Algorithm

In the first step of the algorithm we choose a set of knots

$$\{x_{1,1}, \ldots, x_{k,1}\}$$

such that

$$a = x_{0,1} < x_{1,1} < \cdots < x_{k,1} < x_{k+1,1} = b.$$

By applying the algorithm of Remez (Section 3.4 of Chapter I), we compute the values

$$d_{i,1} = d(f, P_m, [x_{i,1}, x_{i+1,1}]), \qquad i = 0, \ldots, k.$$

We set

$$a_1 = 10^{\alpha_1} = \min_{i=0,\ldots,k} d_{i,1}$$

and

$$b_1 = 10^{\beta_1} = \max_{i=0,\ldots,k} d_{i,1}.$$

For $p \geq 1$, we proceed by induction as follows. We set

$$\delta_{p+1} = \frac{\alpha_p + \beta_p}{2}$$

and

$$d_{p+1} = 10^{\delta_{p+1}}.$$

Then as described below, we determine a set of knots

$$\{x_{1,p+1}, \ldots, x_{k,p+1}\}$$

such that

$$a = x_{0,p+1} < x_{1,p+1} < \cdots < x_{j_{p+1},p+1} < x_{j_{p+1}+1,p+1} = \cdots = x_{k+1,p+1} = b$$

and

$$d(f, P_m, [x_{i,p+1}, x_{i+1,p+1}]) = d_{p+1}, \qquad i = 0, \ldots, j_{p+1} - 1.$$

Here we choose the maximal index j_{p+1} such that these properties are satisfied. Moreover, we compute

$$c_{p+1} = d(f, P_m, [x_{k,p+1}, x_{k+1,p+1}]).$$

Then we set

$$a_{p+1} = 10^{\alpha_{p+1}} = \max\{a_p, \min\{d_{p+1}, c_{p+1}\}\}$$

and

$$b_{p+1} = 10^{\beta_{p+1}} = \min\{b_p, \max\{d_{p+1}, c_{p+1}\}\}.$$

Computation of the Knots

We compute the knots $x_{1,p+1}, \ldots, x_{k,p+1}$ successively as follows. For computing the first knot $x_{1,p+1}$, we apply the algorithm of Remez and determine points y_1 and y_2 such that

$$d(f, P_m, [x_{0,p+1}, y_1]) \leq d_{p+1} < d(f, P_m, [x_{0,p+1}, y_2]).$$

We now apply the regula falsi method for the exponents $\gamma_1, \delta_{p+1}, \gamma_2$ by setting

$$y_3 = y_1 + (y_2 - y_1) \frac{\delta_{p+1} - \gamma_1}{\gamma_2 - \gamma_1},$$

where $d(f, P_m, [x_{0,p+1}, y_1]) = 10^{\gamma_1}$ and $d(f, P_m, [x_{0,p+1}, y_2]) = 10^{\gamma_2}$.
Then either

$$d(f, P_m, [x_{0,p+1}, y_1]) \leq d_{p+1} < d(f, P_m, [x_{0,p+1}, y_3])$$

or

$$d(f, P_m, [x_{0,p+1}, y_3]) \leq d_{p+1} < d(f, P_m, [x_{0,p+1}, y_2]).$$

In the first (respectively second) case we compute γ_4 analogously to above by regula falsi for the exponents $\gamma_1, \delta_{p+1}, \gamma_3$ (respectively $\gamma_3, \delta_{p+1}, \gamma_2$), where $d(f, P_m, [x_{0,p+1}, y_3]) = 10^{\gamma_3}$, and proceed by induction. By continuing this method, we obtain a sequence (y_n) which converges to the knot $x_{1,p+1}$. (In practice we stop after a few steps.) Having computed the knot $x_{1,p+1}$, we determine the other knots $x_{2,p+1}, \ldots, x_{k,p+1}$ analogously.

We now discuss some details concerning this algorithm.

Remark 4.23. (i) By statement (i) in Theorem 4.22 we obtain inclusions of the minimal deviation $d(f, PP_{m,k})$. In the first step we have

$$a_1 \leq d(f, PP_{m,k}) \leq b_1.$$

Then by definition of the sequences (a_p) and (b_p) we get

$$a_p \leq d(f, PP_{m,k}) \leq b_p$$

for all $p = 2, 3, \ldots$. This shows that in general it suffices to compute the knot sets

$$\{x_{1,p}, \ldots, x_{k,p}\}$$

only approximately and we still get inclusions of the value $d(f, PP_{m,k})$ in each step (compare Theorem 4.24).

(ii) Also in the computation of the knots we get inclusions in each step. If, for example, we compute the knot $x_{1,p+1}$ as above, then we have

$$y_1 \leq x_{1,p+1} \leq y_2.$$

Then it follows that

$$y_1 \leq x_{1,p+1} \leq y_3$$

or

$$y_3 \leq x_{1,p+1} \leq y_2,$$

where $y_1 \leq y_3 \leq y_2$. We obtain analogous inclusions in the further steps.

(iii) In particular, it follows from (i) that it suffices in practice to compute the values $d(f, P_m, I)$ for the knot interval I which appears only approximately, e.g. by interpolation at the Chebyshev points (compare Section 2 of Chapter I).

The next result shows that the sequence (d_p) converges to the minimal deviation $d(f, PP_{m,k})$.

Theorem 4.24. *For the value $d(f, PP_{m,k}) = 10^\delta$ and the sequence $d_p = 10^{\delta_p}$ $(p = 2, 3, \ldots)$, the following statements hold:*

(i) $|\delta - \delta_p| \leq \dfrac{1}{2} |\alpha_{p-1} - \beta_{p-1}| \leq \cdots \leq \dfrac{1}{2^{p-1}} |\alpha_1 - \beta_1|, \quad p = 2, 3, \ldots$

(ii) $\lim\limits_{p \to \infty} d_p = d(f, PP_{m,k}).$

Proof. (i). It follows from Theorem 4.22, (i) that

$$a_1 \leq d(f, PP_{m,k}) \leq b_1$$

and

$$\min\{d_p, c_p\} \leq d(f, PP_{m,k}) \leq \max\{d_p, c_p\}.$$

Therefore, by the defintion of a_{p-1} and b_{p-1} we have

$$a_{p-1} \leq d(f, PP_{m,k}) \leq b_{p-1}. \tag{4.31}$$

Moreover, by the defintion of d_p we have

$$a_{p-1} \leq d_p \leq b_{p-1}. \tag{4.32}$$

This implies that if $d_p = \min\{d_p, c_p\}$ (respectively $d_p = \max\{d_p, c_p\}$), then $a_p = \max\{a_{p-1}, d_p\} = d_p$ (respectively $b_p = \min\{b_{p-1}, d_p\} = d_p$). Moreover, if $a_p = d_p$ (respectively $b_p = d_p$), then $\beta_p - \alpha_p \leq \beta_{p-1} - \alpha_p = \beta_{p-1} - \delta_p = \beta_{p-1} - \frac{1}{2}(\alpha_{p-1} + \beta_{p-1}) = \frac{1}{2}(\beta_{p-1} - \alpha_{p-1})$ (respectivly $\beta_p - \alpha_p \leq \beta_p - \alpha_{p-1} = \delta_p - \alpha_{p-1} = \frac{1}{2}(\alpha_{p-1} + \beta_{p-1}) - \alpha_{p-1} = \frac{1}{2}(\beta_{p-1} - \alpha_{p-1}))$. This implies that

$$|\alpha_p - \beta_p| \leq \frac{1}{2} |\alpha_{p-1} - \beta_{p-1}|. \tag{4.33}$$

Finally, it follows from (4.31)–(4.33) and the definition of d_p that

$$|\delta - \delta_p| \leq \frac{1}{2} |\alpha_{p-1} - \beta_{p-1}| \leq \cdots \leq \frac{1}{2^{p-1}} |\alpha_1 - \beta_1|.$$

(ii). This statement follows directly from (i). This proves Theorem 4.24.

We now discuss some further details concerning the convergence of the algorithm.

Remark 4.25. (i) Theorem 4.25 shows that the sequence (d_p) converges relatively fast to $d(f, PP_{m,k})$, since the difference of the corresponding exponents $|\delta - \delta_p|$ is monotonely decreasing at least by the factor $1/2$ in each step.

(ii) It is easy to verify that the knot sequence $(x_{1,p}, \ldots, x_{k,p})$ $(p = 2, 3, \ldots)$ converges to a leveled set of knots $(x_1, \ldots, x_k)$ if there exists a unique leveled set of knots for f. A more general convergence result is given in Nürnberger, Sommer & Strauß [1986]. By using Theorem 3.22 of Chapter I, it can be shown that for functions f with $f^{(n+1)} \neq 0$ for all $t \in (a, b)$, there exists a unique optimal set of knots (see e.g. Meinardus [1967]).

In the following we describe a combination of the above algorithm and the algorithm in Section 4.2 which yields good spline approximations for free knots. (Note, that the above algorithm only yields discontinuous piecewise polynomials, in general.)

We denote by $S_{m,k}$ the set of splines of degree m with k free knots. Roughly speaking, this is the set of all splines of degree m with k knots, where the knots may be chosen arbitrarily. (A precise definition is given in Section 1 of the Appendix.)

Best approximation by splines from $S_m(x_1, \ldots, x_k)$ is a linear problem. However, $S_{m,k}$ is a nonconvex set and therefore, best approximation by splines from $S_{m,k}$ leads to a nonlinear problem. The standard algorithms (Newton methods and tangent methods), if they converge, in general only yield local best approximations (compare Cromme [1976] and Hettich & Zencke [1982]). At present, there is no algorithm for computing global best approximations from $S_{m,k}$.

We will describe a method for computing good global approximations from $S_{m,k}$ (see Nürnberger [1986] and Meinardus, Nürnberger, Sommer & Strauß [1988]). In particular, the subsequent numerical results show that by this method we obtain approximations which in general are much better than best approximations from $S_m(x_1, \ldots, x_k)$ for equidistant knots.

Algorithm for Good Spline Approximations
with Free Knots

Let a function $f \in C[a, b]$ be given. We construct the spline approximation in two steps.

In the first step we apply to f the algorithm for computing best uniform approximations from $PP_{m,k}$. In this way we obtain a leveled set of knots $\{\overline{x}_1, \ldots, \overline{x}_k\}$ for f. This set of knots reflects the "critical" parts of f. Because, if the knot interval $[\overline{x}_{i+1}, \overline{x}_i]$ is relatively large (respectively small), then f behaves well (respectively badly) in this part.

In the second step we apply to f the algorithm of Section 4.2 for computing best uniform approximations from $S_m(\overline{x}_1, \ldots, \overline{x}_k)$, where $\{\overline{x}_1, \ldots, \overline{x}_k\}$ is the above leveled set of knots for f. Then we obtain a spline which in general is a good or nearly best approximation with respect to $S_{m,k}$. (Note, that for $m \geq 2$ the resulting spline is differentiable.)

By using the estimate

$$d(f, PP_{m,k}) \leq d(f, S_{m,k}) \leq d(f, S_m(\overline{x}_1, \ldots, \overline{x}_k)),$$

we can control how efficient the resulting spline approximation is. Since, in general, $d(f, PP_{m,k})$ is strictly smaller than $d(f, S_{m,k})$ (because no continuity is required for the functions from $PP_{m,k}$), the relevant error

$$d(f, S_m(\overline{x}_1, \ldots, \overline{x}_k)) - d(f, S_{m,k})$$

is strictly smaller than the value

$$d(f, S_m(\overline{x}_1, \ldots, \overline{x}_k)) - d(f, PP_{m,k}),$$

which can be computed.

Moreover, by comparing the value $d(f, S_m(\overline{x}_1, \ldots, \overline{x}_k))$ with the value $d(f, S_m(x_1, \ldots, x_k))$, we can control how much the improvement is, if we use the knots $\overline{x}_1, \ldots, \overline{x}_k$ instead of equidistant knots $x_1, \ldots, x_k$ which would be the simplest choice.

Numerical Examples 4.26. We give numerical results on best uniform approximation by cubic splines and piecewise polynomials with free knots. In particular, we compare the minimal deviations $d(f, PP_{m,k})$, $d(f, S_m(\overline{x}_1, \ldots, \overline{x}_k))$ and $d(f, S_m(x_1, \ldots, x_k))$, where $\{\overline{x}_1, \ldots, \overline{x}_k\}$ is a leveled set of knots for f w.r.t. $PP_{m,k}$ and $\{x_1, \ldots, x_k\}$ is the set of equidistant knots.

Table 1. Minimal deviations for $f(t) = t^{1/2}$ on $[0, 2]$

k	$d(f, PP_{3,k})$	$d(f, S_3(\overline{x}_1, \ldots, \overline{x}_k))$ (optimal knots for $PP_{3,k}$)	$d(f, S_3(x_1, \ldots, x_k))$ (equidistant knots)
0	6.50e-2	6.50e-2	6.50e-2
2	4.56e-3	9.14e-3	4.42e-2
4	9.89e-4	2.87e-3	3.45e-2
6	3.34e-4	1.11e-3	2.92e-2
8	1.41e-4	5.07e-4	2.58e-2
10	7.10e-5	2.62e-4	2.33e-2
12	3.91e-5	1.48e-4	2.14e-2

Table 2. Minimal deviations for $f(t) = 1/(1 + t^2)$ on $[-5, 5]$ (*Runge's function*)

k	$d(f, PP_{3,k})$	$d(f, S_3(\overline{x}_1, \ldots, \overline{x}_k))$ (optimal knots for $PP_{3,k}$)	$d(f, S_3(x_1, \ldots, x_k))$ (equidistant knots)
0	3.23e-1	3.23e-1	3.23e-1
2	1.42e-2	1.97e-1	2.32e-1
4	2.86e-3	8.11e-2	1.62e-1
6	4.54e-4	1.23e-2	1.06e-1
8	2.08e-4	3.89e-3	6.74e-2
10	7.02e-5	9.83e-4	4.19e-2
12	3.66e-5	4.09e-4	2.59e-2

Table 3. Leveled set of knots for best uniform approximation of $f(t) = t^{1/2}$ by $PP_{3,k}$ on $[0,2]$

k	2	4	6	8	10	12
	9.76e-3	4.68e-4	5.32e-5	9.74e-6	2.41e-6	7.35e-7
	2.30e-1	1.10e-2	1.26e-3	2.30e-4	5.68e-5	1.73e-5
		9.58e-2	1.09e-2	1.99e-3	4.93e-4	1.51e-4
		5.09e-1	5.80e-2	1.06e-2	2.62e-3	7.99e-4
			2.28e-1	4.16e-2	1.03e-2	3.14e-3
			7.28e-1	1.33e-1	3.29e-2	1.00e-2
				3.66e-1	9.04e-2	2.76e-2
				8.95e-1	2.21e-1	6.75e-2
					4.95e-1	1.51e-1
					1.03e-0	3.13e-1
						6.10e-1
						1.13e-0

Table 4. Leveled set of knots for best uniform approximation of $f(t) = 1/(1+t^2)$ by $PP_{3,k}$ on $[-5,5]$

k	2	4	6	8	10	12
	-0.6670	-1.660	-2.4667	-2.7591	-3.1578	-3.372
	0.6670	-0.405	-0.9792	-1.5172	-2.0484	-2.335
		0.405	-0.3543	-0.7429	-1.0719	-1.567
		1.660	0.1600	-0.2014	-0.6318	-0.935
			0.8120	0.2093	-0.1774	-0.586
			2.4502	0.7468	0.1340	-0.171
				1.5296	0.6249	0.094
				2.7773	1.0620	0.586
					2.0493	0.935
					3.1588	1.569
						2.337
						3.375

By comparing the errors $d(f, S_3(\bar{x}_1, \ldots, \bar{x}_k))$ and $d(f, S_3(x_1, \ldots, x_k))$, we see that there is an improvement up to a factor 144 if we use optimal knots for $PP_{3,k}$ instead of equidistant knots. In addition, by taking into consideration that $d(f, PP_{3,k})$ is strictly less than $d(f, S_{3,k})$, we may conclude that the value $d(f, S_3(\bar{x}_1, \ldots, \bar{x}_k))$ is relatively near to the optimal value $d(f, S_{3,k})$. Finally, if we compare for $f(t) = t^{1/2}$, $t \in [0,2]$, the values $d(f, S_3(\bar{x}_1, \ldots, \bar{x}_k))$ with the best results in the book of de Boor [1978], p.187-217 obtained by interpolation and quasi–interpolation using good knot placement algorithms, we see that best approximation methods yield an improvement up to the factor 23.

We close this section by discussing further possibilities for an appropriate choice of knots. By the above algorithm we compute for an arbitrary function

$f \in C[a, b]$, a set of knots $a = x_0 < x_1 < \cdots < x_k < x_{k+1} = b$ such that

$$d(x_{i-1}, x_i) = d(x_i, x_{i+1}), \quad i = 1, \ldots, k,$$

where $d(x_i, x_{i+1}) = d(f, P_m, [x_i, x_{i+1}])$, $i = 0, \ldots, k$. In the case of functions f which satisfy certain differentiability properties, we may use the following approach. It follows from Theorem 3.30 in Chapter I that

$$d(f, P_m, [x_i, x_{i+1}]) \leq K (x_{i+1} - x_i)^j \|f^{(j)}\|_{[x_i, x_{i+1}]} \|_\infty, \quad i = 0, \ldots, k,$$

for functions $f \in C^j[a, b]$ $(1 \leq j \leq m)$. For this reason, Dodson [1972], de Boor [1973], [1974b] and Burchard [1974] (see also de Boor [1978]) suggested (aside from other methods) to use knots $a = x_0 < x_1 < \cdots < x_k < x_{k+1} = b$ such that

$$d(x_{i-1}, x_i) = d(x_i, x_{i+1}), \quad i = 1, \ldots, k,$$

where

$$d(x_i, x_{i+1}) = (x_{i+1} - x_i)^j \|f^{(j)}\|_{[x_i, x_{i+1}]} \|_\infty, \quad i = 0, \ldots, k,$$

as fixed knots for the approximation of f by splines. It was shown by Meinardus, Nürnberger, Sommer & Strauß [1988] that a variant of the above algorithm for piecewise polynomials can be applied to solve such nonlinear systems. These authors also developed various methods for computing good knots for spline approximation.

We also note that the algorithm for piecewise polynomials in this section can be applied to general *segment approximation* problems (see Nürnberger, Sommer & Strauß [1986]). Further segment approximation methods were developed by Lawson [1964], Pavlidis & Maika [1974], Kioustelidis [1980] and McLaughlin & Zacharski [1980].

4.4. Approximation Power of Splines

In this section we give a result on the approximation power of splines which, in particular, shows that the minimal deviation of a function $f \in C^{m+1}[a, b]$ from $S_m(x_1, \ldots, x_k)$ is up to some constant less or equal to the $(m+1)$-th power of the mesh size $\delta = \max\{|x_{i+1} - x_i| : i = 0, \ldots, k\}$.

The minimal deviation of a function $f \in C[a, b]$ from the space of splines $S_m(x_1, \ldots, x_k)$ in the uniform norm is defined by

$$d_\infty(f, S_m(x_1, \ldots, x_k)) = \inf_{s \in S_m(x_1, \ldots, x_k)} \|f - s\|_\infty.$$

Theorem 4.27. *Let integers $m \geq 1$ and $j \in \{0, \ldots, m\}$ be given. Then there exists a constant $K > 0$ (depending only on m and j) such that for every function $f \in C^j[a, b]$ and every set of knots $\{x_1, \ldots, x_k\}$, we have*

$$d_\infty(f, S_m(x_1, \ldots, x_k)) \leq K \delta^j \omega(f^{(j)}; \delta), \tag{4.34}$$

where $\delta = \max\{|x_{i+1} - x_i| : i = 0, \ldots, k\}$. In particular, if $f \in C^{j+1}[a, b]$, then

$$\omega(f^{(j)}; \delta) \leq \delta \|f^{(j+1)}\|_\infty. \tag{4.35}$$

Proof. We first prove the following

Claim. For every $f \in C[a, b]$, we have

$$d_\infty(f, S_m(x_1, \ldots, x_k)) \leq \frac{1}{2}(m+1)\omega(f; \delta). \qquad (4.36)$$

For proving the claim, we set for every index $i \in \{-m, \ldots, k\}$,

$$t_i = \begin{cases} \frac{1}{2}(x_i + x_{i+m+1}), & \text{if } \frac{1}{2}(x_i + x_{i+m+1}) \in [a, b] \\ a, & \text{if } \frac{1}{2}(x_i + x_{i+m+1}) < a \\ b, & \text{if } \frac{1}{2}(x_i + x_{i+m+1}) > b \end{cases}.$$

Let a function $f \in C[a, b]$ be given. We define $s \in S_m(x_1, \ldots, x_k)$ by

$$s(t) = \sum_{i=-m}^{k} f(t_i) N_i^m(t), \qquad t \in [a, b],$$

where for all $i \in \{-m, \ldots, k\}$ the function N_i^m is the normalized B–spline of degree m with support (x_i, x_{i+m+1}) (see Definition 2.13). It suffices to show that

$$\|f - s\|_\infty \leq \frac{1}{2}(m+1)\omega(f; \delta).$$

Let $t \in [a, b]$ be given. Since by Theorem 2.15

$$\sum_{i=-m}^{k} N_i^m(t) = 1, \qquad (4.37)$$

it follows that

$$f(t) - s(t) = \sum_{i=-m}^{k} (f(t) - f(t_i)) N_i^m(t).$$

Now, let $i \in \{-m, \ldots, k\}$ be given. If $t \notin (x_i, x_{i+m+1})$, then

$$(f(t) - f(t_i)) N_i^m(t) = 0.$$

If $t \in (x_i, x_{i+m+1})$, then by definition of the point t_i we have

$$|t - t_i| \leq \frac{1}{2}(x_i + x_{i+m+1}) \leq \frac{1}{2}(m+1)\delta$$

and therefore,

$$|f(t) - f(t_i)| |N_i^m(t)| \leq \omega(f; \frac{1}{2}(m+1)\delta) \cdot N_i^m(t) = \frac{1}{2}(m+1)\omega(f; \delta) \cdot N_i^m(t).$$

Then it follows from (4.37) that for all $t \in [a, b]$,

$$|f(t) - s(t)| \leq \sum_{i=-m}^{k} |f(t) - f(t_i)| |N_i^m(t)| \leq \frac{1}{2}(m+1)\omega(f; \delta)$$

which proves the claim.

We now proceed by induction as follows. Suppose that (4.34) holds for $j - 1 \le m - 2$. Let a function $f \in C^j[a, b]$ be given. Then it follows that for all $s \in S_m(x_1, \ldots, x_k)$,

$$
\begin{aligned}
d_\infty(f, S_m(x_1, \ldots, x_k)) &= d_\infty(f - s, S_m(x_1, \ldots, x_k)) \\
&\le K_{j-1}\,\delta^{j-1}\,\omega(f^{(j-1)} - s^{(j-1)}; \delta) \\
&\le K_{j-1}\,\delta^j\,\|f^{(j)} - s^{(j)}\|_\infty,
\end{aligned}
$$

where $K_{j-1} = K$ denotes the constant in (4.34). Since

$$
S_{m-j}(x_1, \ldots, x_k) = \{s^{(j)} : s \in S_m(x_1, \ldots, x_k)\},
$$

we get

$$
\begin{aligned}
d_\infty(f, S_m(x_1, \ldots, x_k)) &\le K_{j-1}\,\delta^j\,d_\infty(f^{(j)}, S_{m-j}(x_1, \ldots, x_k)) \\
&\le K_{j-1}\,\delta^j\,\frac{1}{2}(m+1)\omega(f^{(j)}; \delta).
\end{aligned}
$$

Finally, we consider the case when $f \in C^m[a, b]$. We have already shown that

$$
d_\infty(f, S_m(x_1, \ldots, x_k)) \le K_{m-2}\,\delta^{m-1}\,d_\infty(f^{(m-1)}, S_1(x_1, \ldots, x_k)).
$$

By using arguments similar to above, if we consider each interval $[x_i, x_{i+1}]$, $i = 0, \ldots, k$, separately, it is easy to verify that

$$
d_\infty(f^{(m-1)}, S_1(x_1, \ldots, x_k)) \le \frac{1}{2}\delta\,\omega(f^{(m)}; \delta).
$$

This proves Theorem 4.27.

A detailed study of the approximation power of splines with fixed or free knots is given in the book of Schumaker [1981].

5. Continuity of the Set Valued Metric Projection for Spline Spaces

The operator of best approximation, called metric projection, is a set valued mapping, in general. We investigate various continuity concepts of this mapping. In particular, it is shown that w.r.t the uniform norm, the set valued metric projection onto a given spline space is always upper semicontinuous, never lower semicontinuous, and that there exists a continuous selection if and only if the number of knots is less than or equal to the degree of splines plus one.

5.1. Upper Semicontinuity

It is shown that the metric projection onto a given finite–dimensional subspace of a normed linear space is upper semicontinuous.

We first recall the definition of the metric projection. Let G be a subset of a real normed linear space E. The set valued mapping $P_G : E \to POW(G)$ which associates to every $f \in E$, the set $P_G(f)$ of the best approximations of f from G is called the *metric projection* onto G, where $POW(G)$ denotes the set of all subsets of G.

In the following we consider continuity properties of the metric projection. Since it is a set valued mapping, in general, we have to say what we mean by continuity. There are various concepts which will be investigated.

We begin with the notion of upper semicontinuity.

Definition 5.1. The metric projection $P_G : E \to POW(G)$ is called *upper semicontinuous* if the set

$$\{f \in E : P_G(f) \cap A \neq \emptyset\} \tag{5.1}$$

is closed for every closed subset A of G.

If $P_G(f)$ is a singleton for every $f \in E$, then P_G is upper semicontinuous if and only if $P_G : E \to G$ is continuous in the usual sense.

The next well-known result shows that the metric projection onto a finite–dimensional subspace is upper semicontinuous (see Singer [1970]).

Theorem 5.2. *Let G be a finite–dimensional subspace of a real normed linear space E. Then the metric projection $P_G : E \to POW(G)$ is upper semicontinuous.*

Proof. Let A be a closed subset of G. We will show that

$$B = \{f \in E : P_G(f) \cap A \neq \emptyset\}$$

is a closed subset of E. Therefore, let (f_n) be a sequence in B converging to some element $f \in E$. Since $(f_n) \subset B$, we may choose an element $g_n \in P_G(f_n) \cap A$ for all n. The space G is finite–dimensional and therefore, by passing to a subsequence, we may assume that (g_n) converges to some element $g_0 \in G$. By Theorem 3.7 in Chapter I the mapping $f \to d(f, G)$ $(f \in E)$ is continuous. Thus, it follows that

$$\|f_n - g_n\| = d(f_n, G) \to d(f, G)$$

and

$$\|f_n - g_n\| \to \|f - g_0\|.$$

This implies that $\|f - g_0\| = d(f, G)$, i.e. $g_0 \in P_G(f)$. Since A is closed, $(g_n) \subset A$ and $g_n \to g_0$, it follows that $g_0 \in P_G(f) \cap A$ which implies that $f \in B$. This proves Theorem 5.2.

5.2. Lower Semicontinuity

It is shown that the metric projection onto spline spaces is never lower semicontiuous w.r.t. the uniform norm.

We begin with the definition of lower semicontinuity.

Definition 5.3. The metric projection $P_G : E \to POW(G)$ is called *lower semicontinuous* if the set

$$\{f \in E : P_G(f) \cap A \neq \emptyset\} \tag{5.2}$$

is open for every open subset A of G.

In contrast to Theorem 5.2, the metric projection is not lower semicontinuous, in general. The next result actually shows that the metric projection onto spline spaces is never lower semicontinuous.

Theorem 5.4. *The metric projection* $P_{S_m(x_1,...,x_k)} : C[a,b] \to POW(S_m(x_1, ..., x_k))$ *is not lower semicontinuous (where $C[a,b]$ is endowed with the uniform norm).*

Proof. We will construct a function f and a sequence (f_n) in $C[a,b]$ converging to f such that $0, s_0 \in P_{S_m(x_1,...,x_k)}(f)$, where $s_0 \neq 0$, and for all n,

$$P_{S_m(x_1,...,x_k)}(f_n) = \{0\}.$$

Then it is easy to see that $P_{S_m(x_1,...,x_k)}$ is not lower semicontinuous. Let $f \in C[a,b]$ be a function with the following properties: $\|f\|_\infty = 1$; f has $m+2$ alternating extreme points in each interval $[x_0, x_1], ..., [x_{k-1}, x_k]$; $f(x_k) = 1$, $f(x_{k+1}) = -1$ and f is linear between these alternating extreme points and on $[x_k, x_{k+1}]$. We now define $s_0 \in S_m(x_1, ..., x_k)$ by

$$s_0(t) = -\frac{1}{(x_{k+1} - x_k)^m} (t - x_k)_+^m, \qquad t \in [a,b].$$

Then by Theorem 4.2 we get that $0 \in P_{S_m(x_1,...,x_k)}(f)$ and, since $\|f - s_0\|_\infty = 1 = \|f\|_\infty$, that $s_0 \in P_{S_m(x_1,...,x_k)}(f)$. Moreover, it is easy to verify that by small perturbations of f we obtain a sequence (f_n) converging to f such that for sufficiently large n, $\|f_n\|_\infty = 1$ and f_n has the same alternating extreme points as f, except that the extreme point x_k is shifted to $x_k + \frac{1}{n}$. Then condition (ii) in Theorem 4.4 is satisfied which implies that $P_{S_m(x_1,...,x_k)}(f_n) = \{0\}$ for all n. This proves Theorem 5.4.

In contrast to Theorem 5.4, it was proved by Berens & Nürnberger [1987] that there exists a dense, open subset of $C[a,b]$ on which $P_{S_m(x_1,...,x_k)}$ is lower semicontinuous. Blatter [1967] proved that the metric projection $P_G : C[a,b] \to POW(G)$ is not lower semicontinuous w.r.t. the uniform norm for every finite–dimensional non–Chebyshev subspace G of $C[a,b]$. For more results on the lower semicontinuity, see e.g. Blatter, Morris & Wulbert [1968], Brown [1964], Brosowski & Wegman [1973], Blatter & Schumaker [1982], and the survey article of Nürnberger & Sommer [1985].

5.3. Continuous Selections

We give a characterization of those spline spaces which admit a continuous selection for the corresponding metric projection w.r.t. the uniform norm.

Definition 5.5. A continuous mapping $F : E \to G$ is called a *continuous selection* for $P_G : E \to POW(G)$ if $F(f) \in P_G(f)$ for all $f \in C[a,b]$.

It follows from the selection theorem of Michael [1956] for general set valued mappings that lower semicontinuity is a sufficient condition for the existence of a continuous selection for the metric projection. However, Theorem 5.4 shows that Michael's theorem is not applicable to spline spaces. Nevertheless, the following characterization due to Nürnberger & Sommer [1978b] shows that there may exist continuous selections.

Let G be an n-dimensional subspace of $C[\alpha, \beta]$ and $f \in C[\alpha, \beta]$. A best uniform approximation $g_f \in G$ of f is called an *alternation element*, briefly denoted by A-*element*, if $A(f - g_f)|_{[\alpha,\beta]} \geq n + 1$ (compare Definition 4.1.).

Theorem 5.6. *There exists a continuous selection for the metric projection* $P_{S_m(x_1,...,x_k)} : C[a,b] \to POW(S_m(x_1,...,x_k))$ *if and only if* $k \leq m + 1$ *(where* $C[a,b]$ *is endowed with the uniform norm).*

Proof. Necessity. Suppose that $k > m + 1$. We will construct a function f and sequences (f_n) and $(\tilde{f}_n)$ in $C[a,b]$ converging to f such that

$$P_{S_m(x_1,...,x_k)}(f_n) = \{0\} \qquad \text{for all } n,$$

and

$$P_{S_m(x_1,...,x_k)}(\tilde{f}_n) = \{s_0\} \qquad \text{for all } n,$$

where $s_0 \neq 0$. Then obviously, there does not exist a continuous selection for $P_{S_m(x_1,...,x_k)}$.

We set $s_0 = (1/\|B_1^m\|_\infty) B_1^m$, where B_1^m is the B-spline with support (x_1, x_{m+2}). By Theorem 2.5 the spline s_0 has a unique maximum in (x_1, x_{m+2}) which we denote by $\tilde{x}$. It is easy to construct a function $f \in C[a,b]$ with the following properties: $\|f\|_\infty = 1$; f has $m + 2$ alternating extreme points in each interval $[x_0, x_1], [x_{m+2}, x_{m+3}], \ldots, [x_k, x_{k+1}]$; f is linear between these alternating extreme points, except on $[x_1, x_{m+2}]$; $f(x_1) = 1$ and $f(x_{m+2}) = -1$. Moreover, we define f on $[x_1, x_{m+2}]$ as follows: $f(\tilde{x}) = 0$; f is linear on $[x_1, \tilde{x}]$ and $f(t) = s_0(t) - 1$ for all $t \in [\tilde{x}, x_{m+2}]$. It follows from Theorem 4.2 that $0 \in P_{S_m(x_1,...,x_k)}(f)$ and, since $\|f - s_0\|_\infty = 1 = \|f\|_\infty$, that $s_0 \in P_{S_m(x_1,...,x_k)}(f)$. Moreover, it is easy to verify that by small perturbations of f we obtain a sequence (f_n) converging to f such that for sufficiently large n, $\|f_n\|_\infty = 1$ and f_n has the same alternating extreme points as f, except that the extreme point x_1 (respectively x_{m+2}) is shifted to $x_1 + \frac{1}{n}$ (respectively $x_{m+2} - \frac{1}{n}$). Analogously, we obtain a sequence $(\tilde{f}_n)$ such that for sufficiently large n, $\|\tilde{f}_n - s_0\|_\infty = 1$ and $\tilde{f}_n - s_0$ has the same alternating extreme points as $f - s_0$, except that the extreme point x_1 (respectively x_{m+2})

is shifted to $x_1 + \frac{1}{n}$ (respectively $x_{m+2} - \frac{1}{n}$). Now, it is easy to see that for sufficiently large n, the error $f_n - 0$ and $\tilde{f}_n - s_0$ satisfy condition (ii) in Theorem 4.4 which implies that

$$P_{S_m(x_1,\dots,x_k)}(f_n) = \{0\}$$

and

$$P_{S_m(x_1,\dots,x_k)}(\tilde{f}_n) = \{s_0\}.$$

Sufficiency. We assume that $k \le m+1$, and show that there exists a continuous selection $F : C[a, b] \to S_m(x_1, \dots, x_k)$ in this case.

Construction of the Selection F

Let $f \in C[a, b]$ be given and $g_0 \in S_m(x_1, \dots, x_k)$ be a best uniform approximation of f. By Theorem 4.3 there exists an interval $[x_p, x_{p+1}]$ such that all best uniform approximations of f from $S_m(x_1, \dots, x_k)$ coincide with g_0 on $[x_p, x_{p+1}]$. The construction of the spline $F(f)$ is based on local alternation elements. For this, we define the spaces

$$G_j = \{B^m_{p+j}, \dots, B^m_k\}, \qquad j = 1, \dots, k - p.$$

By Theorem 2.8 all these spaces are weak Chebyshev. Therefore, by Theorem 1.8 there exists a best uniform approximation g_1 of $f - g_0$ from G_1 on $[x_{p+1}, x_{k+1}]$ which is an A–element. We now define further functions $g_2, \dots, g_{k-p}$ inductively as follows. For each $j \in \{2, \dots, k - p\}$, let g_j be a best uniform approximation of $f - g_0 - \cdots - g_{j-1}$ from G_j on $[x_{p+j}, x_{k+1}]$ which is an A–element. Moreover, we define the spaces

$$G_j = \{B^m_{-m}, \dots, B^m_{p-m+j}\}, \qquad j = -p, \dots, -1.$$

Analogously to above, let g_{-1} be a best uniform approximation of $f - g_0$ from G_{-1} on $[x_0, x_p]$ which is an A–element. We now define $g_{-2}, \dots, g_{-p}$ inductively as follows. For each $j \in \{-2, \dots, -p\}$, let g_j be a best uniform approximation of $f - g_0 - g_{-1} - \cdots - g_{j-1}$ from G_j on $[x_0, x_{p+j+1}]$. Having constructed these functions, we now set

$$F(f) = g_{-p} + g_{-p+1} + \cdots + g_{-1} + g_0 + g_1 + \cdots + g_{k-p}.$$

It is easy to verify that $F(f)$ is a best uniform approximation of f from $S_m(x_1, \dots, x_k)$. There is some freedom in the construction of $F(f)$: the choice of the interval $[x_p, x_{p+1}]$, the best approximation g_0 and the local A–elements g_j. We will show that in spite of this freedom, the function $F(f)$ is uniquely determined by the above construction.

We first show that any local A–element g_1 above is uniquely determined on $[x_{p+1}, x_{p+2}]$. Let $\tilde{g}_1 \in G_1$ be an arbitrary A–element of $f - g_0$ on $[x_{p+1}, x_{k+1}]$. We will show that $\tilde{g}_1 = g_1$ on $[x_{p+1}, x_{p+2}]$. By the definition of g_1 and $\tilde{g}_1$ there exist $\sigma, \tilde{\sigma} \in \{-1, 1\}$, $u_1 < \cdots < u_{k-p+1}$ and $v_1 < \cdots < v_{k-p+1}$ in $[x_{p+1}, x_{k+1}]$ such that

$$\sigma(-1)^i \left(f(u_i) - g_0(u_i) - g_1(u_i) \right) = \|f - g_0 - g_1\|_\infty, \quad i = 1, \dots, k - p + 1, \quad (5.3)$$

and

$$\tilde{\sigma}(-1)^i \left(f(v_i) - g_0(v_i) - \tilde{g}_1(v_i) \right) = \|f - g_0 - \tilde{g}_1\|_\infty, \quad i = 1, \ldots, k-p+1. \quad (5.4)$$

Then it follows from (5.3) that

$$\sigma(-1)^i \left(f(u_i) - g_0(u_i) - g_1(u_i) \right) = \|(f - g_0 - g_1)|_{[x_{p+1}, x_{k+1}]}\|_\infty$$
$$= \|(f - g_0 - \tilde{g}_1)|_{[x_{p+1}, x_{k+1}]}\|_\infty \geq \sigma(-1)^i \left(f(u_i) - g_0(u_i) - \tilde{g}_1(u_i) \right),$$

which implies that

$$\sigma(-1)^i \left(\tilde{g}_1(u_i) - g_1(u_i) \right) \geq 0, \qquad i = 1, \ldots, k-p+1. \quad (5.5)$$

Analogously, it follows from (5.4) that

$$\tilde{\sigma}(-1)^i \left(\tilde{g}_1(v_i) - g_1(v_i) \right) \leq 0, \qquad i = 1, \ldots, k-p+1. \quad (5.6)$$

Case 1. $u_i = v_i$ for all $i \in \{1, \ldots, k-p+1\}$.
Then it follows from (5.5) and (5.6) that $u_2, \ldots, u_{k-p+1}$ are $k-p$ distinct zeros of $\tilde{g}_1 - g_1$ in $(x_{p+1}, x_{k+1}]$. By Theorem 3.3 this is only possible if $\tilde{g}_1 - g_1$ vanishes on some subinterval $[x_i, x_{i+1}]$ of $[x_{p+1}, x_{k+1}]$. Since $k \leq m+1$ and $\tilde{g}_1 - g_1 = 0$ on $[x_p, x_{p+1}] \cup [x_i, x_{i+1}]$, it follows from Remark 2.4 that $\tilde{g}_1 - g_1 = 0$ on $[x_p, x_{i+1}]$. In particular, $\tilde{g}_1 = g_1$ on $[x_{p+1}, x_{p+2}]$.
Case 2. $u_j \neq v_j$ for some $j \in \{1, \ldots, k-p+1\}$.
In this case there exist a sign $\delta \in \{-1, 1\}$ and points $w_1 < \cdots < w_{k-p+1}$ in $(x_{p+1}, x_{k+1}]$ such that

$$\delta(-1)^i \left(\tilde{g}_1(w_i) - g_1(w_i) \right) \geq 0, \qquad i = 1, \ldots, k-p+1. \quad (5.7)$$

This is obvious if $x_p < u_i$ or $x_p < v_i$. But, if $x_{p+1} = u_1 = v_1$, then it follows from (5.3) and (5.4) that $\sigma = \tilde{\sigma}$. We may assume that $u_j < v_j$. Then by (5.5) and (5.6), the sign $\delta = \sigma$ and the points $u_2 < \cdots < u_j < v_j < \cdots < v_{k-p+1}$ have the desired property. We now add arbitrary points $w_{-m} < w_{-m+1} < \cdots < w_0$ from $[x_p, x_{p+1}]$ to $w_1 < \cdots < w_{k-p+1}$ and obtain $k-p+m+2$ points. Note, that $\dim S_m(x_1, \ldots, x_k)|_{[x_p, x_{k+1}]} = k-p+m+1$. Analogously to the proof of (i) $\Rightarrow$ (ii) in Theorem 4.2, there exists a subinterval $[x_r, x_{r+q}]$ of $[x_p, x_{k+1}]$, $[x_r, x_{r+q}] \neq [x_p, x_{p+1}]$ which contains points $y_1 < \cdots < y_{q+m+1}$ from $\{w_{-m}, w_{-m+1}, \ldots, w_{k-p+1}\}$ such that for some $\tilde{\delta} \in \{-1, 1\}$,

$$\tilde{\delta}(-1)^i \left(\tilde{g}_1(y_i) - g_1(y_i) \right) \geq 0, \qquad i = 1, \ldots, q+m+1,$$

and for all $i \in \{1, \ldots, q+m+1\}$, the set

$$\{y_1, \ldots, y_{i-1}, y_{i+1}, \ldots, y_{q+m+1}\}$$

is poised with respect to $S_m(x_1, \ldots, x_k)|_{[x_r, x_{r+q}]}$. Since $S_m(x_1, \ldots, x_k)|_{[x_r, x_{r+q}]}$ is an $(q+m)$-dimensional weak Chebyshev space, it follows from Lemma 1.11 that

$\tilde{g}_1 - g_1 = 0$ on $[x_r, x_{r+q}]$. As in Case 1, we finally get that $\tilde{g}_1 = g_1$ on $[x_{p+1}, x_{p+2}]$. This proves the uniqueness of A–elements on $[x_{p+1}, x_{p+2}]$.

We now show that $g_0 + g_1$ as above is uniquely determined on $[x_p, x_{p+2}]$, independent of the choice of g_0. Let $\tilde{g}_0 \in S_m(x_1, \ldots, x_k)$ be an arbitrary best uniform approximation of f and $\tilde{g}_1$ be an A–element of $f - \tilde{g}_0$ from G_1 on $[x_{p+1}, x_{k+1}]$. We will show that

$$\tilde{g}_0 + \tilde{g}_1 = g_0 + g_1 \quad \text{on } [x_p, x_{p+2}].$$

Since $\tilde{g}_1, g_1 \in G_1$ and all best uniform approximations of f coincide on $[x_p, x_{p+1}]$, we have

$$\tilde{g}_0 + \tilde{g}_1 = g_0 + g_1 \quad \text{on } [x_p, x_{p+1}].$$

Moreover, since $\tilde{g}_1 \in G_1$ is an A–element of $f - \tilde{g}_0$ on $[x_{p+1}, x_{k+1}]$ and $f - g_0 - (\tilde{g}_0 + \tilde{g}_1 - g_0) = f - \tilde{g}_0 - \tilde{g}_1$, the function $(\tilde{g}_0 + \tilde{g}_1 - g_0)|_{[x_{p+1}, x_{k+1}]} \in G_1|_{[x_{p+1}, x_{k+1}]}$ is an A–element of $f - g_0$ on $[x_{p+1}, x_{k+1}]$. Then it follows from the above proved uniqueness of A–elements that $\tilde{g}_0 + \tilde{g}_1 - g_0 = g_1$ on $[x_{p+1}, x_{p+2}]$.

By proceeding with this method, we can prove uniqueness of the local A–elements on $[x_{p+2}, x_{p+3}], \ldots, [x_k, x_{k+1}]$ and then on $[x_{p-1}, x_p] \ldots, [x_0, x_1]$, which implies that the function $F(f)$ is independent of the choice of the best approximation g_0 and the local A–elements g_j.

Finally, we show that $F(f)$ is also independent of the choice of the starting interval $[x_p, x_{p+1}]$. Let $[x_p, x_{p+1}]$ and $[x_q, x_{q+1}]$ be two intervals on which all best uniform approximations of f from $S_m(x_1, \ldots, x_k)$ coincide, where $p + 1 < q$. Moreover, let two best approximations $g_0, \tilde{g}_0 \in S_m(x_1, \ldots, x_k)$ of f be given. Then we have $g_0 - \tilde{g}_0 = 0$ on $[x_p, x_{p+1}] \cap [x_q, x_{q+1}]$. Therefore, since $k \leq m + 1$, it follows from Remark 2.4 that $g_0 - \tilde{g}_0 = 0$ on $[x_p, x_{q+1}]$. This implies that there exists a unique maximal knot-interval I_f on which all best uniform approximations of f from $S_m(x_1, \ldots, x_k)$ coincide. Therefore, we may start our construction on any subinterval $[x_r, x_{r+1}]$ of I_f. This shows that the selection F is well defined.

Continuity of the Selection F

Suppose that $F : C[a, b] \to S_m(x_1, \ldots, x_k)$ is not continuous, i.e. there exists a function $f \in C[a, b]$ and a sequence $(f_n) \subset C[a, b]$ converging to f such that the sequence $(F(f_n))$ does not converge to $F(f)$. Since $S_m(x_1, \ldots, x_k)$ is finite–dimensional, we may assume that $(F(f_n))$ converges to a best uniform approximation $s \in S_m(x_1, \ldots, x_k)$, $s \neq F(f)$, of f. We will show that this assumption leads to a contradiction. By passing to a subsequence, we may assume that there exists a knot-interval I such that $I_{f_n} = I$ for all n. We first show that there exists an interval $[x_p, x_{p+1}]$ which is contained in $I \cap I_f$. Again passing to a subsequence, it follows from the proof of (i) $\Rightarrow$ (ii) in Theorem 4.2 that there exists an interval $[x_r, x_{r+q}] \subset I$ such that for all n, $f_n - F(f_n)$ has $m + q + 1$ alternating extreme points $t_{1,n} < \cdots < t_{m+q+1,n}$ in $[x_r, x_{r+q}]$. Thus, there exists a sign $\sigma \in \{-1, 1\}$ such that for all n,

$$\sigma(-1)^i \left(f_n(t_{i,n}) - F(f_n)(t_{i,n})\right) = \|f_n - F(f_n)\|_\infty, \qquad i = 1, \ldots, m + q + 1.$$

We may assume that $t_{i,n} \to t_i$, $i = 1, \ldots, m+q+1$. Then taking limits, it follows that

$$\sigma \, (-1)^i \, (f(t_i) - s(t_i)) = \|f - s\|_\infty, \qquad i = 1, \ldots, m+q+1.$$

Again by (i) $\Rightarrow$ (ii) of the proof of Theorem 4.2 it follows that there exists an interval $[x_p, x_{p+1}] \subset [x_r, x_{r+q}]$ on which all best uniform approximations of f from $S_m(x_1, \ldots, x_k)$ coincide. It follows that $[x_p, x_{p+1}] \subset I \cap I_f$. Therefore, in the construction of the function $F(f)$ and the sequence $(F(f_n))$, we may choose $[x_p, x_{p+1}]$ as the starting interval. In this way, we get

$$F(f) = g_{-p} + \cdots + g_0 + \cdots + g_{k-p}$$

and

$$F(f_n) = g_{-p,n} + \cdots + g_{0,n} + \cdots + g_{k-p,n} \qquad \text{for all } n.$$

By passing to a subsequence, we may assume that for all $j \in \{-p, \ldots, k-p\}$ the sequence $(g_{j,n})$ converges to some function $\tilde{g}_j \in G_j$. Since $(F(f_n))$ converges to s, we have

$$s = \tilde{g}_{-p} + \cdots + \tilde{g}_0 + \cdots + \tilde{g}_{k-p}.$$

Since for all n, the spline $g_{0,n}$ is a best uniform approximation of f_n, it follows that $\tilde{g}_0$ is a best uniform approximation of f. Moreover, the alternation properties of $f_n - F(f_n)$ carry over to $f - s$. Then it follows from the above proved uniqueness of $F(f)$ that $s = F(f)$ which is the desired contradiction. This proves the continuity of F and Theorem 5.6.

In contrast to Theorem 5.6, it was proved by Berens & Nürnberger [1987] that for $k > m+1$, there exists a selection for $P_{S_m(x_1,\ldots,x_k)}$ which is continuous on a dense open subset of $C[a,b]$. Blatt, Nürnberger & Sommer [1981] showed that the selection in Theorem 5.6 is even pointwise–Lipschitz–continuous (i.e. for each $f \in C[a,b]$ there exists a constant $C_f > 0$ such that for all $\tilde{f} \in C[a,b]$, $\|F(f) - F(\tilde{f})\|_\infty \leq C_f \|f - \tilde{f}\|_\infty$). Blatter & Schumaker [1982],[1983] proved that there is never a *unique* continuous selection for the metric projection onto spline spaces.

In a series of papers, Nürnberger & Sommer gave a characterization of those arbitrary finite–dimensional subspaces G of $C[a,b]$ which admit continuous selections for P_G w.r.t. the uniform norm (see the survey paper of Nürnberger & Sommer [1985], where also remarks on the relationship between continuous selections and the convergence of algorithms are given). In particular, Nürnberger [1980] proved that the existence of a continuous selection for P_G implies that G is weak Chebyshev. A more general characterization was given by Li [1986]. Additional results on continuous selections can be found in the survey article of Deutsch [1983].

6. Best L_1–Approximation by Weak Chebyshev Spaces

We investigate best L_1–approximation by weak Chebyshev spaces with special emphasis on spline spaces. It is shown that every continuous function has a unique best L_1-approximation from a given spline space. Moreover, we show that under certain assumptions best L_1-approximations from weak Chebyshev spaces are uniquely determined by Lagrange interpolation at canonical points.

6.1. Unicity of Best L_1-Approximations

We show that every continuous (periodic) function has a unique best L_1-approximation from a given space of (periodic) splines.

In Section 4.1 of Chapter I we have seen that best L_1-approximations from Chebyshev spaces are always unique. This result does not hold for the more general class of weak Chebyshev spaces. On the other hand, the following theorem due to Galkin [1974] and Strauß [1975a] shows that global unicity of best L_1-approximations holds for spline spaces. Extensions of this result were given by Carroll & Braess [1974], Sommer [1979], [1983a], Strauß [1981], Nürnberger, Schumaker, Sommer & Strauß [1985] and Króo [1984] (see also Pinkus [1987]).

Theorem 6.1. *For every function in $C[a, b]$, there exists a unique best L_1-approximation from $S_m(x_1, \ldots, x_k)$.*

Proof. Suppose to the contrary that there exists a function $f \in C[a, b]$ such that 0 and a nontrivial spline $s_0 \in S_m(x_1, \ldots, x_k)$ are best L_1-approximations of f. Then it follows from the proof of Theorem 4.4 in Chapter I that

$$Z(f - \frac{1}{2} s_0) \subset Z(s_0). \tag{6.1}$$

Claim. There exists a nontrivial spline $s \in S_m(x_1, \ldots, x_k)$ such that

$$(f(t) - \frac{1}{2} s_0(t)) s(t) \geq 0, \qquad t \in [a, b], \tag{6.2}$$

and

$$Z(f - \frac{1}{2} s_0) \subset Z(s), \tag{6.3}$$

where (6.3) holds except possibly for finitely many points.

If the claim is true, then it follows from the proof of Theorem 4.4 in Chapter I that $\frac{1}{2} s_0 \in S_m(x_1, \ldots, x_k)$ is not a best L_1-approximation of f which contradicts the fact that 0 and s_0 are best L_1-approximations of f. Therefore, it remains to prove the claim. We distinguish the following cases.

Case 1. s_0 has only finitely many zeros in $[a, b]$.

It follows from Corollary 3.4 that s_0 has at most $k + m$ distinct zeros. Then by (6.1) the function $f - \frac{1}{2} s_0$ has at most $k + m$ sign changes. This implies that there exist points

$$a = t_0 < t_1 < \cdots < t_{r-1} < t_r = b,$$

where $r \in \{1, \ldots, k + m + 1\}$, and a sign $\sigma \in \{-1, 1\}$ such that

$$\sigma (-1)^i (f(t) - \frac{1}{2} s_0(t)) \geq 0, \qquad t \in [t_{i-1}, t_i], \quad i = 1, \ldots, r.$$

By Corollary 1.7 there exists a nontrivial $s \in S_m(x_1, \ldots, x_k)$ such that

$$\sigma\,(-1)^i\, s(t) \geq 0, \qquad t \in [t_{i-1}, t_i], \quad i = 1, \ldots, r.$$

Therefore, the spline s is the desired function.

Case 2. There exists an interval $[x_v, x_w]$ such that $s_0 = 0$ on $[x_v, x_w]$ and s_0 has only finitely many zeros in $[a, b] \setminus [x_v, x_w]$.

We may assume that $v > 0$. (The other case follows analogously.) By Theorem 2.8 the space

$$S_1 = \mathrm{span}\{B^m_{-m}, \ldots, B^m_{v-m-1}\}$$

is a v– dimensional weak Chebyshev space. Since $s_0 = 0$ on $[x_v, x_w]$, the spline s_0 belongs to S_1 on $[x_0, x_w]$. Therefore, by Theorem 3.3 the spline s_0 has at most $v - 1$ distinct zeros on $[x_0, x_v)$. Then by (6.1) the function $f - \frac{1}{2} s_0$ has at most $v - 1$ sign changes on $[x_0, x_v]$. Analogously to Case 1, by Corollary 1.7 there exists a nontrivial function $s \in S_1$ such that

$$(f(t) - \frac{1}{2}\, s_0(t))\, s(t) \geq 0, \qquad t \in [x_0, x_v].$$

Since $s = 0$ on $[x_v, x_{k+1}]$, the spline s is the desired function.

Case 3. There exist intervals $[x_u, x_v]$ and $[x_w, x_z]$, $v < w$, such that $s_0 = 0$ on $[x_u, x_v] \cup [x_w, x_z]$ and s_0 has only finitely many zeros in (x_v, x_w).

We first note that $w - v \geq m + 1$, otherwise by Remark 2.4 $s_0 = 0$ on $[x_v, x_w]$. By Theorem 2.8 the space

$$S_2 = \mathrm{span}\{B^m_v, \ldots, B^m_{w-m-1}\}$$

is a $(w - v - m)$–dimensional weak Chebyshev space. Since $s_0 = 0$ on $[x_u, x_v] \cup [x_w, x_z]$, the spline s_0 belongs to S_2 on $[x_u, x_z]$. Therefore, by Theorem 3.3 the spline s_0 has at most $w - v - m - 1$ distinct zeros on (x_v, x_w). Then by (6.1) the function $f - \frac{1}{2} s_0$ has at most $w - v - m - 1$ sign changes on $[x_v, x_w]$. Analogously to Case 1, by Corollary 1.7 there exists a nontrivial function $s \in S_2$ such that

$$(f(t) - \frac{1}{2}\, s_0(t))\, s(t) \geq 0, \qquad t \in [x_v, x_w],$$

Since $s = 0$ on $[x_0, x_v] \cup [x_w, x_{k+1}]$, the spline s has the desired property. This proves Theorem 6.1.

In connection with Theorem 6.1, we state without proof a further unicity result for periodic splines.

We denote by

$$P_m(x_1, \ldots, x_k) = \{s \in S_m(x_1, \ldots, x_k): \; s^{(j)}(a) = s^{(j)}(b), \; j = 0, \ldots, m-1\}$$

the $(k+1)$–dimensional space of periodic splines of degree m with knots $x_1, \ldots, x_k$. It was proved by Meinardus & Nürnberger [1988] that the following result on unicity of best L_1–approximations from $P_m(x_1, \ldots, x_k)$ holds (although $P_m(x_1, \ldots, x_k)$ is not a weak Chebyshev space for odd k).

Theorem 6.2. *For every periodic function $f \in C[a, b]$ (i.e. $f(a) = f(b)$), there exists a unique best L_1–approximation from $P_m(x_1, \ldots, x_k)$.*

6.2. Interpolation at Canonical Points

The relationship between best L_1–approximation and Lagrange interpolation at canonical points is investigated. In particular, it is shown that for a function from the convexity cone, its best L_1–approximations from a given spline space is uniquely determined by interpolation conditions.

We begin with a result on the existence of canonical points for weak Chebyshev spaces due to Micchelli [1977].

Theorem 6.3. *For every n–dimensional weak Chebyshev subspace of $C[a, b]$, there exists a set of n canonical points.*

Proof. Let $G = \text{span}\{g_1, \ldots, g_n\}$ be an n–dimensional weak Chebyshev subspace of $C[a, b]$. By Theorem 1.3 there exists a sequence

$$G_m = \text{span}\{g_{1,m}, \ldots, g_{n,m}\}$$

of n–dimensional Chebyshev subspaces of $C[a, b]$ such that

$$\lim_{m \to \infty} \|g_i - g_{i,m}\|_\infty = 0, \qquad i = 1, \ldots, n.$$

Moreover, by Therorem 4.7 in Chapter I, for each m, there exists a set $\{t_{1,m}, \ldots, t_{n,m}\}$ of canonical points of G_m, i.e.

$$\sum_{i=0}^{n}(-1)^i \int_{t_{i,m}}^{t_{i+1,m}} g_{j,m}(t)\, dt = 0, \qquad j = 1, \ldots, n,$$

where $t_{0,m} = a$ and $t_{n+1,m} = b$. By passing to a subsequence, we may assume that

$$\lim_{m \to \infty} t_{i,m} = t_i, \qquad i = 0, \ldots, n + 1.$$

Then by taking limits, it follows that

$$\sum_{i=0}^{n}(-1)^i \int_{t_i}^{t_{i+1}} g_j(t)\, dt = 0, \qquad j = 1, \ldots, n,$$

i.e. $\{t_1, \ldots, t_n\}$ is a set of canonical points of G. This proves Theorem 6.3.

Theorem 4.7 in Chapter I says that the canonical points for a given Chebyshev space are uniquely determined. This result does not hold for the more general class of weak Chebyshev spaces. Micchelli [1977] developed sufficient conditions for the uniqueness and poisedness of canonical points. We state the following version due to Sommer [1979].

Theorem 6.4. *Let G be an n–dimensional weak Chebyshev subspace of $C[a, b]$ such that for every function in $C[a, b]$, there exists a unique best L_1–approximation from G. Then there exists a unique set of n canonical points of G which is poised with respect to G.*

Proof. Since G is weak Chebyshev, it follows from Theorem 6.3 that there exists a set $\{t_1, \ldots, t_n\}$ of canonical points of G. We first show that $\{t_1, \ldots, t_n\}$ is poised with respect to G. Suppose the contrary, that this is not true. Then there exists a nontrivial $g_0 \in G$ such that

$$g_0(t_i) = 0, \qquad i = 1, \ldots, n.$$

By replacing g_0 by $-g_0$, if necessary, we may assume that

$$(-1)^j g_0(t) > 0 \quad \text{for some interval } [x_j, x_{j+1}].$$

Let $f \in C[a, b]$ be defined by

$$f(t) = \begin{cases} 2 \max\{g_0(t), 0\}, & t \in [t_i, t_{i+1}], \text{ if } i = 0 \text{ or } i \text{ is even} \\ 2 \min\{g_0(t), 0\}, & t \in [t_i, t_{i+1}], \text{ if } i \text{ is odd} \end{cases}$$

Then it follows that

$$(-1)^i (f(t) - g_0(t)) \geq 0, \qquad t \in [t_i, t_{i+1}], \quad i = 0, \ldots, n,$$

and

$$(-1)^i f(t) \geq 0, \qquad t \in [t_i, t_{i+1}], \quad i = 0, \ldots, n.$$

Then, since $\{t_1, \ldots, t_n\}$ is a set of canonical points of G, for all $g \in G$,

$$\begin{aligned}
\|f - g_0\|_1 &= \int_a^b |f(t) - g_0(t)|\, dt \\
&= \sum_{i=0}^n (-1)^i \int_{t_i}^{t_{i+1}} (f(t) - g_0(t))\, dt \\
&= \sum_{i=0}^n (-1)^i \int_{t_i}^{t_{i+1}} (f(t) - g(t))\, dt \\
&\leq \int_a^b |f(t) - g(t)|\, dt = \|f - g\|_1.
\end{aligned}$$

Analogously, we conclude that for all $g \in G$,

$$\|f\|_1 \leq \|f - g\|_1.$$

This shows that f has two distinct best L_1-approximations g_0 and 0 from G which contradicts the assumption.

Finally, the proof that $\{t_1, \ldots, t_n\}$ is the only set of canonical points of G is completely analogous to the proof of Theorem 4.7 in Chapter I. We only have to use the fact that G is weak Chebyshev and $\{t_1, \ldots, t_n\}$ is poised with respect to G. This proves Theorem 6.4.

The next corollary, due to Micchelli [1977], is an immediate consequence of Theorem 6.1 and Theorem 6.4. An extension of this result was obtained by Nürnberger, Schumaker, Sommer & Strauß [1985].

Corollary 6.5. *For the space $S_m(x_1, \ldots, x_k)$, there exists a unique set of $k + m + 1$ canonical points which is poised with respect to $S_m(x_1, \ldots, x_k)$.*

We have already seen in Section 4.2 of Chapter I that for Chebyshev spaces, there is a special relationship between best L_1-approximation and interpolation. Micchelli [1977] proved that best L_1-approximations for functions from the convexity cone of weak Chebyshev spaces can be obtained by Lagrange interpolation.

If G is a weak Chebyshev subspace of $C[a, b]$, then the set

$$K(G) = \{f \in C[a, b] : \text{span } (G \cup \{f\}) \text{ is a weak Chebyshev subspace of } C[a, b]\}$$

is called the *convexity cone* of G.

Theorem 6.6. *Let G be an n–dimensional weak Chebyshev subspace of $C[a, b]$. If the set $\{t_1, \ldots, t_n\}$ of canonical points of G is poised with respect to G, then every function $f \in K(G)$ has a unique best L_1-approximation g_f from G and g_f is uniquely determined by*

$$g_f(t_i) = f(t_i), \qquad i = 1, \ldots, n.$$

Proof. By using that G is weak Chebyshev, that the set $\{t_1, \ldots, t_n\}$ is poised with respect to G and $f \in K(G)$, the proof of Theorem 4.9 in Chapter I shows that for all $g \in G$,

$$\|f - g_f\|_1 \leq \|f - g\|_1.$$

Moreover, the equality sign holds if and only if

$$\sigma(-1)^i (f(t) - g(t)) \geq 0, \qquad t \in [t_i, t_{i+1}], \quad i = 0, \ldots, n.$$

This implies that

$$g(t_i) = f(t_i) = g_f(t_i), \qquad i = 1, \ldots, n.$$

Since $\{t_1, \ldots, t_n\}$ is poised with respect to G, we have $g = g_f$. Thus, it follows that g_f is the unique best L_1-approximation of f from G. This proves Theorem 6.6.

By combining Theorem 6.1, Theorem 6.4 and Theorem 6.6, we obtain the following corollary on spline spaces due to Micchelli [1977]. Extensions of this result were proved by Sommer [1979] and Nürnberger, Schumaker, Sommer & Strauß [1985].

Corollary 6.7. *For every function in $K(S_m(x_1, \ldots, x_k))$, the best L_1-approximation from $S_m(x_1, \ldots, x_k)$ is uniquely determined by Lagrange interpolation at the canonical points.*

7. Best One–Sided L_1–Approximation by Weak Chebyshev Spaces and Quadrature Formulas

We have already seen in Section 5.2 of Chapter I that there is a close relationship between best one–sided L_1–approximation by Chebyshev spaces and quadrature formulas. It will be shown that similar results hold for the more general class of weak Chebyshev spaces. We first examine the unicity of best one–sided L_1–approximations. Then the existence and uniqueness of Gauss quadrature formulas for weak Chebyshev spaces is investigated. In particular, it is shown that for spline spaces, there exist unique Gauss quadrature formulas. Moreover, it is shown that under certain assumptions best one–sided L_1–approximations from a weak Chebyshev space are uniquely determined by Hermite interpolation at points of Gauss quadrature formulas.

7.1. Unicity of Best One–Sided L_1–Approximations

It is shown that for continuous functions, best one–sided L_1–approximations from weak Chebyshev spaces are not always unique. On the other hand, global unicity holds for best one–sided L_1–approximations from spline spaces if the functions to be approximated are differentiable.

We begin with the following nonunicity result for weak Chebyshev spaces due to Strauß [1982]. (Compare the remark after Theorem 5.3 in Chapter I on the nonunicity for arbitary finite–dimensional spaces.)

Theorem 7.1. *Let $n \geq 2$ and G be an n–dimensional weak Chebyshev subspace of $C[a, b]$ which contains a strictly positive function. Then there exists a function in $C[a, b]$ which has more than one best one–sided L_1–approximation from G.*

Proof. Let $n \geq 2$ and G be an n–dimensional weak Chebyshev subspace of $C[a, b]$ which contains a strictly positive function. By Theorem 7.4 below, there exists a Gauss quadrature formula for G with points $a \leq t_1 < \cdots < t_r < b$, where $r \leq n - 1$. Since $r \leq n - 1$ and G is an n–dimensional space, there exists a nontrivial function $g_1 \in G$ such that

$$g_1(t_i) = 0, \qquad i = 0, \ldots, r.$$

We can now apply the proof of Theorem 5.3 in Chapter I which shows that 0 and g_1 are best one–sided L_1–approximations of $f = |g_1|$ from G. This proves Theorem 7.1.

In order to prove a unicity result for spline spaces, we need the following statement.

Lemma 7.2. *Let G be a n–dimensional weak Chebyshev subspace of $C[a, b]$ and points $a \leq t_1 < \cdots < t_r \leq b$ with index $I(\{t_1, \ldots, t_r\}) < \frac{n}{2}$ be given. Then there exists a nontrivial function $g \in G$ such that $g \geq 0$ and*

$$g(t_i) = 0, \qquad i = 1, \ldots, r.$$

Proof. Let points $a \leq t_1 < \cdots < t_r \leq b$ with

$$I(\{t_1, \ldots, t_r\}) < \frac{n}{2} \qquad (7.1)$$

be given. If $t_i \in \{a, b\}$, then we associate with t_i a sequence $(t_{i,m})$ converging to t_i such that $t_{i,m} \neq t_i$ for all m. If $t_i \in (a, b)$, then we associate with t_i two sequences $(u_{i,m})$ and $(v_{i,m})$ converging to t_i such that $u_{i,m} < t_i < v_{i,m}$ for all m. We denote the resulting sequences by $(w_{j,m})$, $j = 1, \ldots, q$, such that for sufficiently large m,

$$a = w_{0,m} < w_{1,m} < \cdots < w_{q,m} < w_{q+1,m} = b.$$

It follows from (7.1) and the choice of the sequences that $q \leq n - 1$. Then by Corollary 1.7, for each m, there exists a nontrivial $g_m \in G$ such that

$$(-1)^i g_m(t) \geq 0, \qquad t \in [w_{i-1,m}, w_{i,m}], \quad i = 1, \ldots, q + 1. \qquad (7.2)$$

By multiplying each g_m with $1/\|g_m\|_\infty$, we may assume that $\|g_m\|_\infty = 1$. Since G is finite–dimensional, there exists a subsequence of (g_m) converging to a nontrivial function $g \in G$. By taking limits, it follows from (7.2) and the definition of the sequence $(w_{j,m})$ that $g \geq 0$ and

$$g(t_i) = 0, \qquad i = 1, \ldots, r.$$

This proves Lemma 7.2.

In contrast to Theorem 7.1, the following unicity theorem for spline spaces due to Pinkus [1976] holds. Extensions of this result were given by Sommer & Strauß [1981], Strauß [1982] and Nürnberger, Schumaker, Sommer & Strauß [1985].

Theorem 7.3. *If $m \geq 2$, then for every function in $C^1[a, b]$, there exists a unique best one–sided L_1–approximation from $S_m(x_1, \ldots, x_k)$.*

Proof. We will show that condition (ii) in Theorem 5.4 in Chapter I is satisfied, which proves Theorem 7.3. Let a nontrivial spline $s_0 \in S_m(x_1, \ldots, x_k)$ be given. We have to show that there exists a nontrivial spline $s \in S_m(x_1, \ldots, x_k)$ such that $s \geq 0$ and $Z_1(s_0) \subset Z_1(s)$. To do this, we distinguish the following cases.
Case 1. s_0 has only finitely many zeros in $[a, b]$.
It follows from Corollary 3.4 that $I(Z_1(s_0)) < \frac{k+m+1}{2}$. Then by Lemma 7.2 there exists a nontrivial $s \in S_m(x_1, \ldots, x_k)$ such that $s \geq 0$ and $s(t) = 0$, $t \in Z_1(s_0)$. This implies that $Z_1(s_0) \subset Z_1(s)$. Therefore, the spline s is the desired function.
Case 2. There exists an interval $[x_v, x_w]$ such that $s_0 = 0$ on $[x_v, x_w]$ and s_0 has only finitely many zeros in $[a, b] \setminus [x_v, x_w]$.
We may assume that $v > 0$. (The other case follows analogously.) By Theorem 2.8 the space

$$S_1 = \text{span}\{B_{-m}^m, \ldots, B_{v-m-1}^m\}$$

is a v–dimensional weak Chebyshev space. Since $s_0 = 0$ on $[x_v, x_w]$, the spline s_0 belongs to S_1 on $[x_0, x_w]$. Therefore, it follows from Theorem 3.3 that

$$I(Z_1(s_0) \cap [x_0, x_v)) < \frac{v}{2}.$$

Then by Lemma 7.2 there exists a nontrivial function $s \in S_1$ such that $s \geq 0$ and

$$s(t) = 0, \qquad t \in Z_1(s_0) \cap [x_0, x_v).$$

Since $s = 0$ on $[x_v, x_{k+1}]$, we have $Z_1(s_0) \subset Z_1(s)$. Therefore, the spline s is the desired function.

Case 3. There exist intervals $[x_u, x_v]$ and $[x_w, x_z]$, $v < w$, such that $s_0 = 0$ on $[x_u, x_v] \cup [x_w, x_z]$ and s_0 has only finitely many zeros in (x_v, x_w).
We first note that $w - v \geq m + 1$, otherwise by Remark 2.4 $s_0 = 0$ on $[x_v, x_w]$. By Theorem 2.8 the space

$$S_2 = \text{span}\{B_v^m, \ldots, B_{w-m-1}^m\}$$

is a $(w - v - m)$–dimensional weak Chebyshev space. Since $s_0 = 0$ on $[x_u, x_v] \cup [x_w, x_z]$, the spline s_0 belongs to S_2 on $[x_u, x_z]$. Therefore, it follows from Theorem 3.3 that

$$I(Z_1(s_0) \cap (x_v, x_w)) < \frac{w - v - m}{2}.$$

Then by Lemma 7.2 there exists a nontrivial function $s \in S_2$ such that $s \geq 0$ and

$$s(t) = 0, \qquad t \in Z_1(s_0) \cap (x_v, x_w).$$

Since $s = 0$ on $[x_0, x_v] \cup [x_w, x_{k+1}]$, we have $Z_1(s_0) \subset Z_1(s)$. Therefore, the spline s is the desired function. This proves Theorem 7.3

7.2. Gauss Quadrature Formulas for Weak Chebyshev Spaces

It is shown that Gauss quadrature formulas exist for weak Chebyshev spaces which contain a strictly positive function. The uniqueness of such quadrature formulas can be proved under an additional assumption which is satisfied by spline spaces. Moreover, we examine the relationship between best one–sided L_1–approximation and Hermite interpolation. In particular, it is shown that for a differentiable function from the convexity cone, its best one–sided L_1–approximations from spline spaces can be computed by solving such an interpolation problem.

We first investigate the existence of Gauss quadrature formulas for weak Chebyshev spaces. The following existence result is due to Micchelli & Pinkus [1977]. (For the definition of Gauss quadrature formulas see Section 5.2 in Chapter I.)

Theorem 7.4. *Let G be a weak Chebyshev subspace of $C[a, b]$ which contains a strictly positive function. Then there exists a Gauss quadrature formula for G.*

Proof. Let $G = \text{span}\{g_1, \ldots, g_n\}$ be an n–dimensional weak Chebyshev subspace of $C[a, b]$. By Theorem 1.3 there exists a sequence $G_m = \text{span}\{g_{1,m}, \ldots, g_{n,m}\}$ of Chebyshev subspace of $C[a, b]$ such that

$$\lim_{m \to \infty} \|g_i - g_{i,m}\|_\infty = 0, \qquad i = 1, \ldots, n. \tag{7.3}$$

Then it follows from Theorem 5.9 in Chapter I that for each m, there exists a Gauss quadrature formula

$$Q_m(f) = \sum_{i=1}^{r} a_{i,m} f(t_{i,m})$$

of G_m, where $a \le t_{1,m} < \cdots < t_{r,m} < b$ and $a_{1,m}, \ldots, a_{r,m} > 0$. Since there exists a function $g_0 \in G$ with $g_0 > 0$, it follows from (7.3) that for all sufficiently large m, there exists a function $g_{0,m} \in G_m$ with $\lim_{m \to \infty} \|g_0 - g_{0,m}\|_\infty = 0$ and $g_{0,m} \ge C > 0$. This implies that for such m,

$$2 \int_a^b g_0(t)\, dt \ge \int_a^b g_{0,m}(t)\, dt = \sum_{i=1}^{r} a_{i,m}\, g_{0,m}(t_{i,m})$$

$$\ge C \sum_{i=1}^{r} a_{i,m}$$

and it follows that

$$\sum_{i=1}^{r} a_{i,m} \le \frac{2}{C} \int_a^b g_0(t)\, dt.$$

This implies that

$$\lim_{m \to \infty} a_{i,m} = a_i, \qquad i = 1, \ldots, r.$$

By taking subsequences we may assume that

$$\lim_{m \to \infty} t_{i,m} = t_i, \qquad i = 1, \ldots, r.$$

Since for all m,

$$\int_a^b g_{j,m}(t)\, dt = \sum_{i=1}^{r} a_{j,m}\, g_{j,m}(t_{i,m}), \qquad j = 1, \ldots, n,$$

by taking limits, it follows that

$$\int_a^b g_j(t)\, dt = \sum_{i=1}^{r} a_j\, g_j(t_i), \qquad j = 1, \ldots, n.$$

Moreover, we have $a \le t_1 < \cdots < t_r < b$, $a_1, \ldots, a_r > 0$ and $I(\{t_1, \ldots, t_r\}) = n/2$. Suppose that this is not true. If $a_j = 0$ for some $j \in \{1, \ldots, r\}$, then we omit the point t_j. Then by Lemma 7.2 there exists a nontrivial $g \in G$ such that $g \ge 0$ and $g(t_i) = 0$, $i = 1, \ldots, r$. This implies that

$$0 < \int_a^b g(t)\, dt = \sum_{i=1}^{r} a_i\, g(t_i) = 0,$$

which is a contradiction. This shows that

$$Q(f) = \sum_{i=1}^{r} a_i \, f(t_i)$$

is a Gauss quadrature formula for G and proves Theorem 7.4.

We now examine the uniqueness of Gauss quadrature formulas. To do this, we need the following result on interpolation due to Strauß [1982].

Lemma 7.5. *Let G be an n–dimensional weak Chebyshev subspace of $C^1[a, b]$ which contains a strictly positive function such that for every function in $C^1[a, b]$, there exists a unique best one–sided L_1–approximation from G. Moreover, let $a \le t_1 < \cdots < t_r < b$ be the points of a Gauss quadrature formula for G. Then for each $f \in C^1[a, b]$, there exists a unique $g_f \in G$ such that*

$$g_f(t_i) = f(t_i), \qquad i = 1, \ldots, r, \tag{7.4}$$

and

$$g_f'(t_i) = f'(t_i), \qquad t_i \neq a, \quad i = 1, \ldots, r. \tag{7.5}$$

Proof. Suppose to the contrary, that there exists a nontrivial $g_0 \in G$ such that

$$g_0(t_i) = 0, \qquad i = 1, \ldots, r,$$

and

$$g_0'(t_i) = 0, \qquad \text{if } t_i \neq a, \quad i = 1, \ldots, r.$$

We will show that there exists a function $h \in C^1[a, b]$ for which 0 and g_0 are two best one–sided L_1–approximations from G. We set $\tilde{h} = |g_0|$. If $\tilde{h} \in C^1[a, b]$, then we define $h = \tilde{h}$. If $\tilde{h} \notin C^1[a, b]$, then by smoothing $\tilde{h}$ we construct a function $h \in C^1[a, b]$ as follows. Let $t \in [a, b]$ be a point at which $\tilde{h}$ is not differentiable. Then it follows that $t \in \left(Z(\tilde{h}) \setminus Z_1(\tilde{h}) \right) \cap (a, b)$. Since G is weak Chebyshev and G contains a strictly positive function $g_1 \in G$, it follows that $Z(\tilde{h}) \setminus Z_1(\tilde{h})$ is finite. Otherwise, there exists a sufficiently small ε such that $g_0 + \varepsilon \, g_1 \in G$ has more that $n - 1$ sign changes. We set

$$Z(\tilde{h}) \setminus Z_1(\tilde{h}) = \{t_1, \ldots, t_p\}.$$

Then there exist sufficiently small neighborhoods U_i of t_i, $i = 1, \ldots, p$, such that $Z_1(\tilde{h}) \cap \bigcup_{i=1}^{p} U_i = \emptyset$. By smoothing $\tilde{h}$, we obtain a function $h \in C^1[a, b]$ such that

$$h(t) = \tilde{h}(t), \qquad t \in [a, b] \setminus \bigcup_{i=1}^{p} U_i,$$

and

$$h(t) \ge \tilde{h}(t), \qquad t \in \bigcup_{i=1}^{p} U_i.$$

Then we have for all $g \in G$ with $g \leq h$,

$$\int_a^b g(t)\,dt = \sum_{i=1}^r a_i\, g(t_i) \leq \sum_{i=1}^r a_i\, h(t_i) = \sum_{i=1}^r a_i\, g_0(t_i) = \int_a^b g_0(t)\,dt.$$

This proves Lemma 7.5.

Micchelli & Pinkus [1977] showed that under certain assumptions, Gauss quadrature formulas for weak Chebyshev spaces are uniquely determined. We state the following version due to Strauß [1982].

Theorem 7.6. *Let G be an n–dimensional weak Chebyshev subspace of $C^1[a,b]$ which contains a strictly positive function such that for every function in $C^1[a,b]$, there exists a unique best one–sided L_1–approximation from G. Then there exists a unique Gauss quadrature formula for G.*

Proof. Let G be a subspace as in Theorem 7.6. Then it follows from Theorem 7.4 that there exists a Gauss quadrature formula for G. Suppose that there exist two different Gauss quadrature formulas

$$Q_1(f) = \sum_{i=1}^r a_i\, f(t_i)$$

and

$$Q_2(f) = \sum_{i=1}^r b_i\, f(u_i)$$

for G. Then we may assume that $t_1 = u_1, \ldots, t_{p-1} = u_{p-1}$ and $t_p < u_p$. We choose a real number $\varepsilon > 0$ such that $t_p + \varepsilon < u_p$. Then we argue similarly to the proof of Lemma 7.2, except that we replace the point t_p by $t_p + \varepsilon$ and choose only one sequence converging to $t_p + \varepsilon$. Then we obtain a nontrivial function $g \in G$ such that

$$g(t) \geq 0, \qquad t \in [a, t_p + \varepsilon], \tag{7.6}$$

$$g(t) \leq 0, \qquad t \in [t_p + \varepsilon, b], \tag{7.7}$$

$$g(t_i) = 0, \qquad i = 1, \ldots, r, \quad i \neq p, \tag{7.8}$$

$$g'(t_i) = 0, \qquad i = 1, \ldots, r, \quad t_i \neq a. \tag{7.9}$$

Then we have

$$g(t_p) > 0, \tag{7.10}$$

otherwise it follows from (7.6) that $g(t_p) = g'(t_p) = 0$ and from (7.8), (7.9) and Lemma 7.5 that $g = 0$ which is a contradiction. Finally, it follows from (7.6)–(7.10) that

$$
\begin{aligned}
0 &= \int_a^b g(t)\,dt - \int_a^b g(t)\,dt = \sum_{i=1}^r a_i\, g(t_i) - \sum_{i=1}^r b_i\, g(u_i) \\
&= a_p\, g(t_p) - \sum_{i=1}^r b_i\, g(u_i) > 0,
\end{aligned}
$$

which is a contradiction. This proves Theorem 7.6.

The following result on the uniqueness of Gauss quadrature formulas for spline spaces, due to Micchelli & Pinkus [1977], is an immediate consequence of Theorem 7.3 and Theorem 7.6.

Corollary 7.7. *If $m \geq 2$, then there exists a unique Gauss quadrature formula for $S_m(x_1, \ldots, x_k)$.*

We note that further results on quadrature formulas for spline spaces were proved by Schoenberg [1964a], [1965], [1966], [1969], Karlin & Studden [1966], Karlin [1969a], [1971], Sommer & Strauß [1981], Strauß [1979], [1984a] and Nürnberger, Schumaker, Sommer & Strauß [1985].

Micchelli & Pinkus [1977] proved that under certain assumptions for functions in the convexity cone, best one–sided L_1–approximations are unique and can be computed by Hermite interpolation at points of Gauss quadrature formulas. We state the following version due to Strauß [1982].

Theorem 7.8. *Let G be a subspace as in Theorem 7.6 and $a \leq t_1 < \cdots < t_r < b$ the points of the unique Gauss quadrature formula for G. Then for every function $f \in K(G) \cap C^1[a, b]$, the best one–sided L_1–approximation g_f from G is uniquely determined by*

$$g_f(t_i) = f(t_i), \qquad i = 1, \ldots, r, \tag{7.11}$$

and

$$g_f'(t_i) = f'(t_i), \qquad t_i \neq a, \quad i = 1, \ldots, r. \tag{7.12}$$

Proof. Let a function $f \in K(G) \cap C^1[a, b]$ be given. It follows from Lemma 7.5 that there exists a unique function $g_f \in G$ which satisfies (7.11) and (7.12). We can now apply the proof of Theorem 5.10 in Chapter I to show that g_f is a best one–sided L_1–approximation of f which by assumption is uniquely determined. This proves Theorem 7.8.

By combining Theorem 7.3 and Theorem 7.8, we obtain the following result on spline spaces due to Micchelli & Pinkus [1977].

Corollary 7.9. *If $m \geq 2$, then for every function in $K(S_m(x_1, \ldots, x_k)) \cap C^1[a, b]$ the best one–sided L_1–approximation from $S_m(x_1, \ldots, x_k)$ is uniquely determined by Hermite interpolation at the points of the Gauss quadrature formula for $S_m(x_1, \ldots, x_k)$.*

8. Approximation of Linear Functionals and Splines

In this section we consider approximation of linear functionals on spaces of continuous functions by linear combinations of point functionals. We investigate those approximations which provide a minimal error for classes of differentiable functions. It is shown that the coefficients of the point functionals are the values of the linear functional to be approximated, applied to certain splines.

A fundamental problem in Numerical Analysis is to approximate a given linear functional $L : C[a, b] \to \mathbf{R}$ by linear functionals $A : C[a, b] \to \mathbf{R}$ of the form

$$A(f) = \sum_{i=0}^{k+1} a_i f(x_i), \qquad f \in C[a, b],$$

where $a = x_0 < x_1 < \cdots < x_k < x_{k+1} = b$ and $a_0, \ldots, a_{k+1}$ are real numbers.

Example 8.1. (i) If $L(f) = \int_a^b f(t) \, dt$ for all $f \in C[a, b]$, then this leads to the problem of approximating the integral $\int_a^b f(t) \, dt$ by quadrature formulas $\sum_{i=0}^{k+1} a_i f(x_i)$.

(ii) If $L(f) = f(t)$ for all $f \in C[a, b]$, where $t \in [a, b]$, then this leads to the problem to approximate $f \in C[a, b]$ by functions of the form

$$s(t) = \sum_{i=0}^{k+1} a_i(t) f(x_i), \qquad t \in [a, b].$$

If, in addition, $a_i(x_j) = \delta_{ij}$, $i, j = 0, \ldots, k + 1$, then it follows that

$$s(x_i) = f(x_i), \qquad i = 0, \ldots, k + 1,$$

which means that the approximation s is an interpolating function.

In the following, we investigate the approximation of L by functionals A which are exact for all $p \in P_r$, i.e.

$$L(p) = \sum_{i=0}^{k+1} a_i p(x_i), \qquad p \in P_r. \tag{8.1}$$

We assume that the points $x_0, \ldots, x_{k+1}$ are given and that the real numbers $a_0, \ldots, a_{k+1}$ have to be determined in a suitable way.

If $k = r - 1$, then the real numbers $a_0, \ldots, a_{k+1}$ are uniquely determined. This can be seen as follows. For all $j \in \{0, \ldots, k + 1\}$, let the polynomial $l_j \in P_{k+1}$ be defined by

$$l_j(t) = \prod_{\substack{i=0 \\ i \neq j}}^{k+1} \frac{(t - x_i)}{(x_j - x_i)}, \qquad t \in [a, b],$$

(Compare Definition 2.12 of Chapter I). Then we have

$$l_j(x_i) = \delta_{ij}, \qquad i, j = 0, \ldots, k + 1. \tag{8.2}$$

Moreover, it follows from (8.1) and (8.2) that for all $j \in \{0, \ldots, k + 1\}$,

$$a_j = \sum_{i=0}^{k+1} a_i l_j(x_i) = L(l_j). \tag{8.3}$$

This shows that the coefficients are uniquely determined as the values of L applied to the polynomials $l_0, \ldots, l_{k+1}$. In the special case when $L(f) = f(t)$ for all

$f \in C[a, b]$, where $t \in [a, b]$, we are led to Lagrange interpolation (compare Theorem 2.13 in Chapter I).

We now consider the case when $k \geq r$. Then there is some freedom in the choice of the real numbers $a_0, \ldots, a_{k+1}$. In the following we determine these numbers in such a way that the corresponding error

$$\left| L(f) - \sum_{i=0}^{k+1} a_i f(x_i) \right|$$

has a certain optimality property for all functions $f \in C^{r+1}[a, b]$.

We first derive the so-called *Peano representation* of the error.

Theorem 8.2. *Let a linear functional* $L : C[a, b] \to \mathbf{R}$, *points* $a = x_0 < x_1 < \cdots < x_k < x_{k+1} = b$ *and real numbers* $a_0, \ldots, a_{k+1}$ *be given. Suppose that*

$$L(p) - \sum_{i=0}^{k+1} a_i p(x_i) = 0 \tag{8.4}$$

for all $p \in P_r$. *Then for all* $f \in C^{r+1}[a, b]$,

$$L(f) - \sum_{i=0}^{k+1} a_i f(x_i) = \int_a^b K(u) f^{(r+1)}(u) \, du, \tag{8.5}$$

where the Peano kernel $K : [a, b] \to \mathbf{R}$ *is defined by*

$$K(u) = \frac{1}{r!} \left\{ L \left((\cdot - u)_+^r \right) - \sum_{i=0}^{k+1} a_i (x_i - u)_+^r \right\} \tag{8.6}$$

for all $u \in [a, b]$.

Proof. Let a function $f \in C^{r+1}[a, b]$ be given. It follows from Taylor's theorem that for all $t \in [a, b]$,

$$f(t) = \sum_{i=0}^{r} \frac{f^{(i)}(a)}{i!} (t - a)^i + \frac{1}{r!} \int_a^t (t - u)^r f^{(r+1)}(u) \, du$$

$$= \sum_{i=0}^{r} \frac{f^{(i)}(a)}{i!} (t - a)^i + \frac{1}{r!} \int_a^b (t - u)_+^r f^{(r+1)}(u) \, du.$$

Then it follows from (8.4) that

$$L(f) - \sum_{i=0}^{k+1} a_i f(x_i) = \frac{1}{r!} L \left(\int_a^b (\cdot - u)_+^r f^{(r+1)}(u) \, du \right)$$

$$+ \frac{1}{r!} \int_a^b \left(\sum_{i=0}^{k+1} a_i (x_i - u)_+^r f^{(r+1)}(u) \right) du.$$

Since Riemann integrals can be approximated by Riemann sums, and since L is linear, it is easy to verify that

$$L\left(\int_a^b (\cdot - u)_+^r f^{(r+1)}(u)\, du\right) = \int_a^b L\left((\cdot - u)_+^r\right) f^{(r+1)}(u)\, du,$$

which shows that (8.5) holds. This proves Theorem 8.2.

By using (8.6) in Theorem 8.2 and the Schwarz inequality, we get the error estimate

$$\left| L(f) - \sum_{i=0}^{k+1} a_i\, f(x_i) \right| \le \|K\|_2\, \|f^{(r+1)}\|_2 \tag{8.7}$$

for all $f \in C^{(r+1)}[a, b]$, where $\|K\|_2$ is independent of the functions f. In order to get a small error, it is natural to require that $\|K\|_2$ is as small as possible. This leads to the following notion of optimality in the sense of Sard [1963] (see also Golomb & Weinberger [1959]).

Definition 8.3. Let a linear functional $L : C[a, b] \to \mathbf{R}$, points $a = x_0 < x_1 < \cdots < x_k < x_{k+1} = b$ and a positive integer $r \le k$ be given. We consider the following minimization problem. Determine real numbers $a_0, \ldots, a_{k+1}$ with

$$L(p) = \sum_{i=0}^{k+1} a_i\, p(x_i) \qquad \text{for all } p \in P_r$$

which minimize the norm $\|K\|_2$ of the corresponding Peano kernel.

The following result, due to Schoenberg [1964a], shows that the optimal numbers in the sense of Definition 8.3 are the values of the linear functional to be approximated applied to natural splines. This is a further optimality property of natural splines (compare Theorem 3.31 and Theorem 3.32).

Theorem 8.4. *For every integer $j \in \{0, \ldots, k+1\}$ let $s_j \in S_{2r+1}(x_1, \ldots, x_k)$ be the unique natural spline which satisfies*

$$s_j(x_i) = \delta_{ij}, \qquad i, j = 0, \ldots, k+1. \tag{8.8}$$

Then the real numbers

$$a_j = L(s_j), \qquad j = 0, \ldots, k+1, \tag{8.9}$$

are the unique solution of the minimization problem in Definition 8.3 and

$$L(s) = \sum_{i=0}^{k+1} a_i s(x_i) \qquad \text{for all natural splines } s \in S_{2r+1}(x_1, \ldots, x_k). \tag{8.10}$$

Proof. We first show that (8.10) holds. Let a natural spline $s \in S_{2r+1}(x_1, \ldots, x_k)$ be given. It follows from (8.8) and (8.9) that $s = \sum\limits_{i=0}^{k+1} s(x_i)s_i$ and therefore,

$$L(s) = L\left(\sum_{i=0}^{k+1} s(x_i)s_i\right) = \sum_{i=0}^{k+1} L(s_i)s(x_i) = \sum_{i=0}^{k+1} a_i s(x_i).$$

Now, let real numbers $b_0, \ldots, b_{k+1}$ be given such that

$$L(p) - \sum_{i=0}^{k+1} b_i p(x_i) = 0, \quad p \in P_r. \tag{8.11}$$

We define

$$K(u) = \frac{1}{r!}\left\{L\left((\cdot - u)_+^r\right) - \sum_{i=0}^{k+1} b_i(x_i - u)_+^r\right\}$$

and

$$K_0(u) = \frac{1}{r!}\left\{L\left((\cdot - u)_+^r\right) - \sum_{i=0}^{k+1} a_i(x_i - u)_+^r\right\}.$$

Then it follows that

$$K(u) - K_0(u) = \frac{1}{r!}\sum_{i=0}^{k+1}(a_i - b_i)(x_i - u)_+^r, \quad u \in (-\infty, \infty). \tag{8.12}$$

Since for all $i \in \{0, \ldots, k+1\}$,

$$(x_i - u)_+^r = (x_i - u)^r + (-1)^{r+1}(u - x_i)_+^{r+1}, \quad u \in (-\infty, \infty),$$

it follows from (8.10) and (8.11) that

$$K(u) - K_0(u) = (-1)^{r+1}\frac{1}{r!}\sum_{i=0}^{k+1}(a_i - b_i)(u - x_i)_+^r, \quad u \in (-\infty, \infty).$$

This shows that

$$K(u) - K_0(u) = 0, \quad u \in (-\infty, x_0] \cup [x_{k+1}, \infty).$$

Therefore, there exists a natural spline $s \in S_{2r+1}(x_1, \ldots, x_k)$ such that

$$s^{(r+1)}(u) = K(u) - K_0(u), \quad u \in [a, b].$$

It follows from (8.5) and (8.10) that

$$\int_a^b K_0(u)(K(u) - K_0(u))\,du = \int_a^b K_0(u)s^{(r+1)}(u)\,du = 0. \tag{8.13}$$

Since

$$\int_a^b K_0(u)^2\,du = \int_a^b K(u)^2\,du - \int_a^b (K(u) - K_0(u))^2\,du$$

$$-2\int_a^b K_0(u)(K(u) - K_0(u))\,du, \tag{8.14}$$

it follows from (8.13) that

$$\int_a^b K_0(u)^2 \, du \le \int_a^b K(u)^2 \, du$$

and that the equality sign holds, if $K = K_0$ which by (8.13) implies that $a_i = b_i$, $i = 0, \ldots, k+1$. This proves Theorem 8.4.

Example 8.5. If $L(f) = f(t)$ for all $f \in C[a, b]$, where $t \in [a, b]$, then by Theorem 8.4 the natural spline $s \in S_{2r+1}(x_1, \ldots, x_k)$, defined by

$$s(t) = \sum_{i=0}^{k+1} s_i(t) \, f(x_i) \tag{8.15}$$

for all $t \in [a, b]$, is the unique approximation of $f \in C^{r+1}[a, b]$ which is optimal in the sense of Definition 8.3. It is easy to see that $s \in S_{2r+1}(x_1, \ldots, x_k)$ is the unique natural spline which satisfies the interpolation conditions

$$s(x_i) = f(x_i), \qquad i = 0, \ldots, k+1, \tag{8.16}$$

(compare Theorem 3.30).

By using the Peano formula (8.5) for the error, we can consider further minimization problems. It follows from (8.5) and the Hölder inequality that

$$\left| L(f) - \sum_{i=0}^{k+1} a_i \, f(x_i) \right| \le \|K\|_p \, \|f^{(r+1)}\|_q \tag{8.17}$$

for all $f \in C^{(r+1)}[a, b]$, where $1 \le p \le \infty$, $1 \le q \le \infty$ and $\frac{1}{p} + \frac{1}{q} = 1$.

In the following we give a short description of an approach for the case when $L(f) = f(t)$ for all $f \in C[a, b]$, where $t \in [a, b]$, $p = 1$ and $q = \infty$ which was independently developed by Gaffney & Powell [1976] and Micchelli, Rivlin & Winograd [1976].

We use a somewhat different notation than above. Let a function $f \in C[a, b]$ and points $a = t_1 < t_2 < \cdots < t_{k+m} < t_{k+m+1} = b$ be given. We approximate f by functions of the form

$$s(t) = \sum_{i=1}^{k+m+1} s_i(t) \, f(t_i), \qquad t \in [a, b], \tag{8.18}$$

and suppose that $p(t) = \sum_{i=1}^{k+m+1} s_i(t) \, p(t_i)$ for all $p \in P_m$. Then it follows from Theorem 8.2 that for all $f \in C^{m+1}[a, b]$ and all $t \in [a, b]$,

$$f(t) - \sum_{i=1}^{k+m+1} s_i(t) \, f(t_i) = \int_a^b K_t(u) \, f^{(m+1)}(u) \, du,$$

where $K_t : [a, b] \to \mathbf{R}$ is the Peano kernel defined by

$$K_t(u) = \frac{1}{m!} \left\{ (t - u)_+^m - \sum_{i=1}^{k+m+1} s_i(t) \, (t_i - u)_+^m \right\}, \qquad u \in [a, b].$$

Then by the Hölder inequality

$$\left| f(t) - \sum_{i=1}^{k+m+1} s_i(t)\, f(t_i) \right| \le \|K_t\|_1\, \|f^{(m+1)}\|_\infty$$

for all $t \in [a, b]$.

Analogously as in Definition 8.3, we have to determine functions $s_1, \ldots,$ s_{k+m+1} which minimize the norm $\|K_t\|_1$ for all $t \in [a, b]$. This minimization problem will be solved as follows.

We define the space $G = \operatorname{span}\{B_1^m, \ldots, B_k^m\}$, where for all $i \in \{1, \ldots, k\}$, the functions B_i^m is the B-spline of degree m with support (t_i, t_{i+m+1}) (see Definition 2.3). By Theorem 2.8 the space G is a k-dimensional weak Chebyshev subspace of $C[a, b]$.

For this space there hold results analogous to those for spline spaces, and with the same proofs. (Note, that spline spaces also have a B-spline spline basis (see Theorem 2.6).) In particular, for every function $f \in C[a, b]$, there exists a unique best L_1-approximation from G (compare Theorem 6.1). Then it follows from Theorem 6.4 that for G, there exists a unique set of k canonical points $x_1 < \cdots < x_k$ in (a, b). Moreover, by Theorem 6.4 this set of points is poised with respect to G which implies that

$$t_i < x_i < t_{i+m+1}, \qquad i = 1, \ldots, k \tag{8.19}$$

(compare Theorem 3.7). Then it follows from (8.18) and Theorem 3.7 that for each $j \in \{1, \ldots, k+m+1\}$, there exists a unique spline $s_j \in S_m(x_1, \ldots, x_k)$ such that

$$s_j(t_i) = \delta_{ij}, \qquad i, j = 1, \ldots, k + m + 1. \tag{8.20}$$

It was proved by Gaffney & Powell [1976] and Micchelli, Rivlin & Winograd [1976] that these splines $s_1, \ldots, s_{k+m+1}$ minimize the norm $\|K_t\|_1$ for all $t \in [a, b]$. This shows that the approximation of $f \in C^{m+1}[a, b]$ which is optimal in the above sense is the unique spline $s \in S_m(x_1, \ldots, x_k)$ which satisfies the interpolation conditions

$$s(t_i) = f(t_i), \qquad i = 1, \ldots, k + m + 1$$

(compare Theorem 3.7). Moreover, it was shown that for the minimal norm we have

$$\|K_t\|_1 = |s_0(t)|, \qquad t \in [a, b],$$

where $s_0 \in S_m(x_1, \ldots, x_k)$ is the (unique) perfect spline which satisfies $s_0(t_i) = 0$, $i = 1, \ldots, k + m + 1$, and $\|s_0^{(m+1)}\|_\infty = 1$. The uniqueness of the perfect spline was proved by Karlin [1975] and Micchelli [1977].

Numerical results on the computation of the optimal knots $x_1, \ldots, x_k$ with the aid of the Newton method can be found in de Boor [1977], [1978]. He suggested to start the Newton method with the knots

$$x_i = \frac{t_{i+1} + \cdots + t_{i+m}}{m}, \qquad i = 1, \ldots, k,$$

or to use these knots directly for interpolation by splines from $S_m(x_1, \ldots, x_k)$ at the points $t_1, \ldots, t_{k+m+1}$. (Note, that these knots satisfy the condition (iii) in Theorem 3.7).

Approximation methods of the above type which minimize the norm of the Peano kernel are prototypes of so–called *optimal algorithms* or *optimal recovery schemes*. For a general theory of optimal algorithms and optimal recovery, the reader is referred to Traub & Wozniakowski [1980], Micchelli & Rivlin [1985] and the references therein.

9. Spaces of Splines with Multiple Knots

In the above sections we have considered the classical polynomial splines with simple knots. In this section we define spaces of polynomial splines with multiple knots and more general spaces of Chebyshev splines. Most of the standard results on polynomial splines with simple knots have analogs in the case of splines with multiple knots. Moreover, in most cases analogous arguments can be applied for proving the more general results. However, the complete proofs are sometimes much more technical than in the classical case, and would have gone beyond the scope of this book.

We begin with the definition of polynomial splines with multiple knots.

Definition 9.1. Let points $a = x_0 < x_1 \leq \cdots \leq x_k < x_{k+1} = b$ and an integer $m \geq 1$ be given such that at most m of these points coincide. For every index $i \in \{1, \ldots, k\}$, let m_i be the cardinality of the set $\{j \in \{1, \ldots, k\} : x_j = x_i\}$. We call

$$S_m(x_1, \ldots, x_k) = \{s \in C[a,b] : s\big|_{(x_i, x_{i+1})} \in P_m, \, i = 0, \ldots, k, \text{ such that } s$$
$$\text{has } m - m_i \text{ continuous derivatives at } x_i, \, i = 1, \ldots, k\}$$

the space of *polynomial splines of degree m with k fixed knots $x_1, \ldots, x_k$ of multiplicities $m_1, \ldots, m_k$*. (Here we use the convention that a spline s has zero continuous derivatives at a point $x \in (a, b)$, if s is continuous at x.)

The definition of $S_m(x_1, \ldots, x_k)$ in this form is somewhat unusual (compare Definition 1.4 and Remark 1.5 in the Appendix), but we use this notation, since then it is relatively easy to describe the analogous results for simple and multiple knots.

If $a = x_0 < x_1 < \cdots < x_k < x_{k+1} = b$, then $m_1 = \cdots = m_k = 1$ and $S_m(x_1, \ldots, x_k)$ is the space of polynomial splines of degree m with k fixed simple knots $x_1, \ldots, x_k$ as introduced in Definition 1.15.

In the following we always associate to $S_m(x_1, \ldots, x_k)$ further points $x_{-m} < \cdots < x_{-1} < a$ and $b < x_{k+2} < \cdots < x_{k+m+1}$ which may be choosen arbitrarily.

Most of the results on splines with simple knots have obvious analogs for multiple knots. In particular, $S_m(x_1, \ldots, x_k)$ is a $(k + m + 1)$–dimensional weak Chebyshev subspace of $C[a,b]$ (compare Theorem 1.19). The set of functions

$\{1, t, \ldots, t^m, (t-x_1)_+^{m-r_1}, \ldots, (t-x_k)_+^{m-r_k}\}$ forms a basis of $S_m(x_1, \ldots, x_k)$, where $r_j = \max\{i : x_j = \cdots = x_{j-i}\}$, $j = 1, \ldots, k$ (compare Theorem 1.17). B–splines can be defined analogously for multiple knots. If knots $x_i \leq \cdots \leq x_{i+m+1}$ are given, then there exists a unique B–spline $B_i^m : (-\infty, \infty) \to \mathbf{R}$ of degree m with knots $x_i, \ldots, x_{i+m+1}$ of multiplicities $m_i, \ldots, m_{i+m+1}$ and support (x_i, x_{i+m+1}) (compare Theorem 2.2). Moreover, B–splines can be written as divided differences of the truncated power function with respect to the possibly coalescing points $x_i, \ldots, x_{i+m+1}$ (compare Theorem 2.9), and a recurrence relation holds for B–splines (compare Theorem 2.11). Furthermore, the set of functions $\{B_{-m}^m, \ldots, B_k^m\}$ forms a basis of $S_m(x_1, \ldots, x_k)$ (compare Theorem 2.6). In particular, analogous results on Lagrange and Hermite interpolation (whenever the interpolation problems are well defined) (compare Section 3) and on best approximation in various norms hold (compare the Sections 4–7). In particular, the alternation theorems of Chapter 4.1 remain true for multiple knots (replacing $x_1 < \cdots < x_k$ by $x_1 \leq \cdots \leq x_k$ in the alternation conditions).

We now consider more general spaces of splines. In the following definition the space of polynomials is replaced by an arbitrary extended complete Chebyshev space (compare Definition 1.9, Theorem 1.11 and 1.12 and Example 1.13 in Chapter I) which leads to the so–called Chebyshev splines introduced by Karlin [1968].

Definition 9.2. Let points $a = x_0 < x_1 \leq \cdots \leq x_k < x_{k+1} = b$ as in Definition 9.1 be given and let G be an $(m+1)$–dimensional extended complete Chebyshev subspace of $C^m[a, b]$. We call

$$S_G(x_1, \ldots, x_k) = \{s \in C[a, b] : s\big|_{(x_i, x_{i+1})} \in G, \ i = 0, \ldots, k, \ \text{such that } s$$
$$\text{has } m - m_i \text{ continuous derivatives at } x_i, \ i = 1, \ldots, k\}$$

the space of *Chebyshev splines of degree m with k fixed knots* $x_1, \ldots, x_k$ *of multiplicities* $m_1, \ldots, m_k$.

For Chebyshev splines, results analogous to the case of polynomial splines hold, although truncated power functions and B–splines have to be defined in a suitable way and need some more investigation. In particular, the results on interpolation and on best approximation in various norms have analogous versions with similar proofs.

Example 9.3. We first note that an $(m + 1)$–dimensional extended complete Chebyshev subspace of $C^m[a, b]$ can be constructed by using Theorem 1.12 in Chapter I, i.e. we may choose strictly positive functions $w_i \in C^m[a, b]$, $i = 1, \ldots, m + 1$, and define a basis $\{g_1, \ldots, g_{m+1}\}$ of G by the integral relation (1.22) in Theorem 1.12. As shown in Example 1.13, the prototypes of extended complete Chebyshev spaces are spaces of polynomials and exponentials. In particular, if for given real numbers $\alpha_1 < \cdots < \alpha_{m+1}$,

$$E_{m+1} = \text{span}\{e^{\alpha_1 t}, \ldots, e^{\alpha_{m+1} t}\}$$

is the $(m+1)$-dimensional subspace of exponentials, then the resulting spline space $S_{E_{m+1}}(x_1, \ldots, x_k)$ (see Definition 9.2) is called the space of *exponential splines*.

Appendix

The last three sections contain a brief introduction to splines with free knots, splines in two variables and the use of splines in the numerical solution of differential equations. We only give the basic definitions, some typical results without proof and several references to the literature, in particular, to survey papers.

1. Splines with Free Knots

In this section we introduce splines with free knots. Best approximation by splines with free knots is a nonlinear problem. In contrast to splines with fixed knots, the theory and methods in the case of free knots are not fully developed. An algorithm for computing good spline approximations with free knots is available (Section 4.3 in Chapter II). We give an illustration of how far the theory is developed for best uniform approximation.

We begin with the definition of splines with free knots.

Definition 1.1. Let integers $m \geq 1$ and $k \geq 1$ be given. We call

$$
\begin{aligned}
S_{m,k} \;=\; &\{s : [a,b] \to \mathbf{R} : \text{ there exist points} \\
& a = x_0 < x_1 < \cdots < x_r < x_{r+1} = b \text{ and} \\
& \text{integers } m_1, \ldots, m_r \in \{1, \ldots, m+1\} \\
& \text{with } \sum_{i=1}^{r} m_i \leq k \text{ such that} \\
& s\big|_{[x_i, x_{i+1})} \in P_m, \; i = 0, \ldots, r-1, \\
& s\big|_{[x_r, x_{r+1}]} \in P_m \text{ and } s \text{ has } m - m_i \\
& \text{continuous derivatives at } x_i, \; i = 1, \ldots, r\}
\end{aligned}
\tag{1.1}
$$

the set of *polynomial splines of degree m with k free knots*. (Here we use the convention that $s : [a,b] \to \mathbf{R}$ has -1 (respectively 0) continuous derivatives at $x \in (a,b)$, if no continuity of s is required at x (respectively, if s is continuous at x)).

It is easy to see that each spline $s \in S_{m,k}$ can be written in the form

$$
s(t) = \sum_{i=0}^{m} a_i \, t^i + \sum_{i=1}^{r} \sum_{j=1}^{m_i} a_{ij} \, (t - x_i)_+^{m-j+1},
\tag{1.2}
$$

where a_i, $i = 0, \ldots, m$, and a_{ij}, $j = 1, \ldots, m_i$, $i = 1, \ldots, r$, are real numbers.

The problem of best approximation for $S_{m,k}$ w.r.t. the various L_p–norms ($1 \leq p \leq \infty$) is defined analogously as in the sections of Chapter I.

Since $S_{m,k}$ is a nonconvex set, these approximation problems are nonlinear. Therefore, the existence of best approximations from $S_{m,k}$ is a nontrivial problem.

The first result is a general existence theorem on best L_p-approximation (for details see Rice [1969]).

Theorem 1.2. *If $1 \le p \le \infty$, then for every function $f \in L_p[a, b]$, there exists a best L_p-approximation from $S_{m,k}$.*

Schumaker [1968b], [1969b] proved the following result on the existence of continuous, respectively differentiable, best uniform approximations from $S_{m,k}$.

Theorem 1.3. *The following statements hold:*
(i) For every function $f \in C[a, b]$, there exists a best uniform approximation $s_f \in S_{m,k}$ such that $s_f \in C[a, b]$.
(ii) If $m \ge 2$, then for every function $f \in C^1[a, b]$, there exists a best uniform approximation $s_f \in S_{m,k}$ such that $s_f \in C^1[a, b]$.
(iii) If $m \ge 3$, then there exists a function $f \in C^\infty[a, b]$ such that no function in $C^2[a, b]$ is a best uniform approximation of f from $S_{m,m-1}$.

Next, we give alternation conditions for best uniform approximations. In contrast to splines with fixed knots (see Section 4 in Chapter II), most of the results are not complete. Several deep problems are unsolved at present.

We first define the following type of spline spaces.

Definition 1.4. Let points $a = x_0 < x_1 < \cdots < x_r < x_{r+1} = b$ and integers $m_1, \ldots, m_r \in \{1, \ldots, m\}$ be given. We call

$$S_m \left(\begin{array}{c} x_1, \ldots, x_r \\ m_1, \ldots, m_r \end{array} \right) = \left\{ s \in C[a, b] : s \big|_{[x_i, x_{i+1}]} \in P_m, \, i = 0, \ldots, r, \right.$$

$$\text{such that } s \text{ has } m - m_i \text{ continuous} \quad (1.3)$$

$$\left. \text{derivatives at } x_i, \, i = 1, \ldots, r \right\}$$

the space of *polynomial splines of degree m with r fixed distinct knots $x_1, \ldots, x_r$ of multiplicities $m_1, \ldots, m_r$.*

Remark 1.5. If we count each knot from the set $\{x_1, \ldots, x_r\}$ as in Definition 1.4 according to its multiplicity and denote the resulting knots by $y_1 \le \cdots \le y_k$, where $k = \sum_{i=1}^r m_i$, then

$$S_m \left(\begin{array}{c} x_1, \ldots, x_r \\ m_1, \ldots, m_r \end{array} \right) = S_m(y_1, \ldots, y_k)$$

in the sense of Definition 9.1 of Chapter II.

Building on an earlier result of Schumaker [1968b], the following sufficient condition for best uniform approximations from $S_{m,k}$ was given by Braess [1974a]. Moreover, by using the tangent method of G. Meinardus and D. Schwedt [1964] (see Meinardus [1967]) the implication (ii) $\Rightarrow$ (iii) was proved by Braess [1974a] for differentiable splines. The implication (ii) $\Rightarrow$ (iii) was shown by Nürnberger, Schumaker, Sommer & Strauß [1988] for continuous splines by using different arguments. Finally, Braess [1974a] gave an example which shows that the gap between the alternation conditions (i) and (ii) cannnot be closed. (We refer to Definition 4.1 in Chapter II.)

Theorem 1.6. *Let a function* $f \in C[a,b]$, *a spline* $s_f \in S_m \begin{pmatrix} x_1, \ldots, x_r \\ m_1, \ldots, m_r \end{pmatrix}$ *and a natural number* $k \geq \sum_{i=1}^r m_i$ *be given. We consider the following statements:*
(i)   There exists an interval $[x_p, x_{p+q}] \subset [a,b]$, $q \geq 1$, *such that*
$$A(f - s_f)\big|_{[x_p, x_{p+q}]} \geq k + m + 2 + \sum_{i=p+1}^{p+q-1} m_i.$$
(ii) The spline s_f *is a best uniform approximation of* f *from* $S_{m,k}$.
(iii) There exists an interval $[x_p, x_{p+q}] \subset [a,b]$, $q \geq 1$, *such that*
$$A(f - s_f)\big|_{[x_p, x_{p+q}]} \geq q + m + 1 + \sum_{i=p+1}^{p+q-1} m_i.$$
Then (i) $\Rightarrow$ *(ii)* $\Rightarrow$ *(iii).*

Building on earlier work of Schumaker [1968b] and Arndt [1974], a sufficient alternation condition for strongly unique best approximations from $S_{m,k}$ was given by Schaback [1978]. Nürnberger [1987b] developed a weaker sufficient condition and also a necessary condition for strong unicity which are summarized in the next result. The implication (i) $\Rightarrow$ (ii) holds for arbitrary continuous splines.

Theorem 1.7. *Let a function* $f \in C[a,b]$ *and a spline* $s_f \in S_m \begin{pmatrix} x_1, \ldots, x_r \\ m_1, \ldots, m_r \end{pmatrix} \cap$ $C^1[a,b] \setminus S_{m,k-1}$ *be given, where* $k = \sum_{i=1}^r m_i$. *We consider the following statements:*
(i)   For every interval $(y_i, y_{i+m+j}) \subset (y_{-m}, y_{2k+m+1})$, $j \geq 1$, *we have*
$$A(f - s_f)\big|_{(y_i, y_{i+m+j})} \geq j + 1,$$
where $a < y_1 \leq \cdots \leq y_{2k} < b$ *are the knots of* s_f *counting each knot* x_i $2m_i-$*times,* $i = 1, \ldots, r$, *and* $y_{-m} < \cdots < y_{-1} < y_0 = a$, $b = y_{2k+1} < y_{2k+2} < \cdots < y_{2k+m+1}$ *are arbitrary points.*
(ii) The spline s_f *is a strongly unique best uniform approximation of* f *from* $S_{m,k}$.
(iii) For every interval $(y_i, y_{i+m+j}) \subset (y_{-m}, y_{r+k+m+1})$, $j \geq 1$, *we have*
$$A(f - s_f)\big|_{(y_i, y_{i+m+j})} \geq j + 1,$$
where $a < y_1 \leq \cdots \leq y_{r+k} < b$ *are the knots of* s_f *counting each knot* x_i $(m_i + 1)-$*times,* $i = 1, \ldots, r$, *and* $y_{-m} < \cdots < y_{-1} < y_0 = a$, $b = y_{r+k+1} < y_{r+k+2} < \cdots < y_{r+k+m+1}$ *are arbitrary points.*

Then (i) ⇒ (ii) ⇒ (iii).

It can be shown by examples that the gap between the alternation conditions (i) and (ii) of Theorem 1.7 cannot be closed, in general. However, if s_f has k simple knots (i.e. $m_1 = \cdots = m_k = 1$), then these conditions coincide, and we obtain the following characterization of strong unicity. Moreover, it was shown by Nürnberger [1987b] that in this case strongly unique best approximations and strongly unique local best approximations coincide, which is a rare phenomena in nonlinear approximation.

Theorem 1.8. *Let a function $f \in C[a,b]$ and a spline $s_f \in S_m\begin{pmatrix} x_1, \ldots, x_k \\ 1, \ldots, 1 \end{pmatrix} \setminus$ $S_{m,k-1}$ be given, where $m \geq 2$. The following statements are equivalent:*
(i)   The spline s_f is a strongly unique best uniform approximation of f from $S_{m,k}$.
(ii) For every interval $(y_i, y_{i+m+j}) \subset (y_{-m}, y_{2k+m+1})$, $j \geq 1$, we have

$$A(f - s_f)\Big|_{(y_i, y_{i+m+j})} \geq j + 1,$$

where $a < y_1 \leq \cdots \leq y_{2k} < b$ are the knots of s_f counting each knot x_i twice, $i = 1, \ldots, k$, and $y_{-m} < \cdots < y_{-1} < y_0 = a$, $b = y_{2k+1} < y_{2k+2} < \cdots < y_{2k+m+1}$ are arbitrary points.

In addition, Nürnberger [1989] proved that functions from the interior of SU $(S_{m,k})$ can be characterized by alternation properties of the error $f - s_f$ if the best approximation s_f has k simple knots. Here, the set SU $(S_{m,k})$ is defined by

$$\text{SU }(S_{m,k}) = \{ f \in C[a,b] : f \text{ has a strongly unique best} \atop \text{uniform approximation from } S_{m,k} \} \tag{1.4}$$

Theorem 1.9. *Let a function $f \in C[a,b]$ be given and s_f be a best uniform approximation of f from $S_{m,k}$ with k simple knots $a < x_1 < \cdots < x_k < b$, where $m \geq 2$. The following statements are equivalent:*
(i) $f \in \text{int SU }(S_{m,k})$.
(ii) $A(f - s_f)\big|_{[a,b]} \geq 2k + m + 2$ and for every interval $[x_p, x_{p+q}] \subsetneq [a,b]$, $q \geq 1$,
*   we have $A(f - s_f)\big|_{[x_p, x_{p+q}]} < 2q + m$.*

We note that Nürnberger [1989] also showed that the set of splines with simple knots in $S_{m,k}(m \geq 2)$ is a so–called *sun w.r.t. strong unicity,* although it is not a *sun.*

Johnson [1960] proved that a function f with $f^{(m+1)}(t) \neq 0$ for all $t \in (a,b)$ has a unique best uniform approximation from $S_{m,k}$. It was shown by Braess [1974a] that such a function f satisfies condition (ii) of Theorem 1.9 which implies that f is even from int SU $(S_{m,k})$.

Theorem 1.10. *Let $m \geq 2$ and let a function $f \in C^{m+1}[a,b]$ with $f^{(m+1)}(t) \neq 0$ for all $t \in (a,b)$ be given. Then $f \in \text{int SU }(S_{m,k})$ and the best uniform approximation from $S_{m,k}$ has k simple knots.*

Finally, we discuss the uniqueness of best L_p-approximations ($1 \leq p < \infty$) from $S_{m,k}$ for functions as in Theorem 1.10. The following unicity result for the particular function $f(t) = t^{m+1}$ was proved by Jetter & Lange [1978] for $p = 2$, by Jetter [1978] and Strauß [1979] for $p = 1$, and finally by Bojanov [1979] for $1 < p < \infty$.

Theorem 1.11. *For $1 \leq p < \infty$ the function $f(t) = t^{m+1}$, $t \in [a, b]$, has a unique best L_p-approximation from $S_{m,k}$.*

The next result shows that in contrast to Theorem 1.10 and Theorem 1.11, best L_p-approximations ($1 \leq p < \infty$) from $S_{m,k}$ are not always unique for functions f with $f^{(m+1)}(t) \neq 0$ for all $t \in [a, b]$. This was proved by Barrrow, Chui, Smith & Ward [1978] for the case $p = 2$ and $m = 1$ and by Nürnberger & Braess [1981] in the general case.

Theorem 1.12. *For $1 \leq p < \infty$, $m \geq 1$ and $k \geq 1$, there exists a function $f \in C^\infty[a, b]$ with $f^{(m+1)}(t) \neq 0$ for all $t \in [a, b]$ which has more than one best L_p-approximation from $S_{m,k}$.*

Every function m of the form $m(t) = t^{m+1} - s(t)$, $t \in [a, b]$, where $s \in S_{m,k}$, is called a *monospline of degree $m + 1$*. It follows from Theorem 1.10 and Theorem 1.11 that there exists a unique monospline of minimal L_p-norm ($1 \leq p \leq \infty$). On the other hand, Theorem 1.12 shows that this is not true, if $1 \leq p < \infty$ and $f(t) = t^{m+1}$ is replaced by an arbitrary function f with $f^{(m+1)}(t) \neq 0$ for all $t \in [a, b]$.

In this section we have considered only some aspects of splines with free knots concerning mainly best uniform approximation. For more information on approximation by splines with free knots in various norms, especially on monosplines and the relationship to optimal quadrature formulas, we refer to the following literature: Augsburger [1967], Barrar & Loeb [1970], [1974], [1976a], [1976b], [1978], [1980], [1984], [1986], Barrar, Loeb & Werner [1980], Barrow & Smith [1978], Bojanov [1977], [1981], de Boor [1963], [1969], [1973], [1974b], [1978], Braess [1974b], [1975], [1986], Braess & Dyn [1982], [1986], Burchard [1974], Chow [1982], Chui, Smith & Ward [1977], Cromme [1976], [1982a], [1982b], Dodson [1972], Dyn [1985], Karlin [1969b], [1976a], [1976b], Karlin & Micchelli [1972], Karlin & Schumaker [1967], Kioustelidis [1981], Kornejchuk [1984], Loeb [1980], Malcom [1977], Marin [1984], Meinardus [1966a], [1966b], [1967], Micchelli [1972], Micchelli & Pinkus [1977], Nürnberger, Schumaker, Sommer & Strauß [1988], Pinkus [1981], Powell [1968], Schoenberg [1969], Schumaker [1979], Strauß [1983], [1984a], [1984d], Taylor [1979], Töpfer [1982], Wulbert [1973] and Zhensykbaev [1981], [1982]. A detailed treatment of the approximation power of splines with free knots is given in the book of Schumaker [1981] (see the corresponding references therein). Finally, we note that the nonlinear problem of approximation by rational splines was investigated by Braess & Werner [1974], Schaback [1973] and Werner [1974], [1979], [1980].

2. Splines in Two Variables

A brief introduction to bivariate splines is given and some special interpolation methods are discussed. The tensor product method and the blending method can be used for interpolation on rectangular grids. Certain interpolation procedures for triangular partitions can be carried out by applying the finite element method. On the other hand — in contrast to the univariate theory — no general methods for interpolation by functions from spaces of bivariate splines are available at present. Even such fundamental problems as the dimension of spline spaces or the construction of a basis are only solved for uniform partitions. In addition, we discuss some aspects of multivariate B–splines which were intensively studied in the last decade. A general approach to interpolation of scattered data using radial functions includes a generalization of natural splines to several variables. Several references to the recent literature are given, in particular, to survey articles, where the reader can find details of these topics.

2.1. Tensor Product and Blending

We give a brief introduction to interpolation on a rectangular grid by combining univariate approximation schemes with aid of tensor products or blending.

Let a rectangle $[a, b] \times [c, d]$ in $\mathbf{R}^2$ and points $x_1 < \cdots < x_p$ in $[a, b]$, $y_1 < \cdots < y_q$ in $[c, d]$ be given. In the following we descibe a method for interpolating functions in $C([a, b] \times [c, b])$ at the points (x_μ, y_ν), $\mu = 1, \ldots, p$, $\nu = 1, \ldots, q$.

This is done by combining two univariate interpolation methods. Therefore, let $G = \operatorname{span}\{g_1, \ldots, g_p\}$ be a p–dimensional subspace of $C[a, b]$ such that

$$D\left(\begin{array}{c} g_1, \ldots, g_p \\ x_1, \ldots, x_p \end{array} \right) \neq 0.$$

Then for each function $\overline{f} \in C[a, b]$, there exists a unique function $P(\overline{f}) \in G$ such that

$$P(\overline{f})(x_i) = \overline{f}(x_i), \qquad i = 1, \ldots, p.$$

This defines a Lagrange interpolation operator (see Definition 1.1 in Chapter I) $P : C[a, b] \to G$. Moreover, let $H = \operatorname{span}\{h_1, \ldots, h_q\}$ be a q–dimensional subspace of $C[c, d]$ such that

$$D\left(\begin{array}{c} h_1, \ldots, h_q \\ y_1, \ldots, y_q \end{array} \right) \neq 0.$$

The corresponding Lagrange interpolation operator w.r.t. the points $y_1, \ldots, y_q$ is denoted by $Q : C[c, d] \to H$. By combining P and Q, we obtain an interpolation operator $P \otimes Q : C([a, b] \times [c, d]) \to G \otimes H$ which is defined as follows.

Let a function $f \in C([a, b] \times [c, b])$ be given. For a fixed point $y \in [c, d]$, we define the function $f_y \in C[a, b]$ by

$$f_y(x) = f(x, y), \qquad x \in [a, b]. \tag{2.1}$$

Then $P(f_y) \in G$ can be written as

$$P(f_y)(x) = \sum_{i=1}^{p} a_i(y)\, g_i(x), \qquad x \in [a, b], \tag{2.2}$$

where $a_i(y)$, $i = 1, \ldots, p$, are real coefficients depending on y. Moreover, the function $P(f_y)$ satisfies

$$P(f_y)(x_i) = f_y(x_i) = f(x_i, y), \qquad i = 1, \ldots, p. \tag{2.3}$$

Let an index $i \in \{1, \ldots, p\}$ be given. Then the mapping

$$y \to a_i(y), \qquad y \in [c, d], \tag{2.4}$$

defines a function $a_i \in C[c, d]$. The function $Q(a_i) \in H$ can be written as

$$Q(a_i)(y) = \sum_{j=1}^{q} a_{ij}\, h_j(y), \qquad y \in [c, d], \tag{2.5}$$

where a_{ij}, $j = 1, \ldots, q$, are real coefficients. Moreover, the function $Q(a_i)$ satisfies

$$Q(a_i)(y_j) = a_i(y_j), \qquad j = 1, \ldots, q. \tag{2.6}$$

The (pq)–dimensional subspace $G \otimes H$ of $C\left([a, b] \times [c, d]\right)$ defined by

$$G \otimes H = \left\{ \sum_{i=1}^{p} \sum_{j=1}^{q} b_{ij}\, g_i(x)\, h_j(y) : b_{ij} \in \mathbf{R}, \, i = 1, \ldots, p, \, j = 1, \ldots, q \right\} \tag{2.7}$$

is called the *tensor product* of the spaces G and H. We now define

$$P \otimes Q(f)(x, y) = \sum_{i=1}^{p} \sum_{j=1}^{q} a_{ij}\, g_i(x)\, h_j(y), \qquad (x, y) \in [a, b] \times [c, d], \tag{2.8}$$

where a_{ij}, $i = 1, \ldots, p$, $j = 1, \ldots, q$, are the coefficients from (2.5). The corresponding operator

$$P \otimes Q : C\left([a, b] \times [c, b]\right) \to G \otimes H \tag{2.9}$$

is called the *tensor product* of the operators P and Q.

We now show that the function $P \otimes Q(f)$ satisfies the following interpolation conditions. Let integers $\mu \in \{1, \ldots, p\}$ and $\nu \in \{1, \ldots, q\}$ be given. Then it follows from the interpolation properties (2.3) and (2.6) that

$$\begin{aligned}
P \otimes Q(f)(x_\mu, y_\nu) &= \sum_{i=1}^{p} \sum_{j=1}^{q} a_{ij}\, h_j(y_\nu)\, g_i(x_\mu) \\
&= \sum_{i=1}^{p} a_i(y_\nu)\, g_i(x_\mu) = P(f_{y_\nu})(x_\mu) = f(x_\mu, y_\nu).
\end{aligned} \tag{2.10}$$

It is easy to see that the following matrix relations hold:

$$(f(x_\mu, y_\nu))_{\mu,\nu} = D\begin{pmatrix} g_1, \ldots, g_p \\ x_1, \ldots, x_p \end{pmatrix}(a_{ij})_{i,j}\, D\begin{pmatrix} h_1, \ldots, h_q \\ y_1, \ldots, y_q \end{pmatrix}^T \tag{2.11}$$

and therefore,

$$(a_{ij})_{i,j} = D\begin{pmatrix} g_1, \ldots, g_p \\ x_1, \ldots, x_p \end{pmatrix}^{-1}(f(x_\mu, y_\nu))_{\mu,\nu}\left(D\begin{pmatrix} h_1, \ldots, h_q \\ y_1, \ldots, y_q \end{pmatrix}^T\right)^{-1}. \tag{2.12}$$

(Here A^T denotes the transposed of a given matrix A.)

This shows that there exists a unique function in $G \otimes H$ which interpolates f at the points (x_μ, y_ν), $\mu = 1, \ldots, p$, $\nu = 1, \ldots, q$.

Analogous results hold for Hermite interpolation (see Definition 1.5 in Chapter I).

An efficient method for computing the coefficients a_{ij}, $i = 1, \ldots, p$, $j = 1, \ldots, q$, was developed by de Boor [1979] (see also de Boor [1978]).

The tensor product method can be used for bivariate spline interpolation.

Let $G = S_m(\bar{x}_1, \ldots, \bar{x}_k)$, $H = S_n(\bar{y}_1, \ldots, \bar{y}_l)$ and let points $x_1 < \cdots < x_{k+m+1}$ in $[a, b]$, $y_1 < \cdots < y_{l+n+1}$ in $[c, d]$ be given which satisfy

$$x_i < \bar{x}_i < x_{i+m+1}, \qquad i = 1, \ldots, k, \tag{2.13}$$

and

$$y_j < \bar{y}_j < y_{j+n+1}, \qquad j = 1, \ldots, l. \tag{2.14}$$

Then by Theorem 3.7 in Chapter II the corresponding univariate interpolation problems have a unique solution.

Moreover, let $\{N_{-m}^m, \ldots, N_k^m\}$ (respectively $\{\tilde{N}_{-n}^n, \ldots, \tilde{N}_l^n\}$) be a normalized B–spline basis of G (respectively H) (see Theorem 2.6 in Chapter II). By applying the above method, we obtain a tensor product spline

$$P \otimes Q(f)(x, y) = \sum_{i=-m}^{k} \sum_{j=-n}^{l} a_{ij}\, N_i^m(x)\, \tilde{N}_j^n(y)$$

which satisfies the interpolation condition (2.10).

The tensor product approximation $P \otimes Q(f)$ interpolates the function f at the points (x_μ, y_ν), $\mu = 1, \ldots, p$, $\nu = 1, \ldots, q$. In the following we construct an approximation which interpolates the function f at the lines $x = x_\mu$, $\mu = 1, \ldots, p$, and $y = y_\nu$, $\nu = 1, \ldots, q$.

The operator

$$P \oplus Q : C([a, b] \times [c, d]) \to C([a, b] \times [c, d])$$

defined by

$$P \oplus Q = P \otimes Id + Id \otimes Q - P \otimes Q \tag{2.15}$$

is called the *Boolean sum* of the operators P and Q and the resulting approximation scheme is called *blending*.

We now show that the function $P \oplus Q(f)$ satisfies the following conditions:

$$P \oplus Q(f)(x_\mu, y) = f(x_\mu, y), \qquad \mu = 1, \ldots, p, \quad y \in [c, d], \qquad (2.16)$$

and

$$P \oplus Q(f)(x, y_\nu) = f(x, y_\nu), \qquad \nu = 1, \ldots, q, \quad x \in [a, b]. \qquad (2.17)$$

Since $P \oplus Q = Q \oplus P$, we have

$$
\begin{aligned}
&P \oplus Q(f)(x_\mu, y) \\
&= P \otimes Id(f)(x_\mu, y) + Id \otimes Q(f)(x_\mu, y) - Q \otimes P(f)(x_\mu, y) \\
&= P(f_y)(x_\mu) + Q(f_{x_\mu})(y) - Q(f_{x_\mu})(y) \\
&= f_y(x_\mu) = f(x_\mu, y).
\end{aligned}
$$

Property (2.17) follows analogously.

Although the Boolean sum operator $P \oplus Q$ satisfies the strong conditions (2.16) and (2.17), in this form it is not suitable for numerical purposes, since $P \otimes Id(f)$ and $Id \otimes Q(f)$ depend on infinitely many values of f. Therefore, Gordon [1969] used the following variant.

Let $\tilde{G} = \text{span}\{\tilde{g}_1, \ldots, \tilde{g}_{\tilde{p}}\}$ be a $\tilde{p}$–dimensional subspace of $C[a, b]$ and let points $\tilde{x}_1 < \cdots < \tilde{x}_{\tilde{p}}$ in $[a, b]$ be given such that

$$\{x_1, \ldots, x_p\} \subset \{\tilde{x}_1, \ldots, \tilde{x}_{\tilde{p}}\} \qquad (2.18)$$

and

$$D \begin{pmatrix} \tilde{g}_1, \ldots, \tilde{g}_{\tilde{p}} \\ \tilde{x}_1, \ldots, \tilde{x}_{\tilde{p}} \end{pmatrix} \neq 0. \qquad (2.19)$$

The corresponding Lagrange interpolation operator w.r.t. the points $\tilde{x}_1, \ldots, \tilde{x}_{\tilde{p}}$ is denoted by $\tilde{P} : C[a, b] \to \tilde{G}$. Moreover, let $\tilde{H} = \text{span}\{\tilde{h}_1, \ldots, \tilde{h}_{\tilde{q}}\}$ be a $\tilde{q}$–dimensional subspace of $C[c, d]$ and let points $\tilde{y}_1 < \cdots < \tilde{y}_{\tilde{q}}$ in $[c, d]$ be given such that

$$\{y_1, \ldots, y_q\} \subset \{\tilde{y}_1, \ldots, \tilde{y}_{\tilde{q}}\} \qquad (2.20)$$

and

$$D \begin{pmatrix} \tilde{h}_1, \ldots, \tilde{h}_{\tilde{q}} \\ \tilde{y}_1, \ldots, \tilde{y}_{\tilde{q}} \end{pmatrix} \neq 0. \qquad (2.21)$$

The corresponding Lagrange interpolation operator w.r.t. the points $\tilde{y}_1, \ldots, \tilde{y}_{\tilde{p}}$ is denoted by $\tilde{Q} : C[c, d] \to \tilde{H}$. In practice $\tilde{p}$ and $\tilde{q}$ are chosen to be much larger then p and q, respectively.

We now consider instead of $P \oplus Q$ the operator $A : C([a, b] \times [c, d]) \to C([a, b] \times [c, d])$ defined by

$$A = P \otimes \tilde{Q} + \tilde{P} \otimes Q - P \otimes Q. \qquad (2.22)$$

This operator has the following properties. Let an index $\mu \in \{1, \ldots, p\}$ and a point $y \in [c, d]$ be given. Then we have

$$
\begin{aligned}
A(f)(x_\mu, y) &= P \otimes \tilde{Q}(f)(x_\mu, y) + Q(f_{x_\mu})(y) - Q(f_{x_\mu})(y) \\
&= P \otimes \tilde{Q}(f)(x_\mu, y).
\end{aligned}
\tag{2.23}
$$

Therefore, the function $y \to A(f)(x_\mu, y)$, $y \in [c, d]$, is the unique function from $\tilde{H}$ which attains the values $f(x_\mu, \tilde{y}_\nu)$ at the points $\tilde{y}_\nu$, $\nu = 1, \ldots, \tilde{q}$. Analogously, for $\nu \in \{1, \ldots, q\}$ the function $x \to A(f)(x, y_\nu)$, $x \in [a, b]$, is the unique function from $\tilde{G}$ which attains the values $f(\tilde{x}_\mu, y_\nu)$ at points $\tilde{x}_\mu$, $\mu = 1, \ldots, \tilde{p}$.

An error analysis was given by Gordon [1969]. The above method can be applied analogously as the tensor product method to construct bivariate spline approximations.

Up to now, we have only considered linear interpolation operators . Although the metric projection (see Definition 3.1 in Chapter I) is nonlinear in general, it can be also used for a modification of the blending method such that it is suitable for numerical purposes.

We first replace the operator $P \otimes Id$ in the Boolean sum (2.15) by using best approximation at the lines $x = x_\mu$, $\mu = 1, \ldots, p$. The function $P \otimes Id(f)$ can be written as

$$
P \otimes Id(f)(x, y) = \sum_{i=1}^{p} a_i(y) \, g_i(x),
\tag{2.24}
$$

and satisfies

$$
\sum_{i=1}^{p} a_i(y) \, g_i(x_\mu) = f(x_\mu, y), \qquad \mu = 1, \ldots, p.
\tag{2.25}
$$

For each $\mu \in \{1, \ldots, p\}$, we replace the function $y \to f(x_\mu, y)$, $y \in [c, d]$ by its best uniform approximation $\overline{h}_\mu = \sum_{j=1}^{\tilde{q}} c_{\mu j} \, \tilde{h}_j$ from $\tilde{H}$ and consider instead of (2.25) the equations

$$
\sum_{i=1}^{p} \overline{a}_i(y) \, g_i(x_\mu) = \overline{h}_\mu(y), \qquad \mu = 1, \ldots, p.
\tag{2.26}
$$

This system of equations is equivalent to

$$
(\overline{a}_1, \ldots, \overline{a}_p) = D \left(\begin{array}{c} g_1, \ldots, g_p \\ x_1, \ldots, x_p \end{array} \right)^{-1} (\overline{h}_1, \ldots, \overline{h}_q),
\tag{2.27}
$$

which shows that the functions $\overline{a}_1, \ldots, \overline{a}_p$ can be computed as linear combinations of the basis $\{\tilde{h}_1, \ldots, \tilde{h}_{\tilde{q}}\}$ of $\tilde{H}$.

We analogously start with $Id \otimes Q(f)$ in (2.15) and obtain equations

$$
\sum_{j=1}^{q} \overline{b}_j(x) \, h_j(y_\nu) = \overline{g}_\nu(x), \qquad \nu = 1, \ldots, q,
\tag{2.28}
$$

where for each $\nu \in \{1, \ldots, q\}$, $\overline{g}_\nu \in \tilde{G}$ is the best uniform approximation of the function $x \to f(x, y_\nu)$, $x \in [a, b]$. Again, the functions $\overline{b}_j \in \tilde{G}$ can be computed by solving the system (2.28).

Finally, instead of $P \oplus Q(f)$, we use the approximation $B(f)$ of f defined by

$$B(f)(x,y) = \sum_{i=1}^{p} \bar{a}_i(y) \, g_i(x) + \sum_{j=1}^{q} \bar{b}_j(x) \, h_j(y) - P \otimes Q(f)(x,y),$$

$$(x,y) \in [a,b] \times [c,b]. \quad (2.29)$$

In the case when the spaces G, H, $\tilde{G}$ and $\tilde{H}$ are spline spaces, the best uniform approximations of f on the lines $x = x_\mu$, $\mu = 1,\ldots,p$, and $y = y_\nu$, $\nu = 1,\ldots,q$, can be computed by applying the algorithms described in the Sections 4.2 and 4.3 of Chapter II to obtain approximations of high accuracy.

For further information on tensor products and blending, we refer the reader to Cheney [1983], [1986] and Light & Cheney [1985].

2.2. Finite Element Functions

We briefly describe some special interpolation methods for triangular partitions by applying the finite element method.

The following notation will be used. For $\alpha = (\alpha_1, \ldots, \alpha_n)$, where $\alpha_1, \ldots, \alpha_n$ are nonnegative integers, we set $|\alpha| = \sum_{i=1}^{n} \alpha_n$ and denote by

$$D^\alpha f = \frac{\partial^{|\alpha|} f}{\partial x_1^{\alpha_1} \cdots \partial x_n^{\alpha_n}}$$

the corresponding partial derivative of order $|\alpha|$ of a function $f : \mathbf{R}^n \to \mathbf{R}$.

If T is an open subset of $\mathbf{R}^n$, then we define

$$C(T) = C^0(T) = \{f : T \to \mathbf{R} : f \text{ continuous}\}$$

and for a positive integer r,

$$C^r(T) = \{f : T \to \mathbf{R} : D^\alpha f \in C(T) \text{ for all } |\alpha| \le r\}.$$

Moreover, we set

$$C^r(\overline{T}) = \{f \in C(T) : D^\alpha f \text{ can be extended}$$
$$\text{continuously to } \overline{T} \text{ for all } |\alpha| \le r\},$$

where $\overline{T}$ denotes the closure of T.

We now introduce spaces of splines in two variables.

Definition 2.1. We call

$$P_m = \text{span}\left\{x^i y^j : 0 \le i + j \le m\right\} \quad (2.30)$$

the space of *polynomials of degree m in two variables*.

Let a closed subset T of $\mathbf{R}^2$ and open subsets $T_1, \ldots, T_k$ of T be given such that

$$T_i \cap T_j = \emptyset, \quad i \ne j, \quad i,j = 1,\ldots,k, \quad (2.31)$$

and

$$T = \bigcup_{i=1}^{k} \overline{T}_i. \tag{2.32}$$

The sets $T_1, \ldots, T_k$ are called a *partition* of T.

If r and m are integers with $0 \le r < m$, then we call

$$S_m^r(T_1, \ldots, T_k) = \left\{ s \in C^r(T) : s|_{\overline{T}_i} \in P_m, \ i = 1, \ldots, k \right\} \tag{2.33}$$

the space of *polynomial splines* of degree m and smoothness r with respect to the partition $T_1, \ldots, T_k$ of T.

We first note that the dimension of P_m is $(m+1)(m+2)/2$.

If $T_1, \ldots, T_k$ is a partition of T consisting of triangles or rectangles, then the sets $T_1, \ldots, T_k$ are called *finite elements* .

At present there are no complete results on interpolation by splines from $S_m^r(T_1, \ldots, T_k)$. Therefore, in the following we describe some special interpolation methods for triangular partitions which are known from the finite element analysis.

The basic idea is to choose on each triangle T_i, $i = 1, \ldots, k$, points in such a way that the corresponding interpolation problem for polynomials from P_m has a unique solution and that the resulting piecewise polynomial defined on the union of these triangles is from $S_m^r(T_1, \ldots, T_k)$. Typical results which are known in the literature deal with the cases $m = 1, 2, 3, 5$ and $r = 0, 1$.

Let T be a polygon in $\mathbf{R}^2$ and let $T_1, \ldots, T_k$ be a partition of T consisting of triangles such that the vertices of each triangle are not contained in the interior of an edge of any other triangle.

We begin with the simplest case when $m = 1$ and $r = 0$. Let a function $f \in C(T)$ and a triangle T_j, $j = 1, \ldots, k$, with vertices $z_i = (x_i, y_i)$, $i = 1, 2, 3$, be given. Then there exists a unique polynomial $p_j \in P_1$ such that

$$p_j(z_i) = f(z_i), \qquad i = 1, 2, 3. \tag{2.34}$$

Moreover, the resulting function $s : T \to \mathbf{R}$ defined by

$$s|_{\overline{T}_j} = p_j, \qquad j = 1, \ldots, k, \tag{2.35}$$

is from $S_1^0(T_1, \ldots, T_k)$ and satisfies the interpolation conditions

$$s(w_i) = f(w_i), \qquad i = 1, \ldots, n, \tag{2.36}$$

where $w_1, \ldots, w_n$ are the vertices of all triangles $T_1, \ldots, T_k$.

We verify these simple properties by using barycentric coordinates, since a modification of this principle also works in the subsequent cases.

Let a point $z = (x, y) \in T_j$ be given. We show that there exist unique nonnegative numbers $\phi_1(z)$, $\phi_2(z)$, $\phi_3(z)$ such that

$$z = \sum_{i=1}^{3} \phi_i(z) \, z_i \tag{2.37}$$

and

$$1 = \sum_{i=1}^{3} \phi_i(z). \tag{2.38}$$

These real numbers $\phi_1(z)$, $\phi_2(z)$, $\phi_3(z)$ are called the *barycentric coordinates* of the point z. The conditions (2.37) and (2.38) are equivalent to the following system of linear equations with nonzero determinant

$$\left.\begin{aligned}
\phi_1(z)\,x_1 + \phi_2(z)\,x_2 + \phi_3(z)\,x_3 &= x \\
\phi_1(z)\,y_1 + \phi_2(z)\,y_2 + \phi_3(z)\,y_3 &= y \\
\phi_1(z) + \phi_2(z) + \phi_3(z) &= 1
\end{aligned}\right\} \tag{2.39}$$

This shows that for every index $i \in \{1, 2, 3\}$, the *barycentric coordinate function* $z \to \phi_i(z)$ is a polynomial from P_1. Since

$$\phi_i(z_j) = \begin{cases} 1, & \text{if } i = j \\ 0, & \text{if } i \neq j \end{cases}, \qquad i, j = 1, 2, 3, \tag{2.40}$$

the polynomial

$$p_j = \sum_{i=1}^{3} f(z_i)\,\phi_i \tag{2.41}$$

satisfies (2.34).

Moreover, let two triangles T_i and T_j be given with a common edge $a_0 + a_1 x + a_2 y = 0$. Since p_i and p_j are linear polynomials in one variable on this edge, they are uniquely determined by the values of f at the two vertices which shows that $s \in S_1^0(T_1, \ldots, T_k)$.

Finally, we note that the functions $s_j \in S_1^0(T_1, \ldots, T_k)$, $j = 1, \ldots, n$, defined by

$$s_j(w_i) = \begin{cases} 1, & \text{if } i = j \\ 0, & \text{if } i \neq j \end{cases}, \qquad i = 1, \ldots, n, \tag{2.42}$$

form a basis of $S_1^0(T_1, \ldots, T_k)$.

In the subsequent cases, the results are given without proofs. The proofs can be found in books on finite elements e.g. Strang & Fix [1973], Oden & Reddy [1976], Mitchell & Wait [1977], Ciarlet [1978] and Schwarz [1980].

We now consider the case when $m = 2$ and $r = 0$. For a given triangle $\tilde{T} \in \{T_1, \ldots, T_k\}$ with vertices z_1, z_2, z_3, we set

$$z_{ij} = (z_i + z_j)/2, \qquad i < j, \quad i, j = 1, 2, 3. \tag{2.43}$$

Then for each function $f \in C(T)$, there exists a unique polynomial $\tilde{p} \in P_2$ such that

$$\tilde{p}(z_i) = f(z_i), \qquad i = 1, 2, 3, \tag{2.44}$$

and

$$\tilde{p}(z_{ij}) = f(z_{ij}), \qquad i < j, \quad i, j = 1, 2, 3. \tag{2.45}$$

We just note that

$$\phi_q(z_{ij}) = \begin{cases} \frac{1}{2}, & \text{if } q \in \{i,j\} \\ 0, & \text{if } q \notin \{i,j\} \end{cases}, \qquad i < j, \; q,i,j = 1,2,3, \tag{2.46}$$

and that $\tilde{p}$ has the form

$$\tilde{p} = \sum_{i=1}^{3} f(z_i)\, \phi_i\, (2\phi_i - 1) + \sum_{\substack{i,j=1 \\ i<j}}^{3} 4f(z_{ij})\, \phi_i\, \phi_j. \tag{2.47}$$

By applying this method to each of the triangles $T_1, \ldots, T_k$, we obtain an interpolating function $s \in S_2^0(T_1, \ldots, T_k)$.

We now consider the case when $m = 3$ and $r = 0$. For a triangle $\tilde{T}$ as above, we set

$$z_{iij} = (2z_i + z_j)/3, \qquad i \neq j, \quad i,j = 1,2,3, \tag{2.48}$$

and

$$z_{123} = (z_1 + z_2 + z_3)/3. \tag{2.49}$$

Then for each function $f \in C(T)$, there exists a unique polynomial $\tilde{p} \in P_3$ such that

$$\tilde{p}(z_i) = f(z_i), \quad \tilde{p}(z_{iij}) = f(z_{iij}), \quad \tilde{p}(z_{123}) = f(z_{123}). \tag{2.50}$$

We just note that

$$\phi_i(z_{iij}) = \tfrac{2}{3}, \quad \phi_j(z_{iij}) = \tfrac{1}{3}, \quad \phi_q(z_{iij}) = 0, \quad q \notin \{i,j\}, \quad \phi_q(z_{123}) = \tfrac{1}{3}, \tag{2.51}$$

and that $\tilde{p}$ has the form

$$\tilde{p} = \sum_{i=1}^{3} \tfrac{1}{2} f(z_i)\, \phi_i\, (3\phi_i - 1)(3\phi_i - 2) + \sum_{\substack{i,j=1 \\ i \neq j}}^{3} \tfrac{9}{2} f(z_{iij})\, \phi_i\, (3\phi_i - 1)\, \phi_j$$

$$+ 27 f(z_{123})\, \phi_1\, \phi_2\, \phi_3. \tag{2.52}$$

The resulting interpolating function $s : T \to \mathbf{R}$ is from $S_3^0(T_1, \ldots, T_k)$.

In the following we construct piecewise polynomials which are differentiable. In order to obtain such functions, we have to consider partial derivatives and also normal derivatives defined as follows.

Let a triangle V and a point z from the boundary of V be given which is not a vertex. Moreover, let $n = (n_1, n_2)$ be the outer unit vector perpendicular to the edge of V which contains z. Then for $f \in C^1(V)$,

$$f_n(z) = n_1\, f_x(z) + n_2\, f_y(z) \tag{2.53}$$

is called the *normal derivative* of f at the point z. Here f_x and f_y denote the partial derivatives $\frac{\partial f}{\partial x}$ and $\frac{\partial f}{\partial y}$, respectively.

We now continue with the case when $m = 5$ and $r = 1$. Let a triangle $\tilde{T} \in \{T_1, \ldots, T_k\}$ with vertices z_1, z_2, z_3 and the points z_{ij} as in (2.43) be given. It can be shown that for each function $f \in C^2(T)$, there exists a unique polynomial $\tilde{p} \in P_5$ such that

$$\left. \begin{array}{l} \tilde{p}(z_i) = f(z_i), \ \tilde{p}_x(z_i) = f_x(z_i), \ \tilde{p}_y(z_i) = f_y(z_i), \ \tilde{p}_{xx}(z_i) = f_{xx}(z_i), \\ \tilde{p}_{xy}(z_i) = f_{xy}(z_i), \ \tilde{p}_{yy}(z_i) = f_{yy}(z_i), \ \tilde{p}_n(z_{ij}) = f_n(z_{ij}). \end{array} \right\} \quad (2.54)$$

The resulting interpolating function $s : T \to \mathbf{R}$ is from $S_5^1(T_1, \ldots, T_k)$.

We just note that the dimension of P_5 is 21 and that the same number of interpolation conditions is given by (2.54). The function s is continuous, since $\tilde{p}$ is a polynomial of degree 5 at each edge of $\tilde{T}$ which is uniquely determined by the values, the first and the second derivative of f at the endpoints. Moreover, s is differentiable, since $\tilde{p}_n$ is a polynomial of degree 4 at each edge of $\tilde{T}$ which is uniquely determined by the values of f_n at the endpoints together with its midpoint and by the first derivative of f_n at the endpoints.

This shows that to achieve differentiable splines we have to consider polynomials of degree 5 and partial derivatives up to order two. On the other hand, these requirements can be weakened if we divide the triangle $\tilde{T}$ into three subtriangles. This finite element is called the *Clough–Tocher triangle.*

Let $\tilde{z}$ be a point from the interior of $\tilde{T}$ whose lines to the vertices z_1, z_2, z_3 divide $\tilde{T}$ into three subtriangles $\tilde{T}_1$, $\tilde{T}_2$, $\tilde{T}_3$. It can be shown that for each function $f \in C^1(T)$, there exists a unique spline $\tilde{s} \in S_3^1(\tilde{T}_1, \tilde{T}_2, \tilde{T}_3)$ such that

$$\tilde{s}(z_i) = f(z_i), \ \tilde{s}_x(z_i) = f_x(z_i), \ \tilde{s}_y(z_i) = f_y(z_i), \ \tilde{s}_n(z_{ij}) = f_n(z_{ij}). \quad (2.55)$$

The resulting interpolating function $s : T \to \mathbf{R}$ is from $S_3^1(V_1, \ldots, V_{3k})$, where $V_1, \ldots, V_{3k}$ is the collection of all subtriagles of $T_1, \ldots, T_k$.

We just note that the dimension of the space $\{\tilde{s} : \tilde{T} \to \mathbf{R} : \tilde{s}|_{\tilde{T}_i} \in P_3\}$ is 30. On the other hand, 12 interpolation conditions are given by (2.55) and to obtain a differentiable $\tilde{s} : \tilde{T} \to \mathbf{R}$, we have to require common values of the polynomials $\tilde{p}_i = \tilde{s}|_{\tilde{T}_i}$, $i = 1, 2, 3$, and its first partial derivatives at the points $\tilde{z}$, z_i, $i = 1, 2, 3$, and of its normal derivatives at the midpoints between $\tilde{z}$ and z_i, $i = 1, 2, 3$, which yield 18 further conditions. The differentiability of $s : T \to \mathbf{R}$ follows similarly as in the above case when $m = 5$ and $r = 1$.

Finally, we note that "basis functions" corresponding to the above interpolation methods can be constructed by requiring for each of these functions that one value which appears in the interpolation conditions is one and all others are zero. This is done for all possible choices.

For related topics concerning in particular so–called *super splines* and *vertex splines* the reader is referred to Chui & Lai [1987], Schumaker [1987], Chui [1988] and the references therein.

There is a vast literature on the use of finite element functions in solving partial differential equations. We refer to the books on finite elements mentioned above.

2.3. Spline Functions

Some results on the construction of bases for bivariate spline spaces, which in special cases consist of locally supported functions, are given. In addition, various types of splines are defined which extend the univariate B–splines and truncated power functions to several variables. Moreover, we discuss a strategy for interpolation of scattered data by using radial functions. A special case of this multivariate method can be considered as a generalization of interpolation by natural splines. The Bernstein–Bézier representation of polynomials in several variables is a suitable form to study splines on triangular partitions.

We begin with an observation which gives a first insight into the structure of bivariate splines.

Let a spline space $S_m^r(T_1, \ldots, T_k)$ as in Definition 2.1 be given. Suppose that two adjacent sets T_i and T_j are separarted by a line $a_0 + a_1 x + a_2 y = 0$. If $s \in S_m^r(T_1, \ldots, T_k)$ and $p_i = s|_{T_i}$, $p_j = s|_{T_j}$, then there exists a bivariate polynomial $q \in P_{m-r-1}$ such that $p_i - p_j = (a_0 + a_1 x + a_2 y)^{r+1} q$, where $q \in P_{m-r-1}$. In particular, $s = p_i + (a_0 + a_1 x + a_2 y)_+^{r+1} q$ on $T_i \cup T_j$, if $T_i \in \{(x, y) \in \mathbf{R}^2 : a_0 + a_1 x + a_2 y \leq 0\}$. Here, we set for a real number α,

$$\alpha_+ = \begin{cases} \alpha, & \text{if } \alpha \geq 0 \\ 0, & \text{if } \alpha < 0 \end{cases}. \tag{2.56}$$

A more general result was given by Chui & Wang [1983a].

Theorem 1.17 of Chapter II shows that there exists a simple basis of a given univariate spline space. On the other hand, the subsequent results show that it is difficult to construct bases of bivariate spline spaces even for uniform partitions.

Let a rectangle $T = [a, b] \times [c, d]$ in $\mathbf{R}^2$ and points

$$a = x_0 < x_1 < \cdots < x_p < x_{p+1} = b \tag{2.57}$$

$$c = y_0 < y_1 < \cdots < y_q < y_{q+1} = d \tag{2.58}$$

be given. The subrectangles are denoted by

$$T_{ij} = (x_i, x_{i+1}) \times (y_j, y_{j+1}), \quad i = 0, \ldots, p, \quad j = 0, \ldots, q, \tag{2.59}$$

and the spline spaces corresponding to this partition by $S_m^r(R_{pq})$. We note that this space is not a space of tensor product splines (compare Section 2.1).

The first result can be proved directly by induction and is also a immediate consequence of a theorem of Chui & Wang [1983a] on more general partitions.

Theorem 2.2. *The following functions form a basis of* $S_m^r(R_{pq})$:

$$\{x^\mu y^\nu : 0 \leq \mu + \nu \leq m\} \tag{2.60}$$

$$\{x^\mu y^\nu (x - x_i)_+^{r+1} : 0 \leq \mu + \nu \leq m - r - 1\},$$

$$i = 1, \ldots, p, \tag{2.61}$$

$$\{x^\mu y^\nu (y - y_j)_+^{r+1} : 0 \leq \mu + \nu \leq m - r - 1\},$$

$$j = 1, \ldots, q, \qquad (2.62)$$

$$\{x^\mu y^\nu (x - x_i)_+^{r+1} (y - y_j)_+^{r+1} : 0 \le \mu + \nu \le m - 2r - 2\},$$

$$i = 1, \ldots, p, \; j = 1, \ldots, q, \qquad (2.63)$$

where (2.63) is omitted, if $m < 2r + 2$.

An alternate basis of $S_m^r(R_{pq})$ formed by tensor products of certain univariate splines was constructed by Chui & Schumaker [1982].

Now, let T_{ij}, $i = 0, \ldots, p$, $j = 0, \ldots, q$, be a partition of $T = [a, b] \times [c, d]$ such that the sets $\{x_0, \ldots, x_{p+1}\}$ and $\{y_0, \ldots, y_{q+1}\}$ consist of equidistant points (see (2.57)–(2.59)). We add to each subrectangle T_{ij}, the diagonal from (x_i, y_j) to (x_{i+1}, y_{j+1}) and denote the resulting partition of T by R_{pq}^1. We use the following notation.

Let $y + c_i x + \tilde{c}_i = 0$ be the straight lines through (x_i, y_0) and (x_{i+1}, y_1), $i = 0, \ldots, p$, and let $y + d_j x + \tilde{d}_j = 0$ be the straight lines through (x_0, y_j) and (x_1, y_{j+1}), $j = 1, \ldots, q$. For each natural number $n \ge 2$, we set

$$d(n) = \tfrac{1}{2} \left(m - r - \left[\tfrac{r+1}{n-1}\right]\right)_+ \cdot$$
$$\cdot \left((n-1)m - (n+1)r + (n-3) + (n-1)\left[\tfrac{r+1}{n-1}\right]\right), \quad (2.64)$$

where $[\alpha]$ denotes the largest integer less than or equal to a given real number α.

We know that $n = 3$ lines of the partition R_{pq}^1 intersect at every point (x_i, y_j), $i = 1, \ldots, p$, $j = 1, \ldots, q$, and it was shown by Chui & Wang [1983b] that there exist exactly $d(3)$ (see (2.64)) linearly independent splines

$$s_{ij1}, \ldots, s_{ijd(3)} \qquad (2.65)$$

in $S_m^r(R_{pq}^1)$ which are zero outside of $(x_i, b] \times (y_j, d]$. The functions (2.65) can be computed by solving certain systems of linear equations, but are not given in explicit form (see Chui & Wang [1983b]).

With this notation, the next theorem which follows from a general result of Chui & Wang [1983b] shows which type of functions form a basis of $S_m^r(R_{pq}^1)$.

Theorem 2.3. *A basis of $S_m^r(R_{pq}^1)$ is given by the functions (2.60)–(2.62) together with:*

$$\{x^\mu y^\nu (y + c_i x + \tilde{c}_i)_+^{r+1} : 0 \le \mu + \nu \le m - r - 1\},$$

$$i = 0, \ldots, p, \qquad (2.66)$$

$$\{x^\mu y^\nu (y + d_j x + \tilde{d}_j)_+^{r+1} : 0 \le \mu + \nu \le m - r - 1\},$$

$$j = 1, \ldots, q, \qquad (2.67)$$

$$\{s_{ij1}, \ldots, s_{ijd(3)}\}, \quad i = 1, \ldots, p, \quad j = 1, \ldots, q, \qquad (2.68)$$

(see (2.64) and (2.65)).

We now add to the partition R^1_{pq} of $T = [a, b] \times [c, d]$ in each subrectangle T_{ij} the diagonal from from (x_i, y_{j+1}) to (x_{i+1}, y_j) and denote the resulting partition of T by R^2_{pq}.

Again, we need some notations. Let $y + e_i x + \tilde{e}_i = 0$ be the straight line through (x_i, y_0) and (x_{i-1}, y_1), $i = 1, \dots, p+1$, and let $y + f_j x + \tilde{f}_j = 0$ be the straight line through (x_{p+1}, y_j) and (x_p, y_{j+1}), $j = 1, \dots, q$.

We see that $n = 4$ lines of the partition R^2_{pq} intersect at every point (x_i, y_j), $i = 1, \dots, p$, $j = 1, \dots, q$, and it was shown by Chui & Wang [1983b] that there exist exactly $d(4)$ (see (2.64)) linearly independent splines

$$\tilde{s}_{ij1}, \dots, \tilde{s}_{ijd(4)} \tag{2.69}$$

in $S^r_m(R^2_{pq})$ which are zero outside of $(x_i, b] \times (y_j, d]$. Analogously, at every point $\left(\frac{x_i + x_{i+1}}{2}, \frac{y_j + y_{j+1}}{2} \right)$, $i = 0, \dots, p$, $j = 0, \dots, q$, there intersect $n = 2$ lines of the partition R^2_{pq} and there exist exactly $d(2)$ linearly independent splines

$$\tilde{s}_{ij1}, \dots, \tilde{s}_{ijd(2)} \tag{2.70}$$

in $S^r_m(R^2_{pq})$ which are zero outside the cone in $\left\{ (x, y) \in \mathbf{R}^2 : x \geq \frac{x_i + x_{i+1}}{2} \right\}$ with the two boundary lines intersecting at the vertex $\left(\frac{x_i + x_{i+1}}{2}, \frac{y_j + y_{j+1}}{2} \right)$. As before, the functions (2.69) and (2.70) can be computed, but are not given explicitly (see Chui & Wang [1983b]).

The next result which follows from a general result of Chui & Wang [1983b] gives a basis of $S^r_m(R^2_{pq})$.

Theorem 2.4. *A basis of $S^r_m(R^2_{pq})$ is given by the functions (2.60)–(2.62), (2.66) and (2.67) together with:*

$$\{ x^\mu y^\nu (y + e_i x + \tilde{e}_i)^{r+1}_+ : 0 \leq \mu + \nu \leq m - r - 1 \},$$
$$i = 1, \dots, p+1, \tag{2.71}$$

$$\{ x^\mu y^\nu (y + f_j x + \tilde{f}_j)^{r+1}_+ : 0 \leq \mu + \nu \leq m - r - 1 \},$$
$$j = 1, \dots, q, \tag{2.72}$$

$$\{ \tilde{s}_{ij1}, \dots, \tilde{s}_{ijd(4)} \}, \quad i = 1, \dots, p, \quad j = 1, \dots, q, \tag{2.73}$$

$$\{ \tilde{s}_{ij1}, \dots, \tilde{s}_{ijd(2)} \}, \quad i = 0, \dots, p, \quad j = 0, \dots, q, \tag{2.74}$$

(see (2.64),(2.69) and (2.70)).

We know that B–splines play a fundamental role in the univariate spline theory (see Section 2 of Chapter II). This type of spline function was generalized to the multivariate setting in several directions. Here we can only discuss very few aspects of this subject. It was intensively studied in the last decade, but we have to refer to several surveys for further information.

We begin with some notation. For a set A, we denote by $\mathrm{vol}_r(A)$ the r-dimensional Lebesgue measure of A, by $\overline{A}$ the closure of A, by int A the interior of A and by $|A|$ the cardinality of A. The set of integers is denoted by $\mathbf{Z}$.

Let points $x^0, \ldots, x^r \in \mathbf{R}^n$ be given. The set

$$\text{conv}(x^0, \ldots, x^r) = \left\{ \sum_{i=0}^{r} t_i \, x^i : \sum_{i=0}^{r} t_i = 1, t_i \geq 0, \, i = 0, \ldots, r \right\}$$

is called the *convex hull* of the points $x^0, \ldots, x^r$. Moreover, the set $\text{conv}(x^0, \ldots, x^r)$ is called an *r–simplex* if $\text{vol}_r(\text{conv}(x^0, \ldots, x^r)) > 0$.

The *directional derivative* of a function $f \in C^1(\mathbf{R}^n)$ along $z = (z_1, \ldots, z_n) \in \mathbf{R}^n$ is defined by

$$f_z = \sum_{i=1}^{n} z_i \frac{\partial f}{\partial x_i}.$$

The set

$$\text{supp} \, f = \overline{\{x \in \mathbf{R}^n : f(x) \neq 0\}}$$

is called the *support* of a given function $f : \mathbf{R}^n \to \mathbf{R}$. The space of continuous functions on $\mathbf{R}^n$ with compact support is defined by

$$C_0(\mathbf{R}^n) = \{ f \in C(\mathbf{R}^n) : \text{supp} \, f \text{ is compact} \}.$$

In the following we define various types of multivariate splines.

Let integers $n \geq 1$ and $m \geq 0$ be given. Moreover, let $x^0, \ldots, x^{n+m} \in \mathbf{R}^n$ be points which are in *general position*, i.e. the convex hull of any $n + 1$ of them is an n–simplex. The function $S(\cdot | x^0, \ldots, x^{n+m}) : \mathbf{R}^n \to \mathbf{R}$ defined by

$$S(x | x^0, \ldots, x^{n+m})$$

$$= (n+m)! \, \text{vol}_m \left\{ (t_0, \ldots, t_{n+m}) \in S^{n+m} : \sum_{i=0}^{n+m} t_i \, x^i = x \right\}, x \in \mathbf{R}^n, (2.75)$$

where

$$S^{n+m} = \left\{ (t_0, \ldots, t_{n+m}) \in \mathbf{R}^{n+m+1} : \sum_{i=0}^{n+m} t_i = 1, \, t_i \geq 0, \, i = 0, \ldots, n+m \right\},$$

is called a *multivariate simplex spline*.

This spline was introduced by de Boor [1976d] in the following equivalent form: Choose any points $y^0, \ldots, y^{n+m} \in \mathbf{R}^{n+m}$ such that $\text{vol}_{n+m}(\text{conv}(y^0, \ldots, y^{n+m})) > 0$ and $y^i |_{\mathbf{R}^n} = x^i$, $i = 0, \ldots, n+m$. Then we have

$$S(x | x^0, \ldots, x^{n+m})$$

$$= \frac{\text{vol}_m(\{y \in \text{conv}(y^0, \ldots, y^{n+m}) : y |_{\mathbf{R}^n} = x\})}{\text{vol}_{n+m}(\text{conv}(y^0, \ldots, y^{n+m}))}, \quad x \in \mathbf{R}^n. \quad (2.76)$$

By multiplying (2.75) with $f \in C(\mathbf{R}^n)$ and integrating over $\mathbf{R}^n$, we get the identity

$$\int_{\mathbf{R}^n} f(x) \, S(x | x^0, \ldots, x^{n+m}) \, dx$$

$$= (n+m)! \int_{S^{n+m}} f \left(\sum_{i=0}^{n+m} t_i \, x^i \right) dt_1 \cdots dt_{n+m}, \qquad f \in C(\mathbf{R}^n), \quad (2.77)$$

(see Dahmen [1980], Micchelli [1980]) which is useful for analysing the structure of simplex splines and also suitable for generalizations (see deBoor & Höllig [1982a]). The relation (2.77) was already given by Curry & Schoenberg [1966] as a characterization of univariate B–splines (see Definition 2.3 in Chapter II) and therefore, the simplex spline is a generalization of the classical B–spline.

The simplex spline $S(\cdot|x^0,\ldots,x^{n+m})$ is a piecewise polynomial of degree m in $C^{m-1}(\mathbf{R}^n)$ and satisfies

$$S(x|x^0,\ldots,x^{n+m}) = 0, \qquad x \in \mathbf{R}^n \setminus \mathrm{conv}(x^0,\ldots,x^{n+m}), \qquad (2.78)$$

$$S(x|x^0,\ldots,x^{n+m}) > 0, \qquad x \in \mathrm{int}\, \mathrm{conv}(x^0,\ldots,x^{n+m}). \qquad (2.79)$$

Since this spline has optimal smoothness $m-1$, the corresponding cut regions are rather complicated. In particular, it is a polynomial in every region which is not intersected by an $(n-1)$-simplex formed by n points from $\{x^0,\ldots,x^{n+m}\}$ (cf. the surveys of deBoor [1982] and Dahmen & Micchelli [1983a]).

The following result due to Micchelli [1979], [1980] shows that the simplex spline can be computed with aid of recurrence relations (cf. Theorem 2.14 and Theorem 2.16 in Chapter II for the univariate case).

Theorem 2.5. *(i)* *For all* $x = \sum\limits_{i=0}^{n+m} a_i\, x^i$ *with* $\sum\limits_{i=0}^{n+m} a_i = 1$,

$$S(x|x^0,\ldots,x^{n+m})$$
$$= \frac{n+m}{n} \sum_{i=0}^{n+m} a_i\, S(x|x^0,\ldots,x^{i-1}, x^{i+1},\ldots,x^{n+m}). \qquad (2.80)$$

(ii) *For all* $z = \sum\limits_{i=0}^{n+m} b_i\, x^i$ *with* $\sum\limits_{i=0}^{n+m} b_i = 0$,

$$S_z(x|x^0,\ldots,x^{n+m})$$
$$= (n+m) \sum_{i=0}^{n+m} b_i\, S(x|x^0,\ldots,x^{i-1}, x^{i+1},\ldots,x^{n+m}), x \in \mathbf{R}^n. (2.81)$$

We next give the definition of truncated power functions introduced by Dahmen [1980].

Let points $x^1,\ldots,x^{n+m} \in \mathbf{R}^n$ be points such that $0 \notin \mathrm{conv}(x^1,\ldots,x^{n+m})$ and any n points from $\{x^1,\ldots,x^{n+m}\}$ are linearly independent. The function $T(\cdot|x^1,\ldots,x^{n+m}) : \mathbf{R}^n \to \mathbf{R}$ defined by

$$T(x|x^1,\ldots,x^{n+m})$$
$$= \mathrm{vol}_m \left\{ (t_1,\ldots,t_{n+m}) \in \mathbf{R}_+^{n+m} : \sum_{i=1}^{n+m} t_i\, x^i = x \right\}, \quad x \in \mathbf{R}^n, \qquad (2.82)$$

where $\mathbf{R}_+^{n+m} = \{(t_1,\ldots,t_{n+m}) \in \mathbf{R}^{n+m} : t_i \geq 0,\, i = 0,\ldots,n+m\}$, is called a *multivariate truncated power function*.

Analogously to above, this function satisfies

$$\int_{\mathbf{R}^n} f(x)\, T(x|x^1, \ldots, x^{n+m})\, dx$$

$$= \int_{\mathbf{R}^{n+m}_+} f\left(\sum_{i=1}^{n+m} t_i\, x^i\right) dt_1 \ldots dt_{n+m}, \qquad f \in C_0(\mathbf{R}^n). \qquad (2.83)$$

The function $T(\cdot|x^1, \ldots, x^{n+m})$ is a piecewise polynomial of degree m in the space $C^{m-1}(\mathbf{R}^n)$ and satisfies

$$T(x|x^1, \ldots, x^{n+m}) = 0, x \in \mathbf{R}^n \setminus \left\{ \sum_{i=1}^{n+m} t_i\, x^i : t_i \geq 0,\, i = 1, \ldots, n+m \right\}, \qquad (2.84)$$

$$T(x|x^1, \ldots, x^{n+m}) > 0, x \in \left\{ \sum_{i=1}^{n+m} t_i\, x^i : t_i > 0,\, i = 1, \ldots, n+m \right\}. \qquad (2.85)$$

The cut regions of this spline are

$$\left\{ \sum_{i=1}^{n-1} t_i\, x^{j_i} : t_i \geq 0,\, i = 1, \ldots, n-1 \right\}, \qquad (2.86)$$

where $1 \leq j_1 < \cdots < j_{n-1} \leq n+m$ (cf. the survey of Dahmen & Micchelli [1983a]).

In particular, if $n = 1$, then $T(\cdot|x^1, \ldots, x^{n+m})$, where $x^i = 1, i = 1, \ldots, m+1$, is the univariate truncated power function of degree m with knot at zero.

The following recurrence relations were established by Dahmen [1980].

Theorem 2.6. *(i)   For all* $x = \sum\limits_{i=1}^{n+m} a_i\, x^i$,

$$T(x|x^1, \ldots, x^{n+m}) = \frac{1}{m} \sum_{i=1}^{n+m} a_i\, T(x|x^1, \ldots, x^{i-1}, x^{i+1}, \ldots, x^{n+m}). \qquad (2.87)$$

(ii)   For all $z = \sum\limits_{i=1}^{n+m} b_i\, x^i$,

$$T_z(x|x^1, \ldots, x^{n+m})$$

$$= \sum_{i=1}^{n+m} b_i\, T(x|x^1, \ldots, x^{i-1}, x^{i+1}, \ldots, x^{n+m}), \quad x \in \mathbf{R}^n. \qquad (2.88)$$

Finally, we give the definition of multivariate box splines introduced by de Boor & DeVore [1983].

Let points $x^1, \ldots, x^{n+m} \in \mathbf{R}^n \setminus \{0\}$ be given such that $\mathrm{span}\{x^1, \ldots, x^{n+m}\}$ $= \mathbf{R}^n$. The function $B(\cdot | x^1, \ldots, x^{n+m}) : \mathbf{R}^n \to \mathbf{R}$ defined by

$$B(x|x^1, \ldots, x^{n+m})$$

$$= \mathrm{vol}_m \left\{ (t_1, \ldots, t_{n+m}) \in [0,1]^{n+m} : \sum_{i=1}^{n+m} t_i\, x^i = x \right\}, \quad x \in \mathbf{R}^n, \quad (2.89)$$

is called a *multivariate box spline*.

Analogously to above, this function satisfies

$$\int_{\mathbf{R}^n} f(x)\, B(x|x^1, \ldots, x^{n+m})\, dx$$

$$= \int_{[0,1]^{n+m}} f\left(\sum_{i=1}^{n+m} t_i\, x^i \right) dt_1 \ldots dt_{n+m}, \quad f \in C(\mathbf{R}^n). \quad (2.90)$$

The function $B(\cdot | x^1, \ldots, x^{n+m})$ is a piecewise polynomial of degree m in $C^r(\mathbf{R}^n)$, where $r = \min\{|Y| : Y \subset X,\ \mathrm{span}(X \setminus Y) \neq \mathbf{R}^n\} - 2 \leq m - 1$ for $X = \{x^1, \ldots, x^{n+m}\}$ and satisfies

$$B(x|x^1, \ldots, x^{n+m}) = 0,$$

$$x \in \mathbf{R}^n \setminus \left\{ \sum_{i=1}^{n+m} t_i\, x^i : t_i \in [0,1],\ i = 1, \ldots, n+m \right\}, \quad (2.91)$$

$$B(x|x^1, \ldots, x^{n+m}) > 0, x \in \left\{ \sum_{i=1}^{n+m} t_i\, x^i : t_i \in (0,1),\ i = 1, \ldots, n+m \right\}. \quad (2.92)$$

The cut regions of this spline are

$$\left\{ \sum_{i=1}^{n-1} t_i\, x^{j_i} + \sum_{i=n}^{n+m} u_i\, x^{j_i} : t_i \in [0,1],\ i = 1, \ldots, n-1; \right.$$

$$\left. u_i \in \{0,1\},\ i = n, \ldots, n+m \right\}, \quad (2.93)$$

where $1 \leq j_1 < \cdots < j_{n-1} \leq n+m$ and $\{j_n, \ldots, j_{n+m}\} = \{1, \ldots, n+m\} \setminus \{j_1, \ldots, j_{n-1}\}$ (c.f. the surveys of Höllig [1986a], [1986b], Jetter [1987] and the monograph of Chui [1988]).

In particular, if $n = 1$, then $B(\cdot | x^1, \ldots, x^{m+1})$, where $x^i = 1, i = 1, \ldots, m+1$, is the univariate B–spline of degree m with knots $0, \ldots, m+1$ and support $(0, m+1)$.

The following recurrence relations were proved by de Boor & Höllig [1982b].

Theorem 2.7. (i)   For all $x = \displaystyle\sum_{i=1}^{n+m} a_i\, x^i$,

$$B(x|x^1, \ldots, x^{n+m})$$

$$= \frac{1}{m} \sum_{i=1}^{n+m} (a_i\, B(x|x^1, \ldots, x^{i-1}, x^{i+1}, \ldots, x^{n+m})$$

$$+ (1 - a_i)\, B(x - x^i | x^1, \ldots, x^{i-1}, x^{i+1}, \ldots, x^{n+m})), \quad (2.94)$$

if these box splines are continuous.

(ii) For all $z = \sum\limits_{i=1}^{n+m} b_i\, x^i$,

$$B_z(x|x^1, \ldots, x^{n+m})$$

$$= \sum\limits_{i=1}^{n+m} b_i (B(x|x^1, \ldots, x^{i-1}, x^{i+1}, \ldots, x^{n+m})$$

$$- B(x - x^i|x^1, \ldots, x^{i-1}, x^{i+1}, \ldots, x^{n+m})), \quad x \in \mathbf{R}^n. \quad (2.95)$$

The next result which was proved by Dahmen & Micchelli [1983b] and Jia [1984] characterizes the linear independence of translates of box splines.

Theorem 2.8. *If $x^1, \ldots, x^{n+m} \in \mathbf{Z}^n$ and $\mathrm{span}\{x^1, \ldots, x^{n+m}\} = \mathbf{R}^n$, then the following conditions are equivalent:*
(i)   *The box splines $B(\cdot - j|x^1, \ldots, x^{n+m})$, $j \in \mathbf{Z}^n$, are linearly independent.*
(ii)   *For all $\{y^1, \ldots, y^n\} \subset \{x^1, \ldots, x^{n+m}\}$ with $\mathrm{span}\{y^1, \ldots, y^n\} = \mathbf{R}^n$,*

$$|\det(y^1, \ldots, y^n)| = 1. \quad (2.96)$$

As we have seen in Chapter II, splines which have local support are very useful for numerical purposes. In the following we discuss for certain bivariate spline spaces the existence of bases consisting of this type of function.

Let $T = [0, p + 1] \times [0, q + 1]$, $T_{ij} = (i, i + 1) \times (j, j + 1)$, $i = 0, \ldots, p$, $j = 0, \ldots, q$, and let R^1_{pq} (respectively R^2_{pq}) be the corresponding partition obtained by adding one diagonal (respectively two diagonals) to each subrectangle T_{ij} (see the beginning of this section). Moreover, we define the unit vectors $e^1 = (1, 0)$ and $e^2 = (0, 1)$.

At present there are no general results on bases of locally supported functions for $S^r_m(R^i_{pq})$, $i = 1, 2$, when r and m are arbitrary (for details see Chui [1988]). The following results show that even for special spaces certain anomalies appear.

The first theorem, due to Chui & Wang [1984a], shows that certain box splines form a basis of $S^1_2(R^2_{pq})$ (compare Theorem 2.4).

Theorem 2.9.
(i)   *The dimension of $S^1_2(R^2_{pq})$ is $(p + 3)(q + 3) - 1$.*
(ii)   *There exist exactly $(p + 3)(q + 3)$ box splines*

$$B(x - \alpha, y - \beta|e^1, e^2, e^1 + e^2, e^2 - e^1),$$

$$\alpha = -1, \ldots, p + 1, \quad \beta = -1, \ldots, q + 1, \quad (2.97)$$

which are not identically zero on $[0, p + 1] \times [0, q + 1]$.
(iii) *Every choice of $(p + 3)(q + 3) - 1$ of the box splines in (ii) form a basis of $S^1_2(R^2_{pq})$.*

We next give a basis result for the space $S_3^1(R_{pq}^1)$. To do this we define the sub-sets $V_1 = \text{conv}\{(0,0), (1,0), (3,2)(3,3), (1,3), (0,2)\}$ and $V_2 = \text{conv}\{(0,0), (2,0),$ $(3,1)(3,3), (2,3), (0,1)\}$ of $\mathbf{R}^2$. There exist splines B_1 and B_2 in $S_3^1(R_{pq}^1)$ such that supp $B_1 = V_1$ and supp $B_2 = V_2$ (see Chui & Wang [1984b]). We note that B_1 and B_2 are not box splines.

The following result was proved by Chui & Wang [1984b].

Theorem 2.10.
(i)   *The dimension of $S_3^1(R_{pq}^1)$ is $2(p+3)(q+3) - 5$.*
(ii) *There exist exactly $2(p+3)(q+3) - 2$ splines*

$$B_1(x - \alpha, y - \beta), \quad B_2(x - \alpha, y - \beta), \quad \alpha, \beta \in \mathbf{Z}, \qquad (2.98)$$

which are not identically zero on $[0, p+1] \times [0, q+1]$.
(iii) *There exist $2(p+3)(q+3) - 5$ splines as in (ii) which form a basis of $S_3^1(R_{pq}^1)$.*

Chui & Wang [1984b] developed criteria for dropping three of the splines in (ii) of Theorem 2.10 such that a basis of $S_3^1(R_{pq}^1)$ is obtained.

The next result on $S_4^2(R_{pq}^1)$ was given by Chui & Wang [1984b] who in addition showed that this space does not have a basis of locally supported splines.

Theorem 2.11.
(i)   *The dimension of $S_4^2(R_{pq}^1)$ is $(p+6)(q+6) - 18$.*
(ii) *There exist exactly $(p+4)(q+4) - 2$ box splines*

$$B(x - \alpha, y - \beta | e^1, e^1, e^2, e^2, e^1 + e^2, e^1 + e^2), \quad \alpha, \beta \in \mathbf{Z}, \qquad (2.99)$$

which are not identically zero on $[0, p+1] \times [0, q+1]$.
(iii) *A basis of $S_4^2(R_{pq}^1)$ is given by the box splines in (ii) together with:*

$$\left\{(x - i)_+^4 : i = 0, \ldots, p\right\} \qquad (2.100)$$

$$\left\{(y - j)_+^4 : j = 0, \ldots, q\right\} \qquad (2.101)$$

$$\left\{(x - y - k)_+^4 : k = -q - 1, \ldots, p\right\}. \qquad (2.102)$$

If a space $S_m^r(T_1, \ldots, T_k)$ with an arbitrary triangulation $T_1, \ldots, T_k$ is given, then in contrast to the results at the beginning of this section, not even the dimension of such spaces is known, in general. In particular, it was shown that the dimension depends on the geometry of the partition (cf. the survey of Schumaker [1984b]). In this case lower and upper bounds on the dimension were developed by Schumaker [1984a]. Moreover, a basis of $S_m^r(T_1, \ldots, T_k)$ consisting of locally supported functions was given for the case when $m \geq 4r + 1$ by Alfeld, Piper & Schumaker [1987]. Further details on properties of spline spaces can be found in Schumaker [1984b] and Chui [1988].

In the study of splines on triangular partitions a special representation of bivariate polynomials is useful which we briefly describe in the following.

Let a triangle T in $\mathbf{R}^2$ with vertices x^1, x^2, x^3 be given. For every point $x \in \mathbf{R}^2$ we denote by $\phi_1(x), \phi_2(x), \phi_3(x) \in \mathbf{R}$ the barycentric coordinates of x relative to T which are uniquely defined by

$$x = \sum_{i=1}^{3} \phi_i(x)\, x^i \tag{2.103}$$

and

$$1 = \sum_{i=1}^{3} \phi_i(x) \tag{2.104}$$

(compare (2.37)–(2.39)). Since the barycentric coordinate function ϕ_1, ϕ_2, ϕ_3 are bivariate polynomials of degree one, every bivariate polynomial $p \in P_m$ can be written in the so-called *Bernstein–Bézier form*

$$p(x) = \sum_{i+j+k=m} a_{ijk}\, \frac{m!}{i!\, j!\, k!}\, \phi_1(x)^i\, \phi_2(x)^j\, \phi_3(x)^k, \qquad x \in \mathbf{R}^2, \tag{2.105}$$

where $a_{ijk} \in \mathbf{R}$ and $(i, j, k) \in \mathbf{Z}_+^3$.

We note that in the univariate case when $[x^1, x^2] = [0, 1]$ the analogous representation of a polynomial $p \in P_m$ is

$$p(x) = \sum_{i+j=m} a_{ij}\, \frac{m!}{i!\, j!}\, (1 - x)^i\, x^j = \sum_{j=0}^{m} a_j \binom{m}{j} (1 - x)^{m-j}\, x^j, \qquad x \in \mathbf{R}, \tag{2.106}$$

where

$$a_j = a_{m-j,j}, \quad j = 0, \ldots, m.$$

The polynomials

$$p_j^m(x) = \binom{m}{j} (1 - x)^{m-j}\, x^j, \qquad x \in \mathbf{R}, \tag{2.107}$$

$j = 0, \ldots, m$ are called *Bernstein polynomials of degree* m.

In particular, the Bernstein–Bézier form (2.105) is suitable for describing smoothness conditions for piecewise polynomials on triangles. We give a typical example.

Let $T_1 = \mathrm{conv}(x^1, x^2, x^3)$ and $T_2 = \mathrm{conv}(x^1, x^2, x^4)$ be two adjacent triangles in $\mathbf{R}^2$ such that $T_1 \cap T_2 = \mathrm{conv}(x^1, x^2)$ and let a function $s : T_1 \cup T_2 \to \mathbf{R}$ be given such that $p_1 = s|_{T_1} \in P_m$ and $p_2 = s|_{T_2} \in P_m$. The Bernstern–Bézier forms of p_1 and p_2 relative to T_1 and T_2, respectively, are denoted by

$$p_1(x) = \sum_{i+j+k=m} a_{ijk}\, \frac{m!}{i!\, j!\, k!}\, \phi_1(x)^i\, \phi_2(x)^j\, \phi_3(x)^k \tag{2.108}$$

and

$$p_2(x) = \sum_{i+j+k=m} \tilde{a}_{ijk}\, \frac{m!}{i!\, j!\, k!}\, \tilde{\phi}_1(x)^i\, \tilde{\phi}_2(x)^j\, \tilde{\phi}_3(x)^k. \tag{2.109}$$

Then the following result on the smoothness of s holds. For a given integer $r \geq 0$ we have $s \in C^r(T_1 \cup T_2)$ if and only if for all $l \in \{0, \ldots, r\}$,

$$\tilde{a}_{ijl} = \sum_{\lambda+\mu+\nu=l} \frac{l!}{\lambda!\mu!\nu!} a_{i+\lambda,j+\mu,\nu} \, \phi_1(x^4)^\lambda \, \phi_2(x^4)^\mu \, \phi_3(x^4)^\nu, \qquad (2.110)$$

where $i + j + l = m$.

For further information on the use of the Bernstein–Bézier form, we refer the reader to de Boor [1987b], Chui [1988], the surveys of Böhm, Farin & Kahmann [1984], Barnhill [1985] on computer aided design and the references therein.

In the univariate case there is a fully developed theory on interpolation and best approximation by polynomials and splines. The situation is completely different in the bivariate case.

Even the classical problem of *interpolation by polynomials in two variables* is not completely solved at present. Results on interpolation at special point configurations are known in the literature (see e.g. Cheney [1986], Micchelli [1986a], Chui [1988] and the references therein).

We give an example on a typical configuration where interpolation is possible. We choose $m + 1$ distinct lines $\gamma_0, \ldots, \gamma_m$ in $\mathbf{R}^2$ and distinct points

$$x^{i,0}, \ldots, x^{i,i} \in \gamma_i, \qquad i = 0, \ldots, m. \qquad (2.111)$$

Then for all real numbers $y_{i,0}, \ldots, y_{i,i}$, $i = 0, \ldots, m$, there exists a unique bivariate polynomial $p \in P_m$ such that

$$p(x^{i,j}) = y_{i,j}, \qquad j = 0, \ldots, i, \quad i = 0, \ldots, m. \qquad (2.112)$$

We note that the total number of points is equal to the dimension of P_m. The unique solvabiltiy is easy to verify e.g. by showing that any polynomial $p \in P_m$ which satisfies the homogeneous equations (2.112) is a product of the linear polynomials corresponding to the lines $\gamma_1, \ldots, \gamma_m$ and thus $p(x^{0,0}) = 0$ implies that $p = 0$.

Some results on interpolation by splines are known for special spaces $S_1^0(T_1, \ldots, T_k)$ and $S_2^1(T_1, \ldots, T_k)$ (see Chui & He [1987] and Chui [1988]).

There exists a vast literature on topics such as linear independence of multivariate B–splines, quasi–interpolation and cardinal interpolation by linear combinations of multivariate B–splines, its approximation power and subdivision algorithms for the evaluation of B–splines. We refer to the above mentioned surveys (e.g. de Boor [1982], [1987a], Dahmen & Micchelli [1983a], Höllig [1986a], [1986b], Jetter [1987], Chui [1988]) and the references therein.

We close this section with a general approach to *interpolation of scattered data* by radial functions which includes a generalization of univariate natural splines to several variables. However, the multivariate interpolating functions are no longer usual splines.

Let us first recall Theorem 3.30 and Theorem 3.31 of Chapter II which say that if $k \geq m - 2$, then for any real numbers $y_0, \ldots, y_{k+1}$, there exists a unique natural spline $s \in S_{2m-1}(x_1, \ldots, x_k)$ such that

$$s(x_j) = y_j, \qquad j = 0, \ldots, k + 1, \qquad (2.113)$$

which minimizes the integral $\int_a^b |g^{(m)}(x)|^2 \, dx$ among all $g \in C^m[a,b]$ with $g(x_j) = y_j$, $j = 0, \ldots, k+1$.

We now consider the following multivariate interpolation problem for scattered data.

Let points $x^1, \ldots, x^N \in \mathbf{R}^n$, $y_1, \ldots, y_N \in \mathbf{R}$ and an integer $m \geq 0$ be given. Moreover, let $g : [0, \infty) \to \mathbf{R}$ be a function depending on m. The interpolation problem is to determine coefficients $a_1, \ldots, a_N \in \mathbf{R}$ and a polynomial $p \in P_{m-1}$ such that the function

$$G(x) = \sum_{i=1}^{N} a_i \, g(|x - x^i|^2) + p(x), \qquad x \in \mathbf{R}^n, \tag{2.114}$$

satisfies

$$G(x^j) = y_j, \qquad j = 1, \ldots, N, \tag{2.115}$$

and

$$\sum_{i=1}^{N} a_i q(x^i) = 0, \qquad q \in P_{m-1}. \tag{2.116}$$

Here

$$P_{m-1} = \operatorname{span}\{x_1^{\alpha_1} \cdots x_n^{\alpha_n} : 0 \leq \alpha_1 + \cdots + \alpha_n \leq m - 1\}$$

is the space of polynomials of degree m in n variables which has dimension $M = \binom{n+m-1}{n}$. The functions $x \to g(|x - x^i|^2)$, $x \in \mathbf{R}^n$, $i = 1, \ldots, N$, are called *radial functions*.

If $\{p_1, \ldots, p_M\}$ is a basis of P_{m-1}, then the problem can be reformulated as follows. Determine coefficients $a_1, \ldots, a_N \in \mathbf{R}$ and $b_1, \ldots, b_M \in \mathbf{R}$ which satisfy the following system of linear equations.

$$\sum_{i=1}^{N} a_i \, g(|x^j - x^i|^2) + \sum_{i=1}^{M} b_i \, p_i(x^j) = y_j, \qquad j = 1, \ldots, N, \tag{2.117}$$

and

$$\sum_{i=1}^{N} a_i p_j(x^i) = 0, \qquad j = 1, \ldots, M. \tag{2.118}$$

We first note that if this problem has a unique solution, then polynomials in P_{m-1} are recovered by this interpolation method.

To give conditions for the unique solvability of the interpolation problem, we need the following definition.

Definition 2.12. A function $g \in C[0, \infty)$ is called *conditionally positive definite* of order m on $\mathbf{R}^n$ if for all distinct points $x^1, \ldots, x^N \in \mathbf{R}^n$ and all scalars $a_1, \ldots, a_N \in \mathbf{R}$ with

$$\sum_{i=1}^{N} a_i q(x^i) = 0, \qquad q \in P_{m-1}, \tag{2.119}$$

we have

$$\sum_{i=1}^{N} \sum_{j=1}^{N} a_i \, a_j \, g(|x^i - x^j|^2) > 0. \tag{2.120}$$

In the following we always assume that there does not exist a nontrivial polynomial $q \in P_{m-1}$ with $q(x^j) = 0$, $j = 1, \ldots, N$. Then a sufficient condition for the unique solvability of (2.117) and (2.118) is that g is conditionally positive definite of order m on $\mathbf{R}^n$.

In this context, Micchelli [1986b] gave a sufficient condition for functions having this property.

Theorem 2.13. *If $g \in C[0,\infty) \cap C^\infty(0,\infty)$ has the property that for all integers $j \geq m$ and all points $t \in (0,\infty)$,*

$$(-1)^j \, g^{(j)}(t) > 0, \tag{2.121}$$

then g is conditionally positive definite of order m on $\mathbf{R}^n$ for all $n \geq 1$.

A sufficient condition of Micchelli [1986b] for the nonsingularity of the matrix $(g(|x^i - x^j|^2))_{i,j=1}^N$ refers to the interpolation problem in the case when (2.118) and the polynomial term in (2.117) are omitted (i.e. $m = 0$).

Theorem 2.14. *If $g \in C[0,\infty) \cap C^\infty(0,\infty)$ has the property that for all integers $j \geq 1$ and all points $t \in (0,\infty)$,*

$$g(t) > 0, \qquad (-1)^{j+1} \, g^{(j)}(t) > 0, \tag{2.122}$$

then for all $n \geq 1$ and all distinct points $x^1, \ldots, x^N \in \mathbf{R}^n$,

$$(-1)^{N-1} \cdot \left(g\left(\left| x^i - x^j \right|^2 \right) \right)_{i,j=1}^N > 0. \tag{2.123}$$

We now give several examples of radial functions which are used in practice. Duchon [1977] (see also Meinguet [1979]) introduced for $m > \frac{n}{2}$ the class of functions

$$g(t) = \begin{cases} t^{m-\frac{n}{2}} \log t, & \text{if } n \text{ is even} \\ t^{m-\frac{n}{2}}, & \text{if } n \text{ is odd} \end{cases}. \tag{2.124}$$

The corresponding functions G in (2.114) satisfying (2.116) are called *surface splines* or *thin plate splines.*

In the univariate case $n = 1$, the functions in (2.114) with g as in (2.124) are

$$G(x) = \sum_{i=1}^N a_i \, |x - x^i|^{2m-1} + p(x), \qquad x \in \mathbf{R}, \tag{2.125}$$

where $p \in P_{m-1}$. Since the functions $x \to |x - x^i|^{2m-1}$, $x \in \mathbf{R}$, are polynomials in P_{2m-1} on $x \in \mathbf{R} \setminus (x^1, x^N)$, it follows from (2.116) that $G^{(j)}(x^1) = G^{(j)}(x^N) = 0$, $j = m, \ldots, 2m - 2$, i.e. G is a natural spline.

Analogously to the univariate case, there exists a unique surface spline G which satisfies (2.115) and minimizes the integral

$$\sum_{|\alpha|=m} \int_{\mathbf{R}^n} \binom{m}{\alpha} |D^\alpha H(x)|^2 \, dx \tag{2.126}$$

among all H in $C^m(\mathbf{R}^n)$ (even in the corresponding Sobolev space) with $H(x^j) = y_j$, $j = 1, \ldots, N$ (see Duchon [1977]). The unique solvability of the surface spline interpolation problem also follows from Theorem 2.13, since g or $-g$ in (2.124) satisfies (2.121).

Hardy [1971] introduced the radial function, called *Hardy multiquadrics*,

$$g(t) = \left(t + h^2\right)^{-\frac{1}{2}} \tag{2.127}$$

and

$$g(t) = \left(t + h^2\right)^{\frac{1}{2}} \tag{2.128}$$

for the case when $h > 0$, $n = 2$ and $m = 0$. Since g in (2.127) (respectively (2.128)) satisfies (2.121) (respectively (2.122)), the corresponding interpolation problem (2.115) has a unique solution.

Dyn, Levin & Rippa [1986] considered the radial function

$$g(t) = \left(t + h^2\right)^m \log\left(t + h^2\right), \tag{2.129}$$

where $n = 2$, $m \geq 0$ and $m + h^2 > 0$. The unique solvability of (2.115) and (2.116) is given, since g or $-g$ satisfies (2.121).

For further informations on interpolation of scattered data, we refer the reader to the surveys of Schumaker [1976b], Meinguet [1984], Böhm, Farin & Kahmann [1984], Barnhill [1985], Micchelli [1986a],[1986b], Dyn [1987], Franke [1987] and Powell [1987].

3. Spline Collocation and Differential Equations

We give a brief introduction to the use of splines in the numerical solution of ordinary and delay differential equations. The analog of interpolation of functions which are given explicitly is the so-called collocation method for approximating functions which are defined implicitly by differential equations.

We first consider the following ordinary *nonlinear differential equation* with side condition:

$$y^{(p)}(t) = F\left(t, y(t), \ldots, y^{(p-1)}(t)\right), \quad t \in [a, b], \tag{3.1}$$

$$\sum_{j=0}^{p-1} a_{ij}\, y^{(j)}(z_i) = b_i, \quad i = 1, \ldots, p, \tag{3.2}$$

where $z_1 \leq \cdots \leq z_p$ are given points in $[a, b]$, a_{ij} and b_i, $j = 0, \ldots, p-1$, $i = 1, \ldots, p$, are given real numbers. Moreover, we will assume that the function $F : \mathbf{R}^{p+1} \to \mathbf{R}$ is sufficiently smooth.

If $p = 1$ (respectively $p > 1$), then we obtain an initial value problem (respectively a boundary value problem).

Since most of these problems cannot be solved exactly, we try to construct an approximation to some exact solution y (if it exists). There is a vast literature on the construction of approximate solutions defined on finite subsets of $[a, b]$. Here we consider the approach to constuct a spline approximation on the whole interval.

We first define the relevant spline spaces. Let knots $a = x_0 < x_1 < \cdots < x_k < x_{k+1} = b$ and integers $0 \le r < m$ be given. We call

$$S_m^r(x_1, \ldots, x_k) = \left\{ s \in C^r[a, b] : s|_{[x_i, x_{i+1}]} \in P_m, i = 0, \ldots, k \right\} \qquad (3.3)$$

the space of *polynomial splines of degree m and smoothness r with k fixed knots* $x_1, \ldots, x_k$. The dimension of $S_m^r(x_1, \ldots, x_k)$ is $N = m + k(m - r) + 1$.

We now outline the *collocation method* for constructing an approximate solution of (3.1) and (3.2). Analogously to the case of interpolation, we choose points $t_1 < \cdots < t_{N-p}$ in $[a, b]$. Then the *collocation problem* is to determine a spline $s \in S_m^r(x_1, \ldots, x_k)$ such that

$$s^{(p)}(t_i) = F\left(t_i, s(t_i), \ldots, s^{(p-1)}(t_i) \right), \qquad i = 1, \ldots, N - p, \qquad (3.4)$$

and

$$\sum_{j=0}^{p-1} a_{ij}\, s^{(j)}(z_i) = b_i, \qquad i = 1, \ldots, p. \qquad (3.5)$$

We note that this problem is well defined, since the number of boundary conditions in (3.5) added to the number of conditions in (3.4) is equal to the dimension N of $S_m^r(x_1, \ldots, x_k)$.

A suitable choice of collocation points is the following. We set $m = q + p - 1$, $r = p - 1$ and choose as collocation points in each knot–interval (x_i, x_{i+1}) the q zeros of the Legendre polynomial of degree q relative to $[x_i, x_{i+1}]$, $i = 0, \ldots, k$ (see Definition 5.11 in Chapter I). These points are called *Gauss points*.

It was proved by de Boor & Swartz [1973] that under certain assumptions, in particular if F is sufficiently smooth and $\delta = \max\{|x_{i+1} - x_i| : i = 0, \ldots, k\}$ is sufficiently small, then the nonlinear collocation problem (3.4) and (3.5) has a unique solution s near any locally unique solution y of (3.1) and (3.2). The approximation s can be computed by applying Newton type methods. Moreover, the following error estimation holds. If $q \ge p$, then there exists a constant $K > 0$ (independent of $\{x_1, \ldots, x_k\}$) such that

$$\|y - s\|_\infty \le K\, \delta^{q+p} \qquad (3.6)$$

and

$$|y(x_i) - s(x_i)| \le K\, \delta^{2q}, \qquad i = 0, \ldots, k+1. \qquad (3.7)$$

The phenomenon that the order of convergence at the knots is higher (if $q > p$) than on the whole interval is called *superconvergence*. For more details we refer to de Boor & Swartz [1973] and de Boor [1978].

In the following we consider general classes of differential equations. The problem to describe a process with a delayed reaction by a mathematical model and to determine the subsequent behavior leads to the so–called delay equations.

Such type of equations appear in many fields of applications. We describe a few aspects of how to use splines in the numerical solution of these equations.

Let the following initial value problem for *nonlinear delay differential equations* with constant delay be given:

$$y'(t) = F(t, y(t), y(t - \omega)), \qquad t \in [0, \infty), \tag{3.8}$$

$$y(t) = \phi(t), \qquad t \in [-\omega, 0], \tag{3.9}$$

where a sufficiently smooth function $F : \mathbf{R}^3 \to \mathbf{R}$, $\phi \in C[-\omega, 0]$ and the so–called *delay* $\omega > 0$ are given. A solution of this initial value problem is a function $y \in C[-\omega, \infty]$ which satisfies (3.8) and (3.9), where y' means the right–hand derivative.

In order to solve (3.8) and (3.9) numerically, we first consider the problem on $[0, \omega]$ and obtain the ordinary differential equation:

$$y'(t) = F(t, y(t), \phi(t - \omega)), \qquad t \in [0, \omega], \tag{3.10}$$

$$y(0) = \phi(0). \tag{3.11}$$

We now apply the above collocation method to this problem, i.e. we choose knots $0 = x_0 < x_1 < \cdots < x_k < x_{k+1} = \omega$ and collocation points $0 \le t_1 < \cdots < t_{N-1} \le \omega$ and determine a spline $s \in S_m^r(x_1, \ldots, x_k)$ such that

$$s'(t_i) = F(t_i, s(t_i), \phi(t_i - \omega)), \qquad i = 1, \ldots, N - 1, \tag{3.12}$$

$$s(0) = \phi(0), \tag{3.13}$$

where $N = \dim S_m^r(x_1, \ldots, x_k)$ (compare (3.4) and (3.5)). Having obtained an approximate solution on $[0, \omega]$, we continue the method on $[\omega, 2\omega], \ldots, [(n - 1)\omega, n\omega]$ and get an approximate solution of (3.8) and (3.9) on $[-\omega, n\omega]$.

Next, we discuss a few aspects of boundary value problems of a very general form for *nonlinear differential equations with arbitrary functional arguments*:

$$y''(t) = F(t, y'(t), y'(h_1(t)), y(t), y(h_0(t))), \qquad t \in [a, b], \tag{3.14}$$

$$y(t) = \begin{cases} \phi_1(t), & t \in (-\infty, a], \\ \phi_2(t), & t \in [b, \infty), \end{cases} \tag{3.15}$$

where $h_0, h_1 \in C[a, b]$, $\phi_1 \in C^1(-\infty, a]$ and $\phi_2 \in C[b, \infty)$ are given. By applying a simple change of variables, we may assume that $\phi_1(a) = \phi(b) = 0$.

Analogously to above, we choose knots $a = x_0 < x_1 < \cdots < x_k < x_{k+1} = b$ and collocation points $0 \le t_2 < \cdots < t_{N-1} \le \omega$. Then we have to determine a function $s \in C(-\infty, \infty)$ such that $s|_{[a,b]} \in S_m^r(x_1, \ldots, x_k)$ and

$$s''(t_i) = F(t_i, s'(t_i), s'(h_1(t_i)), s(t_i), s(h_0(t_i))), i = 2, \ldots, N - 1, \tag{3.16}$$

$$s(t) = \begin{cases} \phi_1(t), & t \in (-\infty, a], \\ \phi_2(t), & t \in [b, \infty), \end{cases} \tag{3.17}$$

where $N = \dim S_m^r(x_1, \ldots, x_k)$.

Special cases of this problem were investigated by Reddien & Travis [1974] and Bellen & Zennaro [1984] who proved results on the unique solvability of (3.16) and (3.17) (if F is sufficiently smooth and δ is sufficiently small) and on the error resulting from this nonlinear collocation method (see also Bader [1985]).

In the case of interpolation by splines there is a characterization of those points for which the corresponding interpolation problem has a unique solution (see Theorem 3.7 of Chapter II). Having this in mind, we now consider the following question: under which conditions do linear collocation problems have a unique solution (if δ is not required to be sufficiently small)?

Let the linear differential equation

$$
\begin{aligned}
L(y)(t) \;=\; & y''(t) + f_1(t)\,y'(t) + g_1(t)\,y'(h_1(t)) \\
& + f_0(t)\,y(t) + g_0(t)\,y(h_0(t)) = f(t), \qquad t \in [a,b],
\end{aligned} \tag{3.18}
$$

together with the boundary conditions (3.15) be given, where f, f_0, f_1, g_0, g_1, h_0, $h_1 \in C[a,b]$. By applying the collocation method, we have to determine a function $s \in C(-\infty,\infty)$ such that $s|_{[a,b]} \in S_m^r(x_1,\ldots,x_k)$ and

$$
L(s)(t_i) \;=\; f(t_i), \qquad i = 2,\ldots, N-1, \tag{3.19}
$$

$$
s(t) \;=\; \begin{cases} \phi_1(t), & t \in (-\infty, a], \\ \phi_2(t), & t \in [b, \infty), \end{cases} \tag{3.20}
$$

where $N = m + k(m-r) + 1 = \dim S_m^r(x_1,\ldots,x_k)$ (compare (3.16) and (3.17)). We set $\{y_1,\ldots,y_{k(m-r)}\} = \{x_1,\ldots,x_k\}$, where each point y_i appears $m-r$ times in the set $\{x_1,\ldots,x_k\}$, $i = 1,\ldots,k$.

Meinardus & Nürnberger [1985] showed that by applying Theorem 3.7 of Chapter II, one obtains the following density result concerning the unique solvability of the collocation problem (3.19) and (3.20). The requirement on L in this result is satisfied for many standard examples. An analogous result holds for initial value problems:

Suppose that for every function $s \in C(-\infty,\infty)$ with $s(t) = 0$, $t \in (-\infty, a] \cup [b,\infty)$ and $s|_{[a,b]} \in S_m^r(x_1,\ldots,x_k)$, and every index $j \in \{0,\ldots,k\}$, the function $L(s)$ is not identically zero on a subinterval of $[x_j, x_{j+1}]$, if $s|_{[x_j,x_{j+1}]}$ is not the zero function. If points $a = w_1 < w_2 < \cdots < w_{N-1} < w_N = b$ are given such that

$$
w_i < y_i < w_{i+m+1}, \qquad i = 1,\ldots, k(m-r), \tag{3.21}
$$

then for each real number $\varepsilon > 0$ there exist points $a < t_2 < \cdots < t_{N-1} < b$ with $|t_i - w_i| < \varepsilon$, $i = 2,\ldots, N-1$, such that the collocation problem (3.19) and (3.20) has a unique solution.

Finally, we note that a further approach to solving (3.14) and (3.15) approximately is the so-called *Galerkin method*. In this approach we choose a basis $\{s_1,\ldots,s_{N-2}\}$ of the space $\{s \in S_m^r(x_1,\ldots,x_k) : s(a) = s(b) = 0\}$ and have to determine a function $s \in C(-\infty,\infty)$ with $s|_{[a,b]} \in S_m^r(x_1,\ldots,x_k)$ and

$$
s(t) = \begin{cases} \phi_1(t), & t \in (-\infty, a], \\ \phi_2(t), & t \in [b, \infty), \end{cases} \tag{3.22}
$$

such that the following orthogonality relations are satisfied:

$$\int_a^b \left(s''(t) - F\left(t, s'(t), s'\left(h_1(t)\right), s(t), s\left(h_0(t)\right)\right)\right) s_j(t)\, dt = 0$$

$$j = 1, \ldots, N - 2. \tag{3.23}$$

Reddien & Travis [1974] investigated a special case of (3.23) by using cubic splines.

The Galerkin method is a standard approach to the numerical solution of boundary value problems for partial differential equations. In this case the bivariate approximating functions are chosen from finite element spaces (see Section 2.2). The interested reader is referred to the books on the finite element method cited in Section 2.2.

More information about numerical methods for delay equations can be found in the surveys of Cryer [1972] and Bellen [1985], the introductory article of Meinardus & Nürnberger [1985] and the references therein. A general treatment of delay equations is given in the books of Bellman & Cooke [1963], El'sgol'ts & Norkin [1973], Cushing [1977], Driver [1977] and Hale [1977].

Concerning the use of splines in the numerical solution of differential equations, we also refer to the books of Böhmer [1974], Prenter [1975] and Micula [1978], to the recent proceedings of Asher & Russell [1985] on boundary value problems, the survey of Norsett [1984] on initial value problems and the references therein.

References

Achieser N.I.
(1956)   *Theory of Approximation*. Frederick Ungar Publishing Co., New York.

Ahlberg J.H., Nilson E.N. and Walsh J.L.
(1967)   *The Theory of Splines and Their Applications*. Academic Press, New York.

Alfeld P., Piper B. and Schumaker L.L.
(1987)   Minimally supported bases for spaces of bivariate piecewise polynomials of smooth-
          ness r and degree $d \geq 4r + 1$. Computer Aided Geometric Design **3**, 189–198.

Arndt H.
(1974)   On uniqueness of best spline approximations with free knots. J. Approx. Theory
          11, 118–125.

Augsburger W.
(1967)   Segmentapproximation in L_p-Räumen. Dissertation, TH Clausthal.

Asher U.M. and Russell R.D. (eds.)
(1985)   *Numerical Boundary Value ODEs*. Birkhäuser, Basel.

Bader G.
(1985)   Solving boundary value problems for functional differential equations by colloca-
          tion. In: Asher, U.M., Russell, R.D. (eds.) *Numerical Boundary Value ODEs*.
          Birkhäuser, Basel, pp. 227–243.

Barnhill R.E.
(1985)   Surfaces in computer aided geometric design: a survey with new results. Computer
          Aided Geometric Design **2**, 1–17.

Barrar R.B. and Loeb H.L.
(1970)   Existence of best spline approximations with free knots. J. Math. Anal. Appl. **31**,
          383–390.
(1974)   Spline functions with free knots as the limit of varisolvent families. J. Approx.
          Theory **12**, 70–77.
(1976a)   Multiple zeros and applications to optimal linear functionals. Numer. Math. **25**,
          257–262.
(1976b)   On a non–linear characterization problem for monosplines. J. Approx. Theory **18**,
          220–240.
(1978)   On monosplines with odd multiplicities of least norm. J. Analyse Math. **33**, 12–38.
(1980)   Fundamental theorem of algebra for monosplines and related results. SIAM J.
          Numer. Anal. **17**, 874–822.
(1984)   Optimal monosplines with a maximal number of zeros. SIAM J. Math. Anal. **15**,
          1196–1204.
(1986)   The strong uniqueness theorem for monosplines. J. Approx. Theory **46**, 157–169.

Barrar R.B., Loeb H.L. and Werner H.
(1980)   On the uniqueness of best uniform extended totally positive monosplines. J. Approx.
          Theory **28**, 20–29.

Barrow D.L., Chui C.K., Smith P.W. and Ward J.D.
(1978)   Unicity of best L_2-approximation by second order splines with variable knots. Math. Comp. **32**, 1131–1141.

Barrow D.L. and Smith P.W.
(1978)   Asymptotic properties of best $L_2[0, 1]$ approximation by splines with variable knots. Quart. Appl. Math. **36**, 293–304.

Bartelt M.W. and McLaughlin H.W.
(1973)   Characterizations of strong unicity in approximation theory. J. Approx. Theory **9**, 255–266.

Bartelt M.W. and Schmidt D.
(1981)   On Poreda's problem for strong unicity constants. J. Approx. Theory **33**, 69–79.
(1984)   Lipschitz conditions, strong uniqueness and almost Chebyshev subspaces of $C(X)$. J. Approx. Theory **40**, 202–215.

Bastien R. and Dubuc J.
(1976)   Systèmes faibles de Tchebycheff et polynomes de Bernstein. Canad. J. Math. **28**, 653–658.

Bellen A.
(1985)   Constrained mesh methods for functional differential equations. In: Meinardus, G., Nürnberger, G. (eds.) *Delay Equations, Approximation and Application.* ISNM 74, Birkhäuser, Basel, pp. 52–70.

Bellen A. and Zennaro M.
(1984)   A collocation method for boundary value problems of differential equations with functional arguments. Computing **32**, 307–318.

Bellman R. and Cooke K.
(1963)   *Differential Difference Equations.* Academic Press, New York.

Berens H. and Nürnberger G.
(1987)   Nonuniqueness and selections in spline approximation. Constr. Approx., to appear.

Bernstein S.N.
(1926)   *Lecons sur les Propriétés Extrémales et la Meilleure Approximation des Fonctions Analytiques d'une Variable Réelle.* Gauthier–Villars, Paris.
(1931)   Sur la limitation des valeurs d'une polynome $P(x)$ de degré n sur tout un segment par ses valeurs en $(n+1)$ points du segment. Izv. Akad. Nauk SSSR **7**, 1025–1050.

Blatt H.-P.
(1982)   Strenge Eindeutigkeitskonstanten und Fehlerabschätzungen bei linearer Tscheby-scheff–Approximation. In: Collatz, L., Meinardus, G., Werner, H. (eds.) *Numerische Methoden der Approximationstheorie.* ISNM 59, Birkhäuser, Basel, pp. 9–25.
(1984)   Exchange algorithms, error estimations and strong unicity in convex programming and Chebyshev approximation. In: Singh, S.P., Burry, J.H.W., Watson, B. (eds.) *Approximation Theory and Spline Functions.* Reidel, Dodrecht, pp. 1–41.

Blatt H.-P., Nürnberger G. and Sommer M.
(1981)   A characterization of pointwise Lipschitz–continuous selections for the metric projection. Numer. Funct. Anal. and Optimiz. **4**, 101–122.

Blatter J.
(1967)   Zur Stetigkeit von mengenwertigen metrischen Projektionen. Report A16, University Bonn.
(1986)   An algorithm for best uniform approximation by splines with fixed knots. In: Chui, C.K., Schumaker, L.L. and Ward, J.D. (eds.) *Approximation Theory V.* Academic Press, New York, pp. 263–266.

Blatter J., Morris P.D. and Wulbert D.E.
(1968)   Continuity of the set valued metric projection. Math. Annalen **178**, 12–24.

Blatter J. and Schumaker L.L.
(1982)   The set of continuous selections of a metric projection in $C(X)$. J. Approx. Theory **36**, 141–155.

(1983)   Continuous selections and maximal alternators for spline approximation. J. Approx. Theory **38**, 71–80.

Böhm W., Farin G. and Kahmann J.

(1984)   A survey of curve and surface methods in CAGD. Computer Aided Geometric Design **1**, 1–60.

Böhmer K.

(1974)   *Spline–Funktionen.* Teubner, Stuttgart.

Bojanic R. and DeVore R.

(1966)   On polynomials of best one sided approximation. L'Enseignement Math. **12**.

Bojanov B.D.

(1977)   Existence of extended monosplines of least deviation. Serdica **3**, 261–272.

(1979)   Uniqueness of the monosplines of least deviation. In: Hämmerlin, G. (ed.) *Numerical Integration.* ISNM 15, Birkhäuser, Basel, pp. 67–97.

(1981)   Uniqueness of the optimal nodes of quadrature formulae. Math. Comp. **36**, 525–546.

de Boor C.

(1963)   Best approximation properties of spline functions of odd degree. J. Math. Mech. **12**, 747–750.

(1968)   On uniform approximation by splines. J. Approx. Theory **1**, 219–235.

(1969)   On the approximation by γ polynomials. In: Schoenberg, I.J. (ed.) *Approximation with Special Emphasis on Spline Functions.* Academic Press, New York, pp. 157–183.

(1972)   On calculating with B–splines. J. Approx. Theory **6**, 50–62.

(1973)   Good approximation by splines with variable knots. In: Meir, A., Sharma, A. (eds.) *Spline Functions and Approximation Theory.* Birkhäuser, Basel, pp. 57–72.

(1974a)   A remark concerning perfect splines. Bull. Amer. Math. Soc. **80**, 724–727.

(1974b)   Good approximation by splines with variable knots II. In: *Conference on the Numerical Solution of Differential Equations.* Lecture Notes in Mathematics **363**, Springer, Berlin, pp. 12–20.

(1975)   On bounding spline interpolation. J. Approx. Theory **14**, 191–203.

(1976a)   Total positivity of the spline collocation matrix. Ind. Univ. J. Math. **25**, 541–551.

(1976b)   On cubic spline functions that vanish at all knots. Advances in Math. **20**, 1–17.

(1976c)   On "best" interpolation. J. Approx. Theory **16**, 28–42.

(1976d)   Splines as linear combinations of B–splines. In: Lorentz, G.G., Chui, C.K., Schumaker, L.L. (eds.) *Approximation Theory II.* Academic Press, New York, pp. 1–47.

(1977)   Computational aspects of optimal recovery. In: Micchelli, C.A., Rivlin, T.J. (eds.) *Optimal Estimation in Approximation Theory.* Plenum. Publ., New York, pp. 69–91.

(1978)   *A Practical Guide to Splines.* Springer, New York.

(1979)   Efficient computer manipulation of tensor products. ACM Trans. Math. Software **5**, 173–182.

(1982)   Topics in multivariate approximation theory. In: Turner, P.R. (ed.) *Topics in Numerical Analysis.* Lecture Notes in Mathematics **965**, Springer, Berlin, pp. 40–78.

(1987a)   Multivariate approximation. In: Iserles, A., Powell, M. (eds.) *State of Art in Numerical Analysis.* Claredon Press, Oxford, pp. 87–109.

(1987b)   B–form basics. In: Farin, G. (ed.) *Geometric Modelling.* SIAM, Philadelphia, pp. 131–148.

de Boor C. and DeVore R.A.

(1983)   Approximation by smooth multivariate splines. Trans. Amer. Math. Soc. **276**, 775–788.

de Boor C. and Fix G.J.

(1973)   Spline approximation by quasi–interpolants. J. Approx. Theory **8**, 19–45.

de Boor C. and Höllig K.

(1982a)   Recurrence relations for multivariate B–splines. Proc. Amer. Math. Soc. **85**, 397–400.

(1982b)   B–splines from parallelepipeds. J. Analyse Math. **42**, 99–115.

de Boor C., Lyche T. and Schumaker L.L.

(1976)   On calculation with B–splines II. Integration. In: Collatz, L., Meinardus, G. (eds.) *Numerische Methoden der Approximationstheorie.* ISNM 30, Birkhäuser, Basel.

de Boor C. and Pinkus A.

(1977)   Backward error analysis for totally positive linear systems. Numer. Math. **27**, 485–490.

(1978)   Proof of the conjectures of Bernstein and Erdös concerning the optimal nodes for polynomial interpolation. J. Approx. Theory **24**, 289–303.

de Boor C. and Swartz

(1973)   Collocation at Gaussian points. SIAM J. Numer. Anal. **10**, 582–606.

Braess D.

(1974a)   Chebyshev approximation by spline functions with free knots. Numer. Math. **17**, 357–366.

(1974b)   On the nonuniqueness of monosplines with least L_2–norm. J. Approx. Theory **12**, 91–93.

(1975)   On the degree of approximation by spline functions with free knots. Aequationes Math. **12**, 80–81.

(1986)   *Nonlinear Approximation Theory.* Springer, Berlin.

Braess D. and Dyn N.

(1982)   On the uniqueness of monosplines and perfect splines of least L_1– and L_2–norm. J. Analyse Math. **41**, 217–233.

(1986)   On the uniqueness of generalized monosplines of least L_p–norm. Constr. Approx. **2**, 79–99.

Braess D. and Werner H.

(1974)   Tschebyscheff–Approximation mit einer Klasse rationaler Splinefunktionen II. J. Approx. Theory **10**, 379–399.

Brosowski B.

(1968)   *Nicht–lineare Tschebyscheff–Approximation.* Bibliographisches Institut, Mannheim.

(1981)   A refinement of the Kolmogorov–criterion. In: *Constructive Function Theory.* Sofia, pp. 241–247.

(1984)   A refinement of an optimality criterion and its application to parametric programming. J. Optimiz. Theory Appl. **42**, 367–382.

Brosowski B., Deutsch F. and Nürnberger G.

(1980)   Parametric approximation. J. Approx. Theory **29**, 261–277.

Brosowski B. and Nürnberger G.

(1988)   Convex sets and optimization. Preprint.

Brosowski B. and Wegmann R.

(1973)   On the lower semicontinuity of the set–valued metric projection. J. Approx. Theory **8**, 84–100.

Brown A.L.

(1964)   Best n–dimensional approximation to sets of functions. Proc. London Math. Soc. **14**, 577–594.

Brutman L.

(1978)   On the Lebesgue function for polynomial interpolation. SIAM J. Numer. Anal. **15**, 694–704.

Burchard H.G.

(1974)   Splines with optimal knots are better. J. Applicable Anal. **3**, 309–319.

Butzer P.L. and Berens H.
(1967)   *Semi-Groups of Operators and Approximation.* Springer, Berlin.

Carasso C. and Laurent P.J.
(1978)   An algorithm of succesive minimization in convex programming. R.A.I.R.O. Numer.
           Anal. **12**, 377–400.

Carroll M.P. and Braess D.
(1974)   On uniqueness of L_1–approximation for certain families of spline functions. J. Ap-
           prox. Theory **12**, 362–364.

Chebyshev P.L.
(1899)   Sur les questions de minima qui se rattachent à la répr̀esentation approximative des
           fonctions. Oeuvres I, St. Petersbourg, 273–378.

Cheney E.W.
(1966)   *Introduction to Approximation Theory.* McGraw Hill, New York.
(1983)   The best approximation of multivariate functions by combinations of univariate
           ones. In: Chui, C.K., Schumaker, L.L., Ward, J.D. (eds.) *Approximation Theory
           IV.* Academic Press, New York, pp. 1–27.
(1986)   *Multivariate Approximation Theory: Selected Topics.* CBSM 51, SIAM, Philadel-
           phia.

Chow J.
(1982)   On the uniqueness of best $L_2[0,1]$ approximation by piecewise polynomials with
           variable breakpoints. Math. Comp. **39**, 571–585.

Chui C.K.
(1988)   *Multivariate Splines: Theory and Applications.* CBMS, SIAM, Philadelphia.

Chui C.K. and He T.X.
(1987)   On location of sample points in C^1 quadratic bivariate spline interpolation. In: Col-
           latz, L., Meinardus, G., Nürnberger, G. (eds.) *Numerical Methods of Approximation
           Theory.* ISNM 81, Birkhäuser, Basel, pp. 30–43.

Chui C.K. and Lai M.J.
(1987)   On multivariate vertex splines and applications. In: Chui, C.K., Schumaker, L.L.,
           Utreras, F.I. (eds.) *Topics in Multivariate Approximation.* Academic Press, New
           York, pp. 19–36.

Chui C.K and Schumaker L.L.
(1982)   On spaces of piecewise polynomials with boundary conditions, I. Rectangles. In:
           Schempp, W., Zeller, K. (eds.) *Multivariate Approximation Theory II.* Birkhäuser,
           Basel, pp. 69–80.

Chui C.K., Smith P.W. and Ward J.D.
(1977)   On the smoothness of best L_2–approximants from nonlinear spline manifolds. Math.
           Comp. **31**, 17–23.

Chui C.K. and Wang R.H.
(1983a)   On smooth multivariate spline functions. Math. Comp. **41**, 131–142.
(1983b)   Multivariate spline spaces. J. Math. Anal. Appl. **94**, 197–221.
(1984a)   On a bivariate B–spline basis. Scienta Sinica **27**, 1129–1142.
(1984b)   Spaces of bivariate cubic and quartic splines on type–1 triangulations. J. Math.
           Anal. Appl. **101**, 540–554.

Ciarlet P.G.
(1978)   *The Finite Element Method for Elliptic Problems.* North Holland, Amsterdam.

Cline A.K.
(1973)   Lipschitz conditions on uniform approximation operators. J. Approx. Theory **8**,
           160–172.

Collatz L. and Krabs W.
(1973)   *Approximationstheorie.* Teubner, Stuttgart.

Cox M.G.
(1972)   The numerical evaluation of B–splines. J. Inst. Maths. Applics. **10**, 134–149.

Cromme L.J.

(1976)   Eine Klasse von Verfahren zur Ermittlung bester nicht-linearer Tschebyscheff-
            Approximationen. Numer. Math. **25**, 447–459.

(1982a)   Regular C^1-parametrization for exponential sums and splines. J. Approx. Theory
            35, 30–44.

(1982b)   A unified approach to differential characterizations of local best approximations for
            exponential sums and splines. J. Approx. Theory **36**, 294–303.

Cryer C.W.

(1972)   Numerical methods for functional differential equations. In: Schmitt, K. (ed.) *Delay
            and Functional Differential Equations and their Applications.* Academic Press, New
            York, pp. 17–101.

Curry H.B. and Schoenberg I.J.

(1947)   On spline distributions and their limits: the Pólya distribution functions. Bull.
            Amer. Math. Soc. **53**, 1114.

(1966)   On Pólya frequency functions IV: The fundamental spline functions and their limits.
            J. Analyse Math. **17**, 71–107.

Cushing J.M.

(1977)   *Integrodifferential Equations and Delay Models in Population Dynamics.* Lecture
            Notes in Biomathematics **20**, Springer, Berlin.

Dahmen W.

(1980)   On multivariate B-splines. SIAM J. Numer. Anal. **17**, 179–191.

Dahmen W. and Micchelli C.A.

(1983a)   Recent progress in multivariate splines. In: Chui, C.K., Schumaker, L.L., Ward,
            J.D. (eds.) *Approximation Theory IV.* Academic Press, New York, pp. 27–121.

(1983b)   Translates of multivariate splines. Linear Algebra Appl. **52**, 217–234.

Delvos F.-J.

(1987)   Periodic interpolation on uniform meshes. J. Approx. Theory **51**, 71–80.

Demko S.

(1985)   On the existence of interpolating projections onto spline spaces. J. Approx. Theory
            43, 151–156.

DeVore R.A.

(1968)   One-sided approximation of functions. J. Approx. Theory **1**, 11–25.

(1972)   *The Approximation of Continuous Functions by Positive Linear Operators.* Lecture
            Notes in Mathematics **293**, Springer, Berlin.

Deutsch F.

(1983)   A survey of metric selections. In: Sine, R.C. (ed.) *Fixed Points and Nonexpansive
            Mappings.* Contemporary Mathematics **18**, Providence, pp. 49–71.

Deutsch F., Nürnberger G. and Singer I.

(1980)   Weak Chebyshev subspaces and alternation. Pacific J. Math. **89**, 9–31.

Diestel J.

(1975)   *Geometry of Banach Spaces — Selected Topics.* Lecture Notes in Mathematics **485**,
            Springer, Berlin.

Dodson D.S.

(1972)   Optimal order approximation by polynomial spline functions. Ph.D. Thesis, Purdue
            University, West Lafayette.

Driver R.D.

(1977)   *Ordinary and Delay Differential Equations.* Springer, Berlin.

Duchon J.

(1977)   Splines minimizing rotation-invariant semi-norms in Sobolev spaces. In: Schempp,
            W., Zeller, K. (eds.) *Constructive Theory of Functions of Several Variables.* Lecture
            Notes in Mathematics **571**, Springer, Berlin, pp. 85–100.

Dyn N.

(1985)   Generalized monosplines and optimal approximation. Constr. Approx. **1**, 137–154.

(1987)   Interpolation of scattered data by radial functions. In: Chui, C.K., Schumaker,
            L.L., Utreras, F.I. (eds.) *Topics in Multivariate Approximation*. Academic Press,
            New York, pp. 47–62.

Dyn N., Levin D. and Rippa S.
(1986)   Numerical procedures for global surface fitting of scattered data by radial functions.
            SIAM J. Sci. Stat. Comp. **7**, 639–659.

Ehlich H. and Haussmann W.
(1970)   Cebysev–Approximation stetiger Funktionen in zwei Veränderlichen, Math. Z. **117**,
            21–34.

El'sgol'ts L.E. and Norkin S.B.
(1973)   *Introduction to the Theory and Application of Differential Equations with Deviating
            Arguments*. Academic Press, New York.

Erdös P.
(1958)   Problems and results on the theory of interpolation. I, Acta Math. Acad. Sci.
            Hungar. **9**, 381–388.

Esch R.E. and Eastman W.L.
(1969)   Computational methods for best spline approximation. J. Approx. Theory **2**, 85–
            96.

Forsythe G. and Moler C.
(1967)   *Computer Solutions of Linear Algebraic Systems*. Prentice–Hall, Englewood Cliffs,
            N.J.

Franke R.
(1987)   Recent advances in the approximation of surfaces from scattered data. In: Chui,
            C.K., Schumaker, L.L., Utreras, F.I. (eds.) *Topics in Multivariate Approximation*.
            Academic Press, New York, pp. 79–98.

Gaffney P.W. and Powell M.J.D.
(1976)   Optimal interpolation. In: Watson, G.A. (ed.) *Numerical Analysis*. Lecture Notes
            in Mathematics **506**, Springer, Berlin.

Galkin R.V.
(1974)   The uniqueness of the element of the best mean approximation to a continuous
            function using splines with fixed nodes. Math. Notes **15**, 3–8.

Gantmacher V. and Krein M.
(1950)   *Oscillation Matrices and Vibrations of Mechanical Systems*. Moscow.

Golomb M. and Weinberger H.
(1959)   Optimal approximation and error bounds. In: Langer, R.E. (ed.) *On Numerical
            Approximation*. University of Wisconsin Press, pp. 117–190.

Gordon W.J.
(1969)   Distributive lattices and approximation of multivariate functions. In: Schoenberg,
            I.J. (ed.) *Approximations with Special Emphasis on Spline Functions*. Academic
            Press, New York, pp. 223–277.

Haar A.
(1918)   Die Minkowskische Geometrie und die Annäherung an stetige Funktionen. Math.
            Annalen **78**, 294–311.

Hale J.
(1977)   *Theory of Functional Differential Equations*. Springer, Berlin.

Hardy R.L.
(1971)   Multiquadric equations of topograghy and other irregular surfaces. J. Geographical
            Res. **76**, 1905–1915.

Henry M.S. and Roulier J.A.
(1978)   Lipschitz and strong unicity constants for changing dimension. J. Approx. Theory
            22, 85–94.

Hettich R. and Zencke H.
(1982)   *Numerische Methoden der Approximation und semi-infiniten Optimierung.* Teubner, Stuttgart.

Hobby C.R. and Rice J.R.
(1965)   A moment problem in L_1-approximation. Proc. Amer. Math. Soc. **65**, 665–670.

Höllig K.
(1986a)   Box splines. In: Chui, C.K., Schumaker, L.L., Ward, J.D. (eds.) *Approximation Theory V.* Academic Press, New York, pp. 71–95.
(1986b)   Multivariate splines. In: de Boor, C. (ed.) *Approxiamtion Theory.* Proc. Sympos. Appl. Math. **36**, AMS, Providence, Rhode Island, pp. 103–127.

Holladay J.C.
(1957)   A smoothest curve approximation. Math. Tables Aids Computation **11**, 233–243.

Holland A.S. and Sahney B.N.
(1979)   *The General Problem of Approximation and Spline Functions.* Krieger Publ. Company, New York.

Jackson D.
(1930)   *The Theory of Approximation.* AMS Vol. XI, Colloq. Publ. Providence, Rhode Island.

Jetter K.
(1978) L_1-Approximation verallgemeinerter konvexer Funktionen durch Splines mit freien Knoten. Math. Z. **164**, 53–66.
(1987)   A short survey on cardinal interpolation by box splines. In: Chui, C.K., Schumaker, L.L., Utreras, F.I. (eds.) *Topics in Multivariate Approximation.* Academic Press, New York, pp. 125–140.

Jetter K. and Lange G.
(1978)   Die Eindeutigkeit L_2-optimaler Monosplines. Math. Z. **158**, 23–34.

Jia R.Q.
(1984)   On the linear independence of translates of box splines. J. Approx Theory **40**, 158–160.
(1988)   Spline interpolation at knot averages. Constr. Approx. **4**, 1–7.

Johnson R.S.
(1960)   On monosplines of least deviation. Trans. Amer. Math. Soc. **96**, 458–477.

Jones R.C. and Karlovitz L.A.
(1970)   Equioscillation under nonuniqueness in the approximation of continuous functions. J. Approx. Theory **3**, 138–145.

Karlin S.
(1968)   *Total Positvity.* Stanford, Carlifornia.
(1969a)   Best quadrature formulas and interpolation by splines satisfying boundary conditions. In: Schoenberg, I.J. (ed.) *Approximation with Special Emphasis on Spline Functions.* Academic Press, New York, pp. 447–466.
(1969b)   The fundamental theorem of algebra for monosplines satisfying boundary conditions and applications to optimal quadrature formulas. In: Schoenberg, I.J. (ed.) *Approximation with Special Emphasis on Spline Functions.* Academic Press, New York, pp. 467–484.
(1971)   Best quadrature formulas and splines. J. Approx. Theory **4**, 59–90.
(1975)   Interpolation properties of generalized perfect splines and the solution of certain extremal problems. Trans. Amer. Math. Soc. **206**, 25–66.
(1976a)   On a class of best nonlinear approximation problem and extended monosplines. In: Karlin, S., Micchelli, C.A., Pinkus, A., Schoenberg, I.J. (eds.) *Studies in Spline Functions and Approximation Theory.* Academic Press, New York, pp. 16–66.
(1976b)   A global improvement theorem for polynomial monosplines. In: Karlin, S., Micchelli, C.A., Pinkus, A., Schoenberg, I.J. (eds.) *Studies in Spline Functions and Approximation Theory.* Academic Press, New York, pp. 67–82.

Karlin S. and Micchelli C.A.
(1972)   The fundamental theorem of algebra for monosplines satisfying boundary condi-
          tions. Israel J. Math **11**, 405–451.

Karlin S., Micchelli C.A., Pinkus A. and Schoenberg I.J. (eds.)
(1976)   *Studies in Spline Functions and Approximation Theory*. Academic Press, New York.

Karlin S. and Schumaker L.L.
(1967)   The fundamental theorem of algebra for Tchebycheffian monosplines. J. Analyse
          Math. **20**, 233–270.

Karlin S. and Studden W.J.
(1966)   *Tchebycheff Systems: with Applications in Analysis and Statistics*. Interscience,
          New York.

Karlin S. and Ziegler Z.
(1966)   Chebyshevian spline functions. SIAM J. Numer. Anal. **3**, 514–543.

Karon J.M.
(1978)   Computing improved Chebyshev approximations by the continuation method I:
          Description of an algorithm. SIAM J. Numer. Anal. **15**, 1269–1288.

Kilgore T.A.
(1978)   A characterization of the Lagrange interpolation projection with minimal Tcheby-
          cheff norm. J. Approx. Theory **24**, 273–288.

Kioustelidis J.B.
(1980)   Optimal segmented approximations. Computing **24**, 1–8.
(1981)   Optimal segmented polynomial approximations. Computing **26**, 239–246.

Kolmogorov A.N.
(1948)   Eine Bemerkung zu den Polynomen von P.L. Tschebyscheff, die von einer gegebenen
          Funktion am wenigsten abweichen (Russian). Usp. Math. Nauk **3**, 216–221.

Kornejchuk N.P.
(1984)   *Splines in Approximation Theory* (Russian). Moscow.

Krein M.G.
(1951)   The ideas of P.L. Chebyshev and A.A. Markov in the theory of limiting values of
          integrals and their further developments. Amer. Math. Soc. Transl. **12**, 1–122.

Kripke B.R. and Rivlin T.J.
(1965)   Approximation in the metric of $L^1(X, \mu)$. Trans. Amer. Math. Soc. **115**, 101–122.

Kroó A.
(1984)   Some theorems on best L_1-approximation of continuous functions. Acta Math. Sci.
          Hungar. **44**, 409–417.

Lawson C.L.
(1964)   Characteristic properties of the segmented rational minimax approximation prob-
          lem. Numer. Math. **6**, 293–301.

Lawson C.L. and Hanson R.J.
(1974)   *Solving Least Square Problems*. Prentice–Hall, Englewood Cliffs, N.J.

Li W.
(1986)   The characterization of continous selections for metric projections in $C(X)$.
          Preprint.

Light W.A. and Cheney E.W.
(1985)   *Approximation Theory in Tensor Product Spaces*. Lecture Notes in Mathematics
          1169, Springer, Berlin.

Loeb H.
(1980)   The monospline of least norm and related problems. In: Cheney, E.W. (ed.) *Ap-
          proximation Theory III*. Academic Press, New York, pp. 21–39.

Lorentz G.G.
(1966)   *Approximation of Functions*. Holt, Rinehart and Winston, New York.

Lorentz G.G., Jetter K. and Riemenschneider S.D.

(1983)   *Birkhoff Interpolation*. Encycl. of Math. and Appl. **19**, Addison–Wesley, Reading.

Lyche T. and Schumaker L.L.

(1975)   Local spline approximation methods. J. Approx. Theory **15**, 294–325.

Mairhuber J.C.

(1956)   On Haar's theorem concerning Chebyshev approximation problems having unique
           solution. Proc. Amer. Math. Soc. **7**, 609–615.

Makarov V.L. and Hlobystov V.V.

(1983)   *Spline Approximation of Functions* (Russian). Vyssa Skola, Moscow.

Malcom M.A.

(1977)   On the computation of nonlinear spline functions. SIAM J. Numer. Anal. **14**,
           254–282.

Malozemov V.N. and Pevnyi A.B.

(1986)   *Polynomial Splines* (Russian). Leningrad State University, Leningrad.

Marin S.P.

(1984)   An approach to data parametrization in parametric cubic spline interpolation prob-
           lems. J. Approx. Theory **41**, 64–86.

Marsden M.J.

(1974a)   Quadratic spline interpolation. Bull. Amer. Math. Soc. **80**, 903–906.

(1974b)   Cubic spline interpolation of continuous functions. J. Approx. Theory **10**, 103–111.

Marsden M.J. and Schoenberg I.J.

(1966)   On variation diminishing spline approximation methods. Mathematica (Cluj) **8**,
           61–82.

McLaughlin H.W. and Somers K.B.

(1975)   Another characterization of Haar subspaces. J. Approx. Theory **14**, 93–102.

McLaughlin H.W. and Zacharski J.J.

(1980)   Segmented approximation. In: Cheney, E.W. (ed.) *Approximation Theory III*.
           Academic Press, New York, pp. 647–654.

Meinardus G.

(1966a)   Über ein Monotonieprinzip bei linearen Approximationen. ZAMM **46**, 227–238.

(1966b)   Zur Segmentapproximation mit Polynomen. ZAMM **46**, 239–246.

(1967)   *Approximation of Functions: Theory and Numerical Methods*. Springer, Berlin.

(1974)   Bemerkungen zur Theorie der B–Splines. In: Böhmer, K., Meinardus, G., Schempp,
           W. (eds.) *Spline–Funktionen*. Bibliographisches Institut, Mannheim, pp. 165–175.

(1976)   Periodische Spline–Funktionen. In: Böhmer, K., Meinardus, G., Schempp, W.
           (eds.) *Spline Functions*. Lecture Notes in Mathematics **501**, Springer, Berlin, pp.
           177–199.

(1987)   Private communication.

Meinardus G. and Nürnberger G.

(1985)   Approximation theory and numerical methods for delay differential equations. In:
           Meinardus, G., Nürnberger, G. (eds.) *Delay Equations, Approximation and Appli-
           cation*. ISNM 74, Birkhäuser, Basel, pp. 13–40.

(1988)   Uniqueness of best L_1–approximations from periodic spline spaces. J. Approx.
           Theory, to appear.

Meinardus G. , Nürnberger G. , Sommer M. and Strauß H.

(1988)   Algorithms for piecewise polynomials and splines with free knots. Math. Comp.,
           to appear.

Meinardus G. and Schwedt D.

(1964)   Nichtlineare Approximation. Arch. Rat. Mech. Anal. **17**, 297–326.

Meinardus G. and Taylor G.D.

(1978)   Periodic quadratic spline interpolant of minimal norm. J. Approx. Theory **23**,
           137–141.

Meinguet J.
(1979)   Multivariate interpolation at arbitrary points made simple. Z. Angew. Math. Phys.
           30, 292–304.
(1984)   Surface spline interpolation: basic theory and computational aspects. In: Singh,
           S.P., Burry, J.H.W., Watson, B. (eds.) *Approximation Theory and Spline Functions.*
           Reidel, Dodrecht, pp. 127–142.

Merz G.
(1979)   Normen von Spline–Interpolationsoperatoren. In: Meinardus, G. (ed.) *Approxima-*
           tion in Theorie und Praxis. Bibliographisches Institut, Mannheim, pp. 183–208.

Micchelli C.A.
(1972)   The fundamental theorem of algebra for monosplines with multiplicities. In:
           Butzer, P.L., Kahane, J., Nagy, B.S. (eds.) *Linear Operators and Approximation.*
           Birkhäuser, Basel, pp. 419–430.
(1977)   Best L_1–approximation by weak Chebyshev systems and the uniqueness of interpo-
           lating perfect splines. J. Approx. Theory **19**, 1–14.
(1979)   On a numerically efficient method for computing B–splines. In: Schemp, W., Zeller,
           K. (eds.) *Multivariate Approximation Theory.* Birkhäuser, Basel, pp. 211–248.
(1980)   A constructive approach to Kergin interpolation in $\mathbf{R}^k$: Multivariate B–splines and
           Lagrange interpolation. Rocky Mountain J. Math. **10**, 485–497.
(1986a)   Algebraic aspects of interpolation. In: de Boor, C. (ed.) *Approximation Theory,*
           Proc. Sympos. Appl. Math. **36**, AMS, Providence, Rhode Island, pp. 81–102.
(1986b)   Interpolation of scattered data: distance matrices and conditionally positive definite
           functions. Constr. Approx. **2**, 11–22.

Micchelli C.A. and Pinkus A.
(1977)   Moment theory for weak Chebyshev systems with applications to monosplines,
           quadrature formulae and best one–sided L_1–approximation by spline functions with
           fixed knots. SIAM J. Math. Anal. **8**, 206–230.

Micchelli C.A. and Rivlin T.J.
(1985)   Lectures on optimal recovery. In: Dold, A., Eckmann, B. (eds.) *Numerical Analysis.*
           Lecture Notes in Mathematics **1129**, Springer, Berlin, pp. 21–93.

Micchelli C.A., Rivlin T.J. and Winograd S.
(1976)   The optimal recovery of smooth functions. Numer. Math. **26**, 191–200.

Michael E.A.
(1956)   Continuous selections. I. Ann. of Math. **63**, 361–382.

Micula G.
(1978)   *Spline Functions and Applications* (Romanian). Ed. Tehnica, Bucuresti.

Mitchell C.A. and Wait R.
(1977)   *The Finite Element Method in Partial Differential Equations.* Wiley, New York.

Müller M.W.
(1978)   *Approximationstheorie.* Akademische Verlagsgesellschaft, Wiesbaden.

Newman D.J. and Shapiro H.S.
(1963)   Some theorems on Chebyshev approximation. Duke Math. J. **30**, 673–684.

Norsett S.P.
(1984)   Splines and collocation for ordinary initial value problems. In: Singh, S.P., Burry,
           J.H.W., Watson, B. (eds.) *Approximation Theory and Spline Functions.* Reidel,
           Dodrecht, pp. 397–417.

Nörlund N.E.
(1954)   *Vorlesungen über Differenzenrechnung.* Chelsea Publ. Co., New York.

Nürnberger G.
(1980)   Nonexistence of continuous selections of the metric projection and weak Chebyshev
           systems. SIAM J. Math. Anal. **11**, 460–467.
(1982a)   A local version of Haar's theorem in approximation theory. Numer. Funct. Anal.
           and Optimiz. **5**, 21–46.

(1982b)   Strong unicity constants for spline functions. Numer. Funct. Anal. and Optimiz.
           5, 319–347.
(1983)   Strong unicity of best approximations: a numerical aspect. Numer. Funct. Anal.
           and Optimiz. 6, 399–421.
(1985a)   Unicity in one–sided L_1–approximation and quadrature formulae. J. Approx. The-
           ory 45, 271–279.
(1985b)   Global unicity in semi–infinite optimization. Numer. Funct. Anal. and Optimiz.
           8, 173–191.
(1985c)   Unicity in semi–infinite optimization. In: Brosowski, B., Deutsch, F. (eds.) *Para-
           metric Optimization and Approximation.* ISNM 72, Birkhäuser, Basel, pp. 231–247.
(1985d)   Best approximation by spline functions: theory and numerical methods. In: Meinar-
           dus, G., Nürnberger, G. (eds.) *Delay Equations, Approximation and Application.*
           ISNM 74, Birkhäuser, Basel, pp. 180–212.
(1986)   Chebyshev approximation by splines with free knots and computation. In: Chui,
           C.K. , Schumaker, L.L., Ward, J.D. (eds.) *Approximation Theory V.* Academic
           Press, New York, pp. 511–514.
(1987a)   Strong unicity constants in Chebyshev approximation. In: Collatz, L., Meinardus,
           G., Nürnberger, G. (eds.) *Numerical Methods of Approximation Theory.* ISNM 81,
           Birkhäuser, Basel, pp. 144–154.
(1987b)   Strongly unique spline approximations with free knots. Constr. Approx. 3, 31–42.
(1989)   On the structure of nonlinear approximating families and splines with free knots.
           In: Chui, C.K., Schumaker, L.L., Ward, J.D. (eds.) *Approximation Theory VI.*
           Academic Press, New York, to appear.

Nürnberger G. and Braess D.
(1981)   Nonuniqueness of best L_p– approximation for generalized convex functions by
           splines with free knots. Numer. Funct. Anal. and Optimiz. 4, 199–209.

Nürnberger G., Schumaker L.L., Sommer M. and Strauß H.
(1983)   Interpolation by generalized splines. Numer. Math. 42, 195–212.
(1984)   Generalized Tchebycheffian splines. SIAM J. Math. Anal. 15, 790–804.
(1985)   Approximation by generalized splines. J. Math. Anal. Appl. 108, 466–494.
(1988)   Uniform approximation by generalized splines with free knots. J. Approx. Theory,
           to appear.

Nürnberger G. and Singer I.
(1982)   Uniqueness and strong uniqueness of best approximations by spline subspaces and
           other spaces. J. Math. Anal. Appl. 90, 171–184.

Nürnberger G. and Sommer M.
(1978a)   Weak Chebyshev subspaces and continuous selections for the metric projection.
           Trans. Amer. Math. Soc. 238, 129–138.
(1978b)   Characterization of continuous selections of the metric projection for spline func-
           tions. J. Approx. Theory 22, 320–330.
(1983a)   Alternation for best spline approximations. Numer. Math. 41, 207–221.
(1983b)   A Remez type algorithm for spline functions. Numer. Math. 41, 117–146.
(1985)   Continuous selections in Chebyshev approximation. In: Brosowski, B., Deutsch, F.
           (eds.) *Parametric Optimization and Approximation.* ISNM 72, Birkhäuser, Basel,
           pp. 248–263.

Nürnberger G., Sommer M. and Strauß H.
(1986)   An algorithm for segment approximation. Numer. Math. 48, 463–477.

Oden J.D. and Reddy J.N.
(1976)   *An Introduction to the Mathematical Theory of Finite Elements.* Wiley, New York.

Pavlidis T. and Maika A.P.
(1974)   Uniform piecewise polynomial approximation with variable joints. J. Approx. The-
           ory 12, 61–69.

Pinkus A.
(1976)   One–sided L_1–approximation by splines with fixed knots. J. Approx. Theory **18**, 130–135.
(1981)   Bernstein's comparison theorem and a problem of Braess. Aequationes Math. **23**, 98–107.
(1985)   *N–Width in Approximation Theory*. Springer, Berlin.
(1986)   N–width and optimal recovery. In: de Boor, C. (ed.) *Approximation Theory*. Proc. Sympos. Appl. Math. **36**, AMS, Providence, Rhode Island, pp. 51–66.
(1987)   Monograph on L_1–approximation. Preprint.

Pinkus A. and Totik V.
(1984)   One–sided L_1–approximation. Canad. Math. Bull. **29**, 84–90.

Pólya G. and Schoenberg I.J.
(1958)   Remarks on the la Vallée Poussin means and convex conformal maps of the circle. Pacific J. Math. **8**, 295–334.

Pólya G. and Szegö G.
(1972)   *Problems and Theorems in Analysis I*. Springer, New York.

Popov V.A.
(1975)   On approximation of absolutely continuous functions by splines.   Mathematica (Cluj) **8**, 1299–1301.

Popoviciu T.
(1959)   Sur le reste dans certaines formulas linéaires d'approximation de l'analyse. Mathematica (Cluj) **1**, 95–142.

Poreda S.J.
(1976)   Counterexamples in best approximation. Proc. Amer. Mat. Soc. **56**, 167–171.

Powell M.J.D.
(1967)   On the maximum errors of polynomial approximations defined by interpolation and by least square criteria. Computer Journal **9**, 404–407.
(1968)   On best L_2–spline approximations. In: Collatz, L. Meinardus, G., Unger, H. (eds.) *Numerische Mathematik, Differentialgleichungen, Approximationstheorie*. ISNM 9, Birkhäuser, Basel, pp. 317–337.
(1981)   *Approximation Theory and Methods*. Cambridge University Press.
(1987)   Radial basis functions for multivariate interpolation: a review. In: Cox, M.G., Mason, J.C. (eds.) *Algorithms for the Approximation of Functions and Data*. Oxford University Press, pp. 143–168.

Prenter P.M.
(1975)   *Splines and Variational Methods*. John Wiley & Sons, New York.

Quade W. and Collatz L.
(1938)   Zur Interpolationtheorie der reellen periodischen Funktionen.   Sitzungsbericht Preuss. Akad. Wiss. **30**, 383–429.

Reddien G. and Travis C.C.
(1974)   Approximation methods for boundary value problems of differential equations with functional arguments. J. Math. Anal. Appl. **46**, 62–74.

Remez E.J.A.
(1934a)   Sur la détermination des polynômes d'approximation de degré donnée. Comm. Soc. Math. Kharkov **10**, 41–63.
(1934b)   Sur un procédé convergent d'approximations successives pour déterminer les polynômes d'approximation. Compt. Rend. Acad. Sc. **198**, 2063–2065.
(1934c)   Sur le calcul effectiv des polynômes d'approximation des Tschebyscheff. Compt. Rend. Acad. Sc. **199**, 337–340.

Rice J.R.
(1964)   *The Approximation of Functions I*. Addison & Wesley, Reading, Massachusetts.

(1967)   Characterization of Chebyshev approximation by splines. SIAM J. Math. Anal. **4**,
           557–567.
(1969)   *The Approximation of Functions II.* Addison & Wesley, Reading, Massachusetts.

Rivlin T.J.
(1969)   *An Introduction to Approximation of Functions.* Blaisdell, Massachusetts.

Sard A.
(1963)   *Linear Approximation.* Amer. Math. Soc., Providence, R.I.

Sard A. and Weintraub S.
(1971)   *A Book of Splines.* Wiley New York.

Schaback R.
(1973)   Spezielle rationale Splinefunktionen. J. Approx. Theory **7**, 281–292.
(1978)   On alternation numbers in nonlinear Chebyshev approximation. J. Approx. Theory
           23, 379–391.

Schempp W.
(1982)   *Cardinal contour integral representation of cardinal spline functions.* Contemporary
           Math. **7**, AMS, Providence, Rhode Island.

Scherer K.
(1974)   Stetigkeitsmoduli und beste Approximation durch polynomiale Splines. In: Böh-
           mer, K., Meinardus, G., Schempp, W. (eds.) *Spline-Funktionen.* Bibliographisches
           Institut, Mannheim, pp. 289–302.

Schmidt D.
(1978)   On an unboundness conjecture for strong unicity constants. J. Approx. Theory **24**,
           216–223.
(1980)   A characterization of strong unicity constants. In: Cheney, E.W. (ed.) *Approxima-
           tion Theory III.* Academic Press, New York, pp. 805–810.

Schoenberg I.J.
(1946a)   Contributions to the problem of approximation of equidistant data by analytic
           functions, Part A: On the problem of smoothing of graduation, a first class of
           analytic approximation formulae. Quart. Appl. Math. **4**, 45–99.
(1946b)   Contributions to the problem of approximation of equidistant data by analytic
           functions, Part B: On the problem of osculatory interpolation, a second class of
           analytic approximation formulae. Quart. Appl. Math. **4**, 112–141.
(1964a)   Spline interpolation and best quadrature formulae. Bull. Amer. Math. Soc. **70**,
           143–148.
(1964b)   On best approximation of linear operators. Nederl. Akad. Wetensch. Indag. Math.
           26, 155–163.
(1964c)   On interpolation by spline functions and its minimal properties. In: Butzer, P.L.,
           Korevaar, J. (eds.) *On Approximation Theory.* ISNM 5, Birkhäuser, Basel, pp.
           109–129.
(1965)   On monosplines of least deviation and quadrature formulae. SIAM J. Numer. Anal.
           2, 144–170.
(1966)   On monosplines of least square deviation and best quadrature formulae II. SIAM
           J. Numer. Anal. **3**, 321–328.
(1967)   On spline functions. In: Shisha, O. (ed.) *Inequalities.* Academic Press, New York,
           pp. 255–291.
(1969)   Monosplines and quadrature formulae. In: Greville, T.N.E. (ed.) *Theory and
           Applications of Spline Functions.* Academic Press, New York, pp. 157–207.
(1973)   *Cardinal spline interpolation.* CBMS **12**, SIAM, Philadelphia.

Schoenberg I.J. and Whitney A.
(1953)   On Polya frequency functions III. The positivity of translation determinants with
           application to the interpolation problem by spline curves. Trans. Amer. Math.
           Soc. **74**, 246–259.

Schönhage A.
(1971)   *Approximationstheorie.* Walter de Gruyter & Co., Berlin.

Schultz M.H.
(1973)   *Spline Analysis.* Prentice Hall, Englewood Cliffs, New Jersey.

Schumaker L.L.
(1968a)   Uniform approximation by Tchebycheffian spline functions. J. Math. Mech. **18**,
            369–378.
(1968b)   Uniform approximation by Chebyshev spline functions II: free knots. SIAM J.
            Numer. Anal. **5**, 647–656.
(1969a)   Some algorithms for the computation of interpolating and approximating spline
            functions. In: Greville, T.N.E. (ed.) *Theory and Applications of Spline Functions.*
            Academic Press, New York, pp. 87–102.
(1969b)   On the smoothness of best spline approximations. J. Approx. Theory **2**, 410–418.
(1976a)   Zeros of spline functions and applications. J. Approx. Theory **18**, 152–168.
(1976b)   Fitting surfaces to scattered data. In: Chui, C.K., Lorentz, G.G., Schumaker, L.L.
            (eds.) *Approximation Theory II.* Academic Press, New York, pp. 203–268.
(1979) L_2-approximation by splines with free knots. In: Meinardus, G. (ed.) *Approxima-*
            tion in Theorie und Praxis. Bibliographisches Institut, Mannheim, pp. 157–182.
(1981)   *Spline Functions: Basic Theory.* Wiley-Interscience, New York.
(1984a)   Bounds on the dimension of spaces of multivariate piecewise polynomials. Rocky
            Mountain J. Math. **14**, 251–264.
(1984b)   On spaces of piecewise polynomials in two variables. In: Singh, S.P., Burry, J.H.W.,
            Watson, B. (eds.) *Approximation Theory and Spline Functions.* Reidel, Dodrecht,
            pp. 151–197.
(1987)   On super splines and finite elements. Preprint.

Schurer F. and Cheney E.W.
(1968)   On interpolating cubic splines with equally spaced nodes. Indag. Math. **30**, 517–
            524.

Schwarz H.R.
(1980)   *Methode der finiten Elemente.* Teubner, Stuttgart.

Shapiro H.S.
(1971)   *Topics in Approximation Theory.* Lecture Notes in Mathematics **187**, Springer,
            Berlin.

Sharma A. and Meir A.
(1966)   Degree of approximation by spline interpolation. J. Math. Mech. **15**, 759–768.

Singer I.
(1970)   *Best Approximation in Normed Linear Spaces by Elements of Linear Subspaces.*
            Springer, Berlin.
(1974)   *Theory of Best Approximation and Functional Analysis.* SIAM, Philadelphia.

Sommer M.
(1979) L_1-approximation by weak Chebyshev spaces. In: Meinardus, G. (ed.) *Approxi-*
            mation in Theorie und Praxis. Bibliographisches Institut, Mannheim, pp. 85–102.
(1980a)   Characterization of continuous selections for the metric projection for generalized
            splines. SIAM J. Math. Anal. **11**, 23–40.
(1980b)   Nonexistence of continuous selections of the metric projection for a class of weak
            Chebyshev spaces. Trans. Amer. Math. Soc. **260**, 403–409.
(1982)   Characterization of continuous selections of the metric projection for a class of weak
            Chebyshev spaces. SIAM J. Math. Anal. **13**, 280–294.
(1983a)   Weak Chebyshev spaces and best L_1-approximation. J. Approx. Theory **39**, 54–71.
(1983b)   Continuous selections and convergence of best L_p– approximations in subspaces of
            spline functions. Numer. Func. Anal. and Optimiz. **6**, 213–234.

Sommer M. and Strauß H.
(1977)   Eigenschaften von schwach tschebyscheffschen Räumen. J. Approx. Theory **21**,
            257–268.
(1981)   Unicity of best one–sided L_1-approximations for certain classes of spline functions.
            Numer. Funct. Anal. and Optimiz. **4**, 413–435.

Späth H.
(1974)   *Spline Algorithms for Curves and Surfaces.* Utilitas Mathemtical Publ. Inc., Win-
          nipeg, Manitoba.
Stechkin S.B. and Subbotin Yu.N.
(1976)   *Splines in Computational Mathematics* (Russian). Izd. Nauka, Moscow.
Strang G. and Fix G.
(1973)   *An Analysis of The Finite Element Method.* Prentice Hall, Englewood Cliffs, New
          Jersey.
Strauß H.
(1975a) L_1-Approximation mit Splinefunktionen. In: Collatz, L., Meinardus, G. (eds.)
          Numerische Methoden der Approximationstheorie. ISNM 26, Birkhäuser, Basel,
          pp. 151–162.
(1975b)   Eindeutigkeit bei der gleichmäßigen Approximation mit Tschebyscheffschen Spline-
          funktionen. J. Approx. Theory **15**, 78–82.
(1979)   Optimale Quadraturformeln und Perfektsplines. J. Approx. Theory **27**, 203–226.
(1981)   Eindeutigkeit in der L_1-Approximation. Math. Z. **176**, 63–74.
(1982)   Unicity in best one–sided L_1-approximation. Numer. Math. **40**, 229–243.
(1983)   Comparision theorems for monosplines and best one–sided approximation. Numer.
          Funct. Anal. and Optimiz. **6**, 423–445.
(1984a)   Monotonicity of quadrature formulae of Gauss type and comparision theorems for
          monosplines. Numer. Math. **44**, 337–347.
(1984b)   Characterization of strict approximations in subspaces of spline functions. J. Ap-
          prox. Theory **41**, 309–328.
(1984c)   An algorithm for the computation of strict approximations in subspaces of spline
          functions. J. Approx. Theory **41**, 329–344.
(1984d)   On Tchebycheffian monosplines and approximation by splines with free knots. Appl.
          Anal. **17**, 283–293.
(1984e)   Best L_1-approximation. J. Approx. Theory **41**, 297–308.

Taylor G.D.
(1979)   Data fitting: some adaptive methods. In: Meinardus, G. (ed.) *Approximation in
          Theorie und Praxis.* Bibliographisches Institut, Mannheim, pp. 291–304.
Töpfer H.J.
(1982)   Models for smooth curve fitting. In: Collatz, L., Meinardus, G., Werner, H. (eds.)
          Numerical Methods of Approximation Theory. ISNM 59, Birkhäuser, Basel, pp.
          209–224.
Traub J.F. and Wozniakowski H.
(1980)   *A General Theory of Optimal Algorithms.* Academic Press, New York.
Vasilenko V.A.
(1983)   *Spline Functions: Theory, Algorithms, Programs* (Russian). Nauka, Novosibirsk.
Varga R.S.
(1971)   *Functional Analysis and Approximation Theory in Numerical Analysis.* CBMS 3,
          SIAM, Philadelphia.
Veidinger L.
(1960)   On the numerical determination of the best approximations in the Chebyshev sense.
          Numer. Math. **2**, 99–105.
Ward J.D.
(1986)   Some constrained approximation problems. In: Chui, C.K., Schumaker, L.L., Ward,
          J.D. (eds.) *Approximation Theory V.* Academic Press, New York, pp. 211–230.
Watson G.A.
(1980)   *Approximation Theory and Numerical Methods.* Wiley–Interscience, Chichester.
Werner H.
(1974)   Tschebyscheff–Approximation mit einer Klasse rationaler Splinefunktionen. J. Ap-
          prox. Theory **10**, 74–92.

(1979)   An introduction to non–linear splines. In: Sahney, B.N. (ed.) *Polynomial and Spline Approximation*. D. Reidel Publishing Company, Boston, pp. 247–306.
(1980)   The development of nonlinear splines and their applications. In: Cheney, E.W. (ed.) *Approximation Theory III*. Academic Press, New York, pp. 125–150.

Wulbert D.E.
(1971)   Uniqueness and differential characterization of approximations from manifolds. Amer. J. Math. **18**, 350–366.
(1973)   A note on polynomial splines with free knots. Numer. Math. **21**, 181–184.

Zalik R.A.
(1975)   Existence of Tchebycheff extensions. J. Math. Anal. Appl. **51**, 68–75.

Zavjalov Yu.S., Kvasov B.I. and Mirosnicenko V.L.
(1980)   *Methods of Spline Functions* (Russian). Izd. Nauka, Moscow.

Zavjalov Yu.S., Leus V.A. and Skorospelov V.A.
(1985)   *Splines in Engineering Geometry* (Russian). Moscow.

Zhensykbaev A.A.
(1981)   Monosplines of minimal norm and the best quadrature formulae. Russ. Math. Surv. **36**, 121–180.
(1982)   Extremal properties of monosplines and best quadrature formulas. Math. Notes **31**, 145–154.

Zielke R.
(1979)   *Discontinuous Cebysev Systems*. Lecture Notes in Mathematics **707**, Springer, Berlin.

Zwick D.
(1986)   The generalized convexity cone of splines with multiple knots. Numer. Funct. Anal. and Optimiz. **3**, 245–260.
(1987)   Strong uniqueness of best spline approximation for a class of piecewise n–convex functions. Numer. Funct. Anal. and Optimiz. **9**, 371–380.

Index

J. Stoer, University of Würzburg; **R. Bulirsch,**
Technische Universität, Munich, FRG

Introduction to Numerical Analysis

Translated from the German by R. Bartels,
W. Gautschi, C. Witzgall

1st ed. 1980. Corr. 2nd printing 1987. IX, 609 pp.
30 figs. ISBN 3-540-90420-4

Contents: Error Analysis. – Interpolation. – Topics in Integration. – Systems of Linear Equations. – Finding Zeros and Minimum Points by Iterative Methods. – Eigenvalue Problems. – Ordinary Differential Equations. – Iterative Methods for the Solution of Large Systems of Linear Equations. Some Further Methods. – General Literature on Numerical Methods. – Index.

This is the corrected second printing of a popular textbook on numerical analysis. It is on the advanced undergraduate/beginning graduate level and is addressed to mathematicians and computer scientists. This combination of modern mathematical standards with an understanding of the needs of the applied computer scientist is one of the characteristic features of the text. Other features are careful discussions of various algorithms, fully worked out examples, and many carefully formulated and selected problems. Though the authors went to great lengths to make the book accessible to the student, it contains an enormous amount of information, such as an extensive discussion of minimization methods and direct methods for the solution of large systems of equations, much of which is not found in the standard text book literature. The book includes numerous references to contemporary research literature.

Springer-Verlag Berlin
Heidelberg New York London
Paris Tokyo Hong Kong

Springer